T0299985

Electric Utility Resource Planning

Economics, Reliability, and Decision-Making

Electric Utility Resource Planning

Economics, Reliability, and Decision-Making

Steven Sim

CRC Press is an imprint of the
Taylor & Francis Group, an **informa** business

CRC Press
Taylor & Francis Group
6000 Broken Sound Parkway NW, Suite 300
Boca Raton, FL 33487-2742

© 2012 by Taylor & Francis Group, LLC
CRC Press is an imprint of Taylor & Francis Group, an Informa business

No claim to original U.S. Government works

Version Date: 20111021

International Standard Book Number: 978-1-4398-8407-2 (Hardback)

This book contains information obtained from authentic and highly regarded sources. Reasonable efforts have been made to publish reliable data and information, but the author and publisher cannot assume responsibility for the validity of all materials or the consequences of their use. The authors and publishers have attempted to trace the copyright holders of all material reproduced in this publication and apologize to copyright holders if permission to publish in this form has not been obtained. If any copyright material has not been acknowledged please write and let us know so we may rectify in any future reprint.

Except as permitted under U.S. Copyright Law, no part of this book may be reprinted, reproduced, transmitted, or utilized in any form by any electronic, mechanical, or other means, now known or hereafter invented, including photocopying, microfilming, and recording, or in any information storage or retrieval system, without written permission from the publishers.

For permission to photocopy or use material electronically from this work, please access www.copyright.com (http://www.copyright.com/) or contact the Copyright Clearance Center, Inc. (CCC), 222 Rosewood Drive, Danvers, MA 01923, 978-750-8400. CCC is a not-for-profit organization that provides licenses and registration for a variety of users. For organizations that have been granted a photocopy license by the CCC, a separate system of payment has been arranged.

Trademark Notice: Product or corporate names may be trademarks or registered trademarks, and are used only for identification and explanation without intent to infringe.

Visit the Taylor & Francis Web site at
http://www.taylorandfrancis.com

and the CRC Press Web site at
http://www.crcpress.com

Contents

Acknowledgments

I view the time I have spent working in resource planning for Florida Power & Light (FPL) as an ongoing learning process that has never stopped. This learning process has been aided, both directly and indirectly, by many people. Each of the individuals that are mentioned below assisted that learning process in various ways. Some directly shared their knowledge and experience. Others suggested that they tended to look at a specific issue or analysis from a different perspective than I had taken and encouraged me to examine that different perspective as well. Other people asked probing questions that forced a deeper look into analyses that had been performed, and conclusions that had been drawn from those analyses. Still others led me to consider how information could be better explained and presented.

Regardless of the manner in which their contributions were provided, I am very grateful for the contributions all of the folks whose names are about to be mentioned have made to my ongoing education regarding utility planning and operation. These individuals have certainly helped me become a more knowledgeable resource planner than would otherwise have been the case. And, unbeknownst to them (and me) at the time, they have helped make this a better book.

I shall start by thanking some outstanding folks I have had the privilege to work directly for at FPL. These include Rene Silva, Sam Waters, Nelson Hawk, and Enrique Hugues. This is a group of very bright and talented people.

The list of folks I have had the privilege to work with at FPL in resource planning evaluations is a lengthy one. This list includes Dennis Brandt, Daisy Iglesias, Juan Enjamio, Rick Hevia, Jeff Benson, Severino Lopez, John Scott, Richard Brown, Raul Montenegro, Tim Wehnes, Leo Green, Anita Sharma, Rosemary Morley, Richard Feldman, John Gnecco, Hector Sanchez, Bob Schoneck, Kiko Barredo, John Hampp, Ray Butts, Gene Ungar, and Ian Nichols ("on loan" from the Land Down Under). I am sure I learned a lot more from each of you than vice versa.

A very special "thank you" goes to another coworker, Sharon Fischer, who assisted greatly in the preparation of certain figures that appear in this book whenever the author had (another) "stupid author" question.

In the course of working on a large number of regulatory dockets, most involving the Florida Public Service Commission (FPSC), certain FPSC Commissioners and Staff members stand out as having repeatedly asked important and interesting questions, thus leading me over time to improve the quality of both the analyses I presented on behalf of FPL and how the analysis results were presented. Those individuals that most readily come to mind in this regard are Commissioners Susan Clark, Terry Deason, and

Joe Cresse. Among the FPSC Staff members, Jim Dean and Tom Ballinger have been particularly impressive in regard to both the breadth and depth of the questions they have asked in numerous FPSC dockets. Needless to say, these questions by the Commissioners and FPSC Staff members represented valuable input in my ongoing education in resource planning. (However, I can remember a time or two when I was on the witness stand when this input, delivered in the form of difficult questions, was not always immediately "treasured.")

I have also been privileged to work with some outstanding attorneys who helped me prepare for numerous regulatory filings and hearings. Their questions and suggestions regarding many testimonies and analyses have also led to significant improvements in both the analyses themselves and in how the analysis results are presented. These improvements have also led to making this a better book. These outstanding barristers include Charles Guyton, Matt Childs, Bonnie Davis, Bryan Anderson, Jessica Cano, John Butler, and Wade Litchfield. Thanks for all of the help and suggestions. (And, just to make sure that I place in writing the answer to the question you asked me on the first day we worked together on draft testimony: "Yes, Bonnie Davis, I did go to college.")

Finally, I owe a large debt of gratitude to my lovely and talented wife, Sue (a.k.a. "the Goddess"), for her patience and understanding as I spent time in the evenings and on weekends trying to get my thoughts for this book down on paper. (And, "yes, dear," it is now time to take you on that vacation I promised.)

To the aforementioned list of folks, and to the numerous other folks that I have unintentionally omitted, one more set of thanks is called for. Not only have I found the ever-evolving field of electric utility resource planning to be constantly challenging, I found the work to be a *lot* of fun because of these challenges and the great people I have been able to work with. I have been very fortunate in both of these respects. One could not ask for more.*

* OK. One *could* ask for more, but the conclusion of a book writing effort is no time to get greedy.

Author

Dr. Steven Sim has worked in the field of energy analysis since the mid-1970s. After graduating from the University of Miami (Miami, Florida) with bachelor's and master's degrees in mathematics in 1973 and 1975, respectively, he earned a doctorate in environmental science and engineering from the University of California at Los Angeles (UCLA) in 1979 with an emphasis on energy.

During his doctoral work, Dr. Sim also completed an internship of approximately a year and a half at the Florida Solar Energy Center (FSEC), a research arm of Florida's university system located in Merritt Island, Florida, near Cape Canaveral. His work at FSEC involved an examination of consumers' experience with solar water heaters and projections of the potential for renewable energy in the Southeastern United States.

In 1979, Dr. Sim joined Florida Power & Light Company (FPL), a subsidiary of NextEra Energy, Inc. FPL is one of the largest electric utilities in the United States. In the more than 30 years from 1979 to the time this book is being written, FPL has experienced tremendous population growth in the geographic area it serves. As an example, in the years from 1990 through 2007, FPL's average annual growth was approximately 80,000 new customer accounts per year. On average, each account serves about 2.5 people that have newly arrived in FPL's service territory. Therefore, the growth that FPL had to keep up with was roughly equivalent to 200,000 new people per year, or 1 million new people every five years.

Although population growth slowed considerably in 2008, and remains relatively slow as this book is written, the point is that FPL has had to plan for—and meet—extraordinary growth during the last few decades. Relatively few electric utilities have had to face the challenges inherent in meeting continued growth of this magnitude over such an extended time. This extraordinary growth, combined with the planning issues that all utilities face involving changing environmental regulations, fuel decisions, modification/retirement of existing generating units, etc., ensured that FPL's planning efforts have addressed a wide variety of resource planning challenges. During that time, Dr. Sim has had several roles in regard to these resource planning efforts.

Dr. Sim spent approximately the first 10 years of his career at FPL designing a number of FPL's demand side management (DSM) programs. One of these, the Passive Home Program, earned a U.S. Department of Energy (DOE) award for innovation. Among the numerous other DSM programs he either designed or codesigned were the Conservation Water Heating Program (that featured heat pump and solar water heaters) and one of the nation's most successful load management programs (FPL's residential load control program that is known as the On Call Program).

In the course of designing these DSM programs, he became very interested in gaining a better understanding of how DSM programs affect FPL's entire utility system. Of particular interest were how FPL's DSM programs impacted the actual operation of FPL's generation, transmission, and distribution system, including impacts on air emissions for the FPL system as a whole, and fuel usage by all of the power plants in FPL's system.

This interest led to him joining FPL's System Planning Department in 1991. This department (since renamed the Resource Assessment and Planning Department) is charged with determining both when new resources should be added to the FPL system and what the best resources are for FPL to add to meet the continued growth in customers (and their demand for electricity), changing environmental laws, and other regulatory requirements. Dr. Sim has found this work to be challenging, continually evolving, and always interesting.

Over this time, Dr. Sim has conducted, supervised, and/or collaborated on thousands of analyses designed to examine how FPL can best serve its customers given these changing circumstances, laws, and regulations. In addition to these analyses, Dr. Sim has also served as an FPL witness in numerous hearings before the Florida Public Service Commission (FPSC) and in hearings/meetings with other Florida governmental agencies and organizations. A partial list of the subjects addressed in these hearings included: the economic and non-economic impacts to the FPL system of proposed new nuclear, coal, and natural gas power plants; the economic and non-economic impacts to the FPL system of FPL's DSM programs that are alternatives to new power plants; the air emission and fuel use impacts to the FPL system from each of these resource options; the reliability of the utility system; and the planning processes that can be used by a utility to plan for new resources.

In these numerous regulatory hearings, Dr. Sim has provided both written and oral testimony. In so doing, he has had the opportunity to respond to a wide variety of inquiries from regulators, administrative law judges, environmental organizations, fuel suppliers, etc., regarding a host of issues.

Dr. Sim has also participated in various collaborative efforts with other electric utility organizations and individual utilities, including advisory groups for the international Electric Power Research Institute (EPRI), the Southeast Electric Exchange's (SEE) Integrated Resource Planning Task Force, and the Florida Reliability Coordinating Council (formerly the Florida Coordinating Group). These activities have further broadened his perspective of the challenges facing other electric utilities across the country and how these challenges are being addressed.

Over the years, he has spoken at a number of electric utility conferences, both national and international. At one of those conferences, the *6th National Demand-Side Management Conference*, Dr. Sim was awarded the Outstanding

Speaker Award. He has also enjoyed speaking with, and to, a variety of individuals and groups that were interested in energy issues, particularly those issues that directly involve electric utilities. He has seen this interest grow in recent years as issues such as energy prices, climate change, renewable energy, and federal/state energy policy are regularly in the news.

In light of this growing public interest, Dr. Sim has written this book to share the insight he has gained in over 30 years of directly analyzing energy-related issues for one of the biggest electric utilities in the United States, and in collaborating with others in the electric utility industry regarding the challenges they have faced. His hope is that this insight will help to facilitate discussions on a variety of energy-related issues, leading to better informed decisions.

He hopes that you will not only find this book informative, but enjoyable to read as well.

1

Introduction

Why Write This Book?

In my career at Florida Power & Light Company (FPL), I have been fortunate to have opportunities to interact with numerous individuals, both on a one-to-one basis and in group situations, in discussions that address a variety of issues pertaining to electric utilities. These individuals generally have two things in common. First, they are not electric utility resource planners. Second, despite this unfortunate choice in their life's work, they have a genuine interest in energy issues, particularly those relating to electric utilities.

These discussions frequently focus on the issue of what "resource options" FPL was considering, or had already chosen, in order to continue to provide electricity reliably and at a reasonable cost. These resource options could include any (or all) of the following:

- The utility building a new power plant that would be fueled by either natural gas, coal, or nuclear fuel;
- The utility purchasing power from new or existing power plants owned by a neighboring utility or a non-utility organization;
- The utility lowering its customers' demand for, and usage of, electricity by implementing demand side management (DSM) programs; and,
- The utility building, or purchasing energy from, renewable energy (i.e., solar, wind, etc.) facilities.

In regard to the discussions I have had regarding these resource options, and the decisions that were being made about them, most of the discussions were characterized by what I will term a sincere intellectual curiosity. In such cases, the individuals who were discussing the topic with me seemed truly interested in learning what specific resource options had been considered and why a particular decision was made. In other cases, the discussions could be characterized as outright disbelief, and/or open hostility, on the part of the individual who was certain that he/she already knew what the "Correct Decision" was. More than a few of these conversations were similar to the following hypothetical exchange:

Questioner: "Did the utility really decide to go with Option A (*insert appropriate name of new power plant, new energy conservation program, etc.*) instead of Option B (*insert the name of an alternative favored by the Questioner*)?"

Me: "Yes."

Questioner: "What were you _____* thinking of? It is obvious that the decision should have been Option B!" (*Appropriate facial expressions, tone of voice, and/or hand gestures may also be supplied by the reader's imagination.*)

However, contrary to this hypothetical exchange, most of the discussions I have had have been very pleasant. One such exchange took place a number of years ago with a women's civic group in South Florida that was sponsoring a panel discussion regarding energy-related issues. I was invited to represent FPL as a panelist. I single out this particular discussion because it influenced this book in two ways.

Both of these ways are related to a question I was asked at one point in the proceedings. During the discussions, one issue that emerged was whether an individual could generate one's own electricity rather than purchasing it from the utility.† The questioning up to that point had actually been a bit intense when one lady in the audience raised her hand and then asked: "What do I have to do to become an electric utility?"

Blindsided by this unexpected question, and attempting to buy time until I could think of how best to respond to her question, I blurted out: "Well, first you've got to get a smokestack."

Fortunately for me, the response resulted in a few giggles that seemed to lighten the mood considerably. The discussion then continued in a much more relaxed atmosphere and actually proved to be quite productive. Grateful for that fact, I have remembered that exchange and actually used my stumbling response to her question as the initial working title for this book (until sage advice from reviewers convinced me to use a more "proper" title).

The second, and much more important, item I took away from this particular meeting was a perception of ***why*** a number of questions were being asked. During the discussion I realized, perhaps for the first time, that many (if not most) people have a number of misconceptions regarding: (i) how electric utilities actually operate, and (ii) how utilities plan for the future. Since that time, this perception has been reinforced in virtually every subsequent meeting I have participated in that discussed utility-related energy issues.

It was apparent that a basic understanding of how a utility actually operates, particularly in regard to the utility's existing generating units,

---

* Please insert an appropriate reference such as "people," "doofuses," "buffalo heads," etc.

† I seem to recall that the phrase "Evil Empires" came up in reference to electric utilities in the course of discussing this particular issue.

would be very useful if one is to understand how utilities plan for the future. With that in mind, Chapter 2 of this book is designed to provide this basic understanding of how a utility actually operates its many generating units.

This basic understanding of utility operation of its generating units in Chapter 2 sets the stage for the remaining chapters of the book. In those chapters the other subject of the book is discussed: how do electric utilities plan for the future? We will refer to this issue as "resource planning" for a utility.

Who Is This Book Written For?

The information presented in this book is primarily intended for two distinct audiences which I view as equally important.

The first audience is comprised of a wide variety of individuals who do not work for electric utilities, but who are interested in energy issues, especially issues related to electric utilities. From my experience, a number of different types of individuals would be included in this audience. These include, but are not limited to, the following:

- Individuals working in municipal, state, and federal governmental organizations, including those who regulate various aspects of electric utility operations and planning;
- State and federal legislators who make laws that impact electric utilities and their customers;
- Journalists who report on utility and energy matters;
- Environmentalists, especially including representatives of numerous environmental organizations that frequently intervene in utility regulatory hearings;
- Educators and students involved in energy and environmental education; and,
- Others who simply wish to know more about energy issues.

A better understanding of how electric utilities actually operate and plan their systems, and just as important, why they do so, should help to eliminate—or at least minimize—numerous misconceptions that get in the way of meaningful discussions about electric utilities and their plans for the future. In turn, this understanding should result in more informed discussions that lead to better decisions regarding energy issues.

The second audience this book is intended for is a variety of individuals who work for electric utilities in many different areas. The information in the book should prove especially useful for utility engineers and analysts who are relatively new to utility resource planning. In addition, I believe the book's information will also prove useful to people who have other jobs in utilities: executives, attorneys, designers of utility DSM programs, individuals working in the power plant business units, individuals working in the power purchase business units, etc. Such individuals may seek a quick education or refresher on a concept/issue that is either new to them, or is one which they have not addressed in some time.

This utility audience also includes individuals in various utilities around the United States (as discussed a bit more in Chapter 2) that are turning their attention back to the concepts discussed in this book after being away from them for a decade or more due to decisions in their respective states to change the fundamental structure of the electric utility business. In recent years, I have received a number of calls from individuals in such utilities. The conversations often proceed with my cell phone ringing at an inconvenient time with the following type of exchange:

> Caller: "Help! We used to do utility resource planning like you guys do, but we have been away from it so long that we don't remember how to do it. Can you help us?"
>
> Me: "Yes. I will be glad to help, but right now I'm on the golf course lining up my putt for a score of 8 on this par 3 hole. I will call you back when this athletic crisis has passed."

If this book helps minimize such tragic distractions to my golf game, then I may benefit as much from this book as I hope the two types of intended audiences will.

An Overview of the Book

The remaining eight chapters of this book present information in the following sequence:

Chapter 2 lays the foundation for much of the rest of the book by explaining certain fundamental concepts of electric utilities. The difference between "regulated" and "unregulated" utilities is discussed, along with an explanation of why this book focuses on regulated utilities. Next, the major operating areas of an electric utility, including those areas that are the primary focus of this book, are presented. The different types of electric generating units (i.e., power plants) a utility typically utilizes are then presented along with an explanation of how a utility decides which generating units to operate at a given time. The chapter concludes by "creating" a hypothetical utility system. This hypothetical utility system will then be used throughout the remainder of the book to illustrate a variety of topics.

In Chapter 3, we introduce the three key questions that utility resource planning must continually answer. We also introduce different types of analytical approaches that are used to answer these three key questions.

Chapter 4 then utilizes the hypothetical utility system we have created to illustrate how analytical approaches designed to answer the first and second of the three key questions are actually implemented. The answers to these two questions are then used in subsequent chapters as we develop answers to the more complex (and, for most people, more interesting) third question.

Chapters 5 through 7 then illustrate how the third key question is answered using our hypothetical utility system and a number of basic types of "resource options." These resource options represent choices a utility has to meet its future resource needs to ensure that it can maintain a reliable and economic supply of electricity.

Chapter 5 focuses on one of two basic types of resource options. This type of resource option is new electric generating units (or "Supply" options). The list of Supply options that could potentially be selected by the hypothetical utility to meet its resource needs includes new generating units that could be fueled by natural gas, oil, coal, nuclear energy, or renewable energy. In this chapter we will select a few generating unit examples and perform the economic analyses to see which of these would be the best economic choice for our hypothetical utility if it chose to meet its resource needs with new generating units.

The focus of Chapter 6 is on the second basic type of resource option. This type of resource option does not supply additional electricity, but instead reduces the demand that a utility's customers have for electricity and the amount of energy that customers use. These options are traditionally referred to as demand side management (DSM) options. Similar to the approach used in Chapter 5, we will select a few DSM examples and perform economic analyses to see which of these selected DSM options would be the best economic choice for our hypothetical utility if it chose to meet its resource needs with DSM options.

In Chapter 7, we bring together the information developed in the previous two chapters that focused on Supply options and DSM options, respectively. We now look at all of the Supply and DSM options examined in the previous two chapters to determine which resource option is really the best economic choice for our hypothetical utility. In addition, we then take a look at these resource options from non-economic perspectives of our hypothetical utility system (such as the types of fuel used, total system air emissions, etc.) Then, armed with the results from the economic and non-economic evaluations, our hypothetical utility will make its selection of the best resource option with which to meet its resource needs.

Once we have completed Chapters 3 through 7, the three key questions for the hypothetical utility will have been answered. The analyses we will have performed to get to these answers for the hypothetical utility are intended to be both illustrative as to the importance of certain fundamental principles of resource planning that will be introduced along the way, and useful in showing how these fundamental principles are applied. However, we are not quite finished yet.

Chapter 8 discusses why the results to the preceding analyses could easily have been much different. The focus of this chapter is on a variety of "constraints" that may apply to a utility in its resource planning work. As we shall discuss, these constraints may significantly change the resource options that a utility selects. Furthermore, as we shall see, these changes may be for the better, or for the worse, from the perspective of the utility's customers.

Chapter 9 then presents some final thoughts in regard to the information presented in the previous chapters. The key points that have been examined in detail in previous chapters are summarized and a few opinions are offered.

In addition, the book contains a number of appendices. The first appendix, Appendix A, contains a list of what I call my four "Fundamental Principles" of electric utility resource planning. Each of these four principles will be introduced at an appropriate point throughout the book. Appendix A contains an easy-to-reference listing of these principles. Appendix B contains a glossary of terms used in the book that are not commonly used by people who are outside of the electric utility and/or energy industries.

The remaining appendices present five "mini-lessons" about various subjects addressed in the book. Rather than include these mini-lessons in the main body of the book, this approach is designed to accomplish a couple of objectives. First, this approach should allow a reader who may already understand the concept in question in a particular chapter to continue reading straight through the chapter without having to wade through an explanation he/she doesn't need.

Second, this approach provides an option for reader who may be unfamiliar with the concept in question. Such a reader can opt to continue reading through the chapter to understand the overall message, and then go back to the appropriate appendix to delve into the concept. Or, the reader could stop when the concept is introduced, go to the appropriate appendix, and then return to the main text. In either case, the appendices will (hopefully) provide a quick and easy-to-understand explanation of that concept.

Are We Keeping It Simple?

To many (if not most) folks who have read to this point, the entire subject matter of electric utility operation and planning may seem complex and a bit daunting. A reader may be questioning whether the subject can be discussed in simple enough terms to make it understandable. The answer is, in a word, "yes." In addition, I hope to make the discussion interesting.

As mentioned earlier, one of the book's two intended audiences is a variety of individuals who do not work in the electric utility industry. The objective is to inform these individuals, but to do so without bogging down the discussion with unnecessary utility industry jargon or to go into an unwarranted level of detail.

Therefore, the book purposefully attempts to simplify a number of complex subjects while maintaining the basic correctness of the concepts that are discussed. The simplifications used are those that have proven useful in my past discussions with various individuals and groups outside of the electric utility industry.

Individuals who are well schooled in the principles discussed in the book, particularly those folks working for electric utilities who are intimately familiar with resource planning work, will certainly recognize where these simplifications have been made. It is hoped that these folks will understand why the simplifications have been made.*

A Few Words Regarding Assumptions Used in the Book

When discussing how electric utilities operate, and how they plan for the future, it is most helpful to do so by means of examples. For that reason the book is full of examples, especially example calculations involving various utility costs and other values. In order to perform those calculations, a number of assumptions needed to be made regarding the cost and operating characteristics of existing generating units and new resource options, fuel costs, environmental compliance costs, etc. Therefore, in our discussions we shall first make a number of assumptions, then we shall use the assumptions in a variety of calculations.

A few words about these assumptions are in order. First, the values for the assumptions used in the book are generally representative of actual and/or forecasted values that have been used by various utilities in the years immediately preceding (and during) the writing of this book. The different individual utility forecasts varied considerably, but when the forecasts are viewed as a group, the values we are assuming typically fell within the range of values encompassed by those forecasts. Moreover, the assumed values typically are not necessarily "tied to" any particular year but are more general, representative values.

Second, the reality for an electric utility is that the assumptions it is using at any point in time regarding costs and operating characteristics are not

* If not, it is hoped that they will "get over it" with the passage of an appropriate amount of time.

only subject to change, but are usually continually changing. This necessitates that analyses are typically reworked when new assumption values become known and are accepted as being an improved assumption value. As a consequence of these new assumptions, the results from the new analysis may or may not agree with the previous analysis results.

Therefore, in regard to this book, virtually any assumption that we make, and use in our calculations, may or may not be "accurate" at the time the book is read if one were to attempt to compare those values to current values. In fact, it is safe to say that the more years between the time this book is written and the time it is read, the more likely it is that a given assumption value used in the book will no longer reflect the then-current value.

However, this fact should not concern the reader. This is because the goal of the book is ***not*** to definitively demonstrate that a specific resource option A is better than a specific resource option B. Instead, the objective of the book is to explain and demonstrate ***how resource planning in electric utilities is (or should be) performed*** in order to determine whether a resource option A is better than a resource option B. With this in mind, it is clear that the assumptions used throughout the book are basically "placeholders" that allow us to discuss resource planning concepts and analytical approaches using quantitative examples. These concepts and analytical approaches will remain valid regardless of any subsequent change in assumption values in the future.

In other words, the focus should be on the resource planning concepts and analytical approaches that will be discussed. The assumed values are a necessary tool with which to discuss these concepts and analytical approaches.

A Couple of Disclaimers

This book can be thought of as having two types of information. The first type of information is factual information. An example of this type of information is the fact that utilities perform economic analyses. The second type of information can be considered as "the author's opinion." An example of this type of information is when I express my opinion as to why certain constraints to utility planning imposed by legislatures and/or regulators are unwise. Especially in regard to the second type of information, it is prudent to offer a disclaimer.

This disclaimer is that the thoughts and opinions expressed in this book are solely mine. As such, they do not necessarily reflect those of my long-term employer, FPL. In addition, any errors or omissions that appear in this book are solely my responsibility.*

* Unless, of course, a suitable scapegoat can be found.

Now that we have dispensed with the obligatory disclaimers, it is time to return to the objective of understanding how electric utilities operate and plan for the future. So we will now dive into the electric utility pool, shallow end first, and paddle over to Chapter 2 where we will learn some basic facts about how electric utilities actually work.

2

How Does an Electric Utility Actually "Work"?

Two "Types" of Electric Utilities

Generally speaking, until a relatively short time ago, virtually all electric utilities were structured to operate in a similar way. This utility operational structure is characterized by having all three main functions associated with electricity production and delivery all "under the same roof" of a single electric utility company. These three main functions are: (i) the generation or procurement of electricity (the operation of the utility's own electric generating units/power plants and/or the purchase of power from other entities' power plants); (ii) the transmission of bulk quantities of electricity from these power plants throughout the utility's service area (utilizing the large transmission towers and electric lines that cross long distances); and (iii) the distribution of electricity from these transmission lines to individual customers' homes and businesses. In such a structure, this one utility company is the customer's sole choice for providing all aspects of electric service.

In this structure, the utility has a monopoly position. In exchange for this monopoly position, virtually all aspects of the utility's retail business (i.e., sales to the ultimate users of electricity) are overseen by a state or local regulatory agency or other authority. (The Federal Energy Regulatory Commission, FERC, has authority over various other aspects of utility business such as transactions between utilities, etc.) For example, the retail business aspects of privately-owned electric utilities (i.e., utilities for which you can purchase shares of its stock) are generally regulated by the state's Public Service Commission, Public Utility Commission, or an equivalent state organization. Municipal-owned utilities are regulated by an arm of the municipal government. Certain other organizations that provide electric service, such as the Tennessee Valley Authority (TVA) and Bonneville Power Authority (BPA), are more directly tied to federal government oversight.

However, in the last decades of the twentieth century, a number of states modified the regulatory structure of privately-owned utilities in an attempt to encourage greater competition in the electric utility industry.

The modifications frequently resulted in the utility breaking up, or "spinning off," one or more of the three functional areas; generation, transmission, or distribution, into separate companies that are now not all owned/controlled by the original utility company. For reasons of practicality (how many sets of wires and poles do you want running down your street?), customers have little or no choice in regard to the distribution company whose wires connect the customer's home or business. However, the customer, either directly or indirectly, will have a choice at least in regard to the sources of the generation of electricity that he or she is served from. In such a case, the transmission function is regulated in a way that attempts to ensure that no one generation company has an unfair advantage over another in regard to delivering its power to the distribution company for subsequent delivery to customers' homes or businesses.

This first (or original) utility structure is often referred to as being a "regulated" environment for utilities while the second structure is commonly referred to as being a "deregulated" or "restructured" environment. Although these terms are commonly used, they are a bit of a misnomer because there is still considerable regulation in either case, although the regulation takes different forms.

For this book, we will be discussing the original or regulated utility structure. There are several reasons for this. First, the vast majority of the generation, transmission, and distribution facilities in operation at the time this book is written came into existence under the original regulated utility structure. At the very least, the focus on the regulated utility structure will aid in understanding how decisions were made that shaped the development of electric utility systems. Second, many states in the United States continue to operate under the original utility regulatory structure.

Third, the basic principles discussed in this book are not only true from the perspective of a regulated utility structure, but many of these principles are also relevant in an unregulated utility structure.* Finally, the track record to-date of the unregulated utility structure "experiment" is a mixed one at the time this book is written. In the last several years, a number of states and utilities have moved back—in varying degrees—toward a regulated utility structure, and toward the type of resource planning issues and analyses that are discussed in this book. (Hence the previously mentioned cellular phone call I unfortunately received on the golf course while I was lining up my putt for an 8.)

* For example, the discussions regarding utility system reliability and system air emissions remain relevant to a significant degree for unregulated utility systems although this analysis is generally performed not by an individual utility alone, but by another party. Conversely, the discussion regarding the economic analyses of Supply options would be significantly different if discussing unregulated utility systems.

Whose Perspective Will Be Taken?

The discussions in this book will take the perspective of an electric utility's customer, not an electric utility's shareholder. Although the perspective of an electric utility's customer is important in both regulated and unregulated utilities, it is arguably more important in a regulated utility. In fact, resource planning in a regulated utility will likely predominately, if not exclusively, focus on the customer perspective.

What Aspects of an Electric Utility Will We Focus On?

We just briefly introduced what are generally thought of as the three main functional areas of an electric utility in a regulated structure: generation, transmission, and distribution. Each of these is a very complex subject area and each area has its own set of fundamental principles, issues, regulations, and analytical approaches.

For the purposes of this book, we will focus on the generation area. This area encompasses all of a utility's existing electrical generating units. It also encompasses new electrical generating units fueled by natural gas, oil, coal, nuclear fuel, or renewable energy (such as solar, wind, etc.). Furthermore, this area also encompasses purchases of power from these types of facilities, regardless of whether such facilities are owned by other utilities or by non-utility organizations.

However, we will broaden the scope from focusing solely on electrical generation options such as these to also include DSM resource options that lower the demand for, and the usage of, electrical energy. These DSM options are often referred to as either energy conservation (or energy efficiency) programs or as load management (or demand response) programs. Collectively, we will refer to these simply as DSM programs/options because they can affect the amount of electricity that customers demand.

Due to this broadening of scope, it can be said that the book's focus is on two types of resource options: (i) Supply options (new power plants/generating units) that provide more electricity to meet the utility customers' demand for electricity, and (ii) DSM options that lower, or otherwise change, the customers' demand for electricity and their usage of electricity.

One has to start somewhere in this discussion of Supply and DSM options. Therefore, we will start with Supply options. Accordingly, only generating units will be discussed in the remainder of this chapter. The choice to first focus on generating units is beneficial because it not only allows one to understand how a utility system is operated, but also is necessary to understand how both Supply and DSM options are evaluated. We will return to the topic of DSM options in subsequent chapters.

Types of Generating Units a Utility May Have

The number of generating units a utility will have may vary considerably depending upon the "size" of the utility; i.e., according to how many customers the utility serves. For example, at the time the writing of this book began, FPL served approximately 4.5 million customer accounts (i.e., electric meters), or about 11 million people. It served these customer accounts and people with 91 generating units. Smaller utilities will typically have a lower number of generating units.

Although the actual number of generating units a utility has may vary significantly from one utility to another, the *types* of existing generating units a utility has typically varies relatively little from one utility to another. For purposes of discussion, we will group the majority of existing generating units into five basic types of generating units that are in existence at the time this book is written. These five basic types of generating units are:

1. *Steam generating units fueled by oil and/or natural gas:* These units are typically units that were placed in-service many years ago. The design of these units is relatively simple. Either oil or natural gas is burned as fuel in a boiler and the heat from the combustion of the fuel is used to convert water into steam. The steam then enters a steam turbine that, in turn, drives an electric generator to produce electricity.
2. *Steam generating units fueled by coal:* These units are similar in many basic design respects to the aforementioned steam generating units that burn oil or natural gas, but these steam units burn pulverized (or fluidized) coal instead. Like their oil- and natural gas-fired counterparts, many of these coal-fired steam units are older units.
3. *Gas/combustion turbine units:* These generating units also burn natural gas and/or oil, but the basic design of this type of generating unit is different from that of the steam units. Fuel is burned in a combustor, and the resulting hot gases rotate a turbine that, in turn, drives an electric generator. This type of generating unit is typically relatively small in size, and the generating unit can be started very quickly. Older units with poor fuel efficiency are commonly referred to as gas turbines, while newer units with design improvements that result in greater fuel efficiency are commonly referred to as combustion turbines.

4. *Combined cycle units:* This newer type of generating unit is basically a combination of steam and combustion turbine technologies. These combined cycle units operate primarily on natural gas, but are also capable of burning oil. (These units are also capable of burning coal that has been converted into a gas by a separate "gasifier" facility. However, these "integrated gasification-combined cycle units" are relatively rare at the time this book is written.) A combined cycle unit uses one or more combustion turbines to produce electricity as described earlier. Then the heat from the gases that exhaust from the turbine is captured in what is termed a "heat recovery steam generator." This heat recovery steam generator operates similarly to steam generating units by using the captured heat to turn water into steam. The steam is then used to produce additional amounts of electricity. As a result, combined cycle units are significantly more fuel-efficient than steam, gas turbine, or combustion turbine generating units.
5. *Nuclear units:* These units are somewhat similar to steam generating units, but only in the sense that heat is used to produce steam by heating water. However, in nuclear units, the heat is not derived from burning a fuel in a boiler. Instead, it is derived from heat that emanates from the nuclear fuel itself.

It should be noted that there are a number of other types of generating units that have emerged relatively recently as mature technologies that are suitable for consideration in large-scale utility use. As a result, these technologies are now being incorporated into various utility systems, and are under active consideration by other utilities. These other types of generating units include, but are not limited to, the following: advanced steam generating units that burn coal, wind turbines, solar thermal technologies that produce steam, and photovoltaic (PV) arrays that produce electricity directly when the sun is shining. In addition, there are other types of generating units, such as hydroelectric facilities, that have been in existence for many years in specific geographic areas where such water storage/electric generation facilities are feasible.

However, the five basic types of generating units discussed earlier are much more likely to found in any given utility system, either as a generating unit owned and operated by the utility, or as the originating source of electrical power that is purchased by a utility from another entity. These five basic types of generating units also comprise the vast majority of generating units currently in operation. For these reasons, our discussion in the early portions of this book will focus on these five basic types of generating units. Furthermore, the basic principles and analytical concepts presented in this book apply to all types of generating units. (One of these "other" types of generating units, PV arrays, will be one of the types of resource options that will be analyzed in later chapters of the book.)

These five basic types of generating units vary greatly in regard to the cost of constructing them, and the cost of operating and maintaining them. Both of these cost considerations are important when a utility decides which type of generating unit to build. However, once a generating unit has been built and is now one of a utility's existing generating units, the operating cost of the unit becomes of primary importance. The cost of constructing a generating unit has now become a moot point and plays no part in deciding when to operate the generating unit.

The primary "operating cost" is the cost of burning fuel to generate electricity from a particular generating unit. This primary operating cost is determined by two factors: (i) the cost of the fuel, and (ii) the efficiency with which the generating unit burns that fuel to produce electricity. The next section focuses on this operating cost. (Later in the book, when the subject of how utilities select new resource options, such as the new types of generating units mentioned earlier, is discussed, we will return to consider the construction and maintenance costs of these generating units and how these costs are captured in the analyses.)

How Does a Utility Decide Which Generating Units to Use?

Because one of the basic assumptions in this book is that we are discussing electric utilities in a regulated structure, it is correct to say that utilities typically operate their generating units to meet two objectives: (1) to provide electricity to all customers at all times (within acceptable electrical voltage limits); and (2) to keep electric rates low by generating electricity at the lowest operating cost.

I realize that the last statement is one that some readers may question, but it is accurate. Utilities in a regulated structure are constantly trying to reliably operate their generating units in a manner that minimizes the total operating costs of the units. But how is the operating cost of an individual generating unit determined? I'm glad you asked.

The operating cost of a generating unit is primarily a simple multiplication of two factors previously mentioned: (i) the cost of the fuel and (ii) the efficiency with which the generating unit burns the fuel. These two factors determine the operating cost of each generating unit on a utility system. Then, once the operating cost for each generating unit on a utility system is known, that information is used to determine when each generating unit is actually used by the utility as it uses (or "dispatches" in utility parlance) its generating units in the order of lowest operating cost to highest operating cost.

The first factor, the cost of the fuel, may vary considerably from 1 year/month/day to the next. This is particularly true for oil and natural gas. This fluctuation in fuel price is often referred to as volatility in the price of the fuel, and we see it in everyday life. We all recognize how the price of gasoline we buy at the pump may vary significantly over a short time frame, and may move in either direction—higher or lower.

Gasoline prices are directly driven in large part by the price of oil. Therefore, the volatility in gasoline prices we regularly see at the pumps is caused in large part by volatility in oil prices that utilities see from day-to-day. Natural gas prices have historically also been relatively volatile over longer time periods. In contrast, the cost of coal and nuclear fuel (that are used, respectively, in coal units and nuclear units) typically varies very little from one year to the next. In utility/energy parlance, these two types of fuel are said to have relatively little cost "volatility" over time compared to oil and natural gas.

All of these fuels typically are discussed in terms of different "units" of measurement. For example, oil is typically spoken about in terms of barrels of oil. Natural gas is spoken about in terms of cubic feet of gas, and coal in terms of tons of coal. In order to be able to meaningfully compare the relative cost of these fuels, a common unit of measurement is needed. This common unit of measurement is the British Thermal Unit (BTU), which represents the amount of heat required to raise the temperature of one pound of water by one degree Fahrenheit at sea level.

The BTU unit of measurement allows the costs of different fuels to be compared in regard to the price per a set amount of heat content (i.e., the price per a set amount of BTUs) of these fuels. This is commonly done by stating the price for each fuel in terms of the cost in dollars to supply one million BTUs, abbreviated as $/mmBTU.

Although these costs, particularly those for oil and natural gas, will vary significantly over time, it is necessary for us to assume cost values for these fuels in order to illustrate the concepts that will be discussed throughout the remainder of this book. Therefore, costs for each of the four types of fuel have been assumed and are presented in Table 2.1. These cost values are representative of actual costs for each of these four types of fuel that occurred or were forecasted at some point during the time this book was written.

The perceptive reader (you, I'm sure) will likely notice that at any time this book is read, the actual costs for one or more of these fuels will be different, and perhaps significantly different, than the values shown in Table 2.1. Do not despair. As explained in Chapter 1, it is recognized that many assumption

TABLE 2.1

Representative Fuel Costs

Type of Fuel	Representative Fuel Cost ($/mmBTU)
Oil	$10.00
Natural gas	$6.00
Coal	$2.00
Nuclear fuel	$0.60

values, including those for fuel costs, will continually change over time. However, the resource planning concepts and analytical approaches that are discussed in the book remain valid regardless of assumption values used to demonstrate those concepts and analytical approaches. Moreover, the assumed fuel cost values we are using are not supposed to represent fuel costs for any particular year.*

If one were to assume fuel costs values that are different than those that appear in the table, then the specific numerical results that illustrate the discussions that follow in the remainder of this book would change. However, and much more importantly, the fundamental utility resource planning concepts and principles that are discussed throughout the book will remain valid.

Furthermore, electric utilities recognize that there is great uncertainty regarding the projected costs of future fuel costs. Utilities recognize that forecasts that attempt to address a number of years in the future are likely to be inaccurate, particularly the more years in the future the forecast attempts to address. This is true of virtually any type of forecast and forecasts of future fuel costs are no exception. As we shall see in later chapters, forecasted fuel prices are an important input in numerous analyses that electric utilities undertake.

For this reason, utilities typically address the uncertainty inherent in any single fuel cost forecast by using multiple fuel cost forecasts in many of their analyses. For example, a utility may first develop a "medium cost" fuel cost forecast. It then may develop at least two other fuel cost forecasts; one with lower projected costs, and one with higher projected costs, than in the "medium" fuel cost forecast.

Then, as appropriate for the type of analysis that will be performed, these multiple fuel cost forecasts may be utilized in the utility's analyses. In practice, the use of multiple fuel cost forecasts increases the number of analyses that the utility will perform. However, this allows a number of perspectives regarding how the analysis results may differ depending upon how actual fuel costs in the future may turn out.

For purposes of this book, the use of any single fuel cost forecast is sufficient to demonstrate the principles and concepts of utility resource planning that will be discussed. Therefore, only one set of fuel cost values will be used throughout the book.

We can now turn our attention to the second factor that determines the operating cost of a generating unit: the efficiency at which a generating unit burns fuel to produce electricity. This efficiency is termed the "heat rate" of the generating unit, and it is expressed in terms of the number of BTUs of fuel that are required to produce one kilowatt-hour (kWh) of electricity. (A kilowatt-hour, or kWh, of electricity is 1,000 watts (W) of electricity delivered

---

* In fact, as will be seen in later chapters, we have conspicuously avoided any mention of a specific year by using terms like "Current Year" and referring to later years as one, two, etc. years from the "Current Year."

or used for one hour. For example, a 100W light bulb burning for one hour consumes 100 watt-hours (Wh) of electricity. Ten such light bulbs burning for one hour would consume 1,000 Wh, or 1 kWh, of electricity.)*

The efficiency of a generating unit is analogous to the miles-per-gallon (mpg) efficiency rating of an automobile, except that the "direction" of the reference is reversed. For an automobile, the "direction" is how much "product" (miles) you get for a standard unit of fuel (gallon of gas). The higher the mpg number is, the more efficient the car. This direction is reversed for the heat rate efficiency value for electrical generating units. Here the direction is how much fuel (BTUs) is needed to produce the product (1 kWh of electricity). Therefore, the *lower* the BTU per kWh (BTU/kWh) heat rate value is, the more efficient the generating unit because it takes less fuel to produce a kWh of electricity.

In contrast to fuel prices that change, often significantly and quickly, over time, a heat rate remains relatively constant over the life of a generating unit. The heat rate is typically at its lowest in the first few years a generating unit operates when the plant is "new and clean." The heat rate typically increases a bit over time until the unit undergoes planned maintenance designed to return the generating unit to optimum efficiency. (This planned maintenance is similar in basic concept to the need years ago with older automobiles to take them in for service for a "tuneup" once the automobile is driven a predetermined number of miles.)

After this planned maintenance, the generating unit's heat rate is lowered so that it is closer to the original heat rate value of the generating unit at the time the generating unit was new. This cycle of increasing, then decreasing heat rates will be repeated numerous times over the 25 year (or longer) life of a generating unit. However, these fluctuations in heat rate values over the life of the unit are generally in a relatively narrow range, and an average heat rate value for a generating unit over its operating life can be safely assumed for each type of generating unit for purposes of utility resource planning.

The assumptions that will be used as representative heat rates for the five different types of generating units a utility is likely to have are presented in Table 2.2 (In order to simplify the discussion at this point, we will ignore—for now—the newer combustion turbine units and list only the older, less efficient gas turbine units.)

As previously mentioned, the fuel portion of the operating cost of individual generating units is determined by multiplying the cost of the type of fuel the generating unit uses or burns ($/mmBTU) by the efficiency at which

* Two concepts should be introduced at this point: electrical demand (or power) and energy. The term "demand" (or "power") refers to amount of electricity being demanded (or produced) at any one instant in time. Demand is often referred to in terms of watts, kilowatts (kW), and megawatts (MW). Energy refers to the amount of electricity being used or produced over a specific amount of time (which is typically measured in terms of hours.) Energy is often referred to in terms of kilowatt-hours (kWh), megawatt-hours (MWh), and gigawatt-hours (GWh). Please refer to Appendix B for further explanation/definition of each of these terms.

TABLE 2.2

Representative Heat Rates for Existing Generating Units

Type of Generating Unit	Representative Heat Rate (BTU/kWh)
Gas turbines	14,000
Steam-oil/gas	10,000
Combined cycle	7,000
Steam–coal	10,000
Nuclear	11,000

TABLE 2.3

Representative Operating Costs for Different Types of Existing Generating Units

(1) Type of Existing Generating Unit	(2) Primary Fuel	(3) Representative Heat Rate (BTU/kWh)	(4) Representative Fuel Prices ($/mmBTU)	(5) Representative Operating Cost ($/MWh)
Gas turbines	Natural gas	14,000	$6.00	$84.00
Steam unit-oil/gas	Oil/gas	10,000	$10.00/$6.00	$100.00/$60.00
Combined cycle	Natural gas	7,000	$6.00	$42.00
Steam unit—coal	Coal	10,000	$2.00	$20.00
Nuclear	Nuclear fuel	11,000	$0.60	$6.60

the generating unit uses this fuel to generate electricity (BTU/kWh). The operating costs for the five basic types of generating units using the aforementioned representative costs for the four fuel types, and the representative heat rates for the five types of generating units, are presented in Table 2.3.

In Table 2.3, Column (1) lists the five basic types of generating units. Column (2) lists the primary fuel the generating unit is assumed to burn. Columns (3) and (4) show, respectively, the representative heat rate and fuel cost assumptions that were presented earlier.

Column (5) uses the information in Columns (3) and (4) to calculate a representative operating cost for these types of generating units. The formula used to derive the values in Column (5) is: Column (3) × Column (4) × (1,000 kWh/MWh) × (1 mmBTU/1,000,000 BTU). The resulting value is presented in terms of dollars per megawatt-hours ($/MWh). A MWh is one thousand kilowatt-hours (i.e., 1 MWh = 1000 kWh). Operating costs of generating units are commonly expressed in terms of $/MWh.*

* Note that the operating cost values in Column (5) account for fuel costs only. Variable operating and maintenance (O&M) costs of the generator are not included in order to keep this discussion simple. Variable O&M costs are typically in the range of approximately $2.00/MWh or less for fossil fueled units. Nuclear units are generally assumed to have zero variable O&M costs. Therefore, the inclusion/omission of variable O&M costs in Column (5) does not significantly change the relative costs shown in this table.

As shown in Table 2.3, the combination of the highest heat rate (i.e., lowest efficiency) and the assumed cost for natural gas, results in gas turbines being the most expensive type of generating unit to operate on a $/MWh basis. Closely following gas turbines in regard to operating cost is a steam unit that burns oil or gas (and, given the assumed cost for natural gas and oil, these units would usually burn natural gas when it is available). Considerably less expensive is the combined cycle unit. It has the lowest heat rate (i.e., the highest efficiency) and provides a much lower operating cost than either gas turbines or steam units that burn oil or gas.

The converse is true of steam units that burn coal and of nuclear units. Although neither of these units has a low heat rate, the cost of coal and nuclear fuel are much lower than the assumed costs of natural gas or oil. Consequently, the cost of generating a MWh of electricity with coal-fired steam units and nuclear units is much lower than with units that burn gas or oil (given the assumptions we are using).

Based on this information, one might be tempted to ask "why would a utility operate anything other than coal or nuclear generating units?" A similar, and directly related, question may also come to mind: "why would a utility build any type of generating unit other than a coal or nuclear unit?"

But if we attempt to answer these questions now, we are getting ahead of ourselves. In order to meaningfully answer questions such as these, one needs to know more about utility systems. Operating on the premise that anything you create yourself is something you will understand better than something else that is just handed to you, we will take that approach to learning more about utility systems.

Let's Create a Hypothetical Utility System

We will begin the process of "creating" a hypothetical utility system by first creating a profile of the utility customers' electricity usage. After all, utility systems are designed to serve their customers in regard to how these customers actually use electricity. Therefore, it is the logical place to begin.

There are many ways to look at customers' electricity usage. One perspective is to look at electricity usage during one 24 hour day. Another perspective is to look at the total electricity usage during one calendar year. A typical year (i.e., a non-leap year) has 365 days of 24 hours each. Therefore, there are 365 days per year × 24 hours per day = 8,760 hours per year. During those 8,760 hours, there will be one hour in which the highest amount of electricity is used, and one hour in which the lowest amount of electricity will be used. There will also be 8,758 other hours in which the electrical demand falls between these highest and lowest electrical demand values.

The hourly "demand" values for electricity are typically discussed in terms of a megawatt (MW) of electricity and the highest level of electricity usage during a calendar year's 8,760 hours is termed the "peak" demand for electricity or the peak load. Note that even within that hour, the actual electrical demand will vary from minute-to-minute. However, for analysis (and electricity billing) purposes, the load over this peak hour is averaged to provide a more convenient unit of measurement.

The demand for electricity for most utilities is driven in large part by the outside temperature. Consequently, the highest or peak demand for electricity during the year is likely to occur on a summer afternoon on a really hot day or on a really cold winter day. In general, the higher electrical loads typically occur in the summer and winter months, with lower loads occurring in spring and fall months. There may be significant differences in the peak load from one month to another, especially as the calendar moves from one season to the next.

There are also significant differences in the electrical loads from one hour to another during a daily 24 hour cycle. The hourly electrical loads over the course of a day are generally driven by two factors: the outside temperature during each hour and what customers are doing at a given hour (i.e., whether they are active at home or at work during the day and evening, or whether they are asleep at night and in the early morning). For these reasons, electrical loads typically increase during the day, then decrease during the night.

We begin to create our hypothetical electric utility system by assuming that our utility system is a summer peaking utility.* We then construct a representative electrical load for a summer day and we select the one day during which our utility's highest/peak load for the year is reached. A peak load of 10,000 MW is assumed for our utility. A graphic depiction of hourly loads for a given period of time is referred to as a "load curve" (or "load profile.") Figure 2.1 presents a daily load curve of our utility system's electrical load during the 24 hours of its summer peak day. The first value on the left represents 1 a.m. and the load for the remaining 23 hours of the day are then shown as one moves from left to right across the figure.†

* The term "utility" is occasionally used instead of "utility system" in places throughout the book. This is done merely for convenience. However, the reader should always keep in mind that an electric utility is truly a complex system of numerous generating units (as well as transmission and distribution lines, etc.). As we shall see in later chapters, as new resources are added to this system, there are numerous impacts to the operation of the existing generating units that are parts of the system.

† To ensure that the assumed loads for our hypothetical utility are representative of actual utility loads, the hourly electrical load values presented in Figure 2.1 are based on a particular summer-peak day load of FPL's. The FPL data was used as a starting point from which to create the values for our hypothetical utility. FPL's highest load that day was slightly greater than 21,000 MW. For our hypothetical utility, this value has been reduced in the graph to a highest load of 10,000 MW and all other hourly loads have been reduced proportionally.

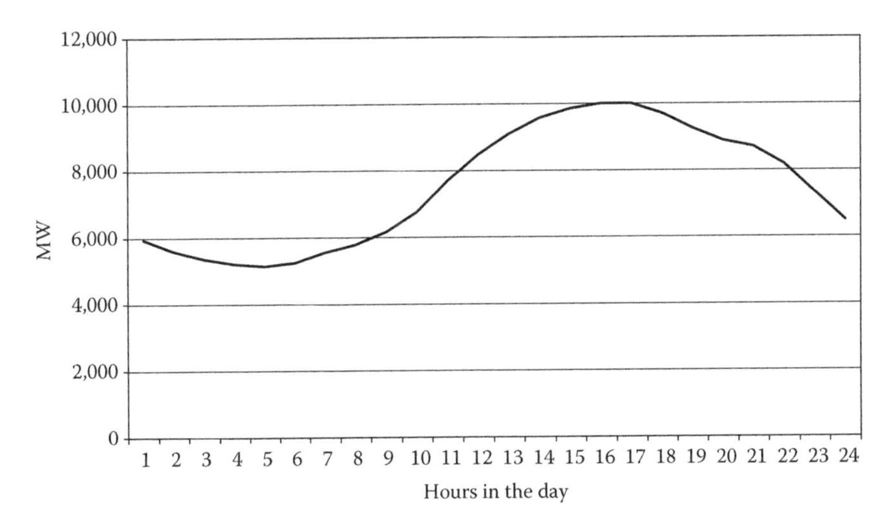

FIGURE 2.1
Representative peak day load curve for a hypothetical utility with a peak load of 10,000 MW.

When viewing this figure starting from the 1 a.m. hour "starting point" on the left-hand side, we see that electrical loads decrease in the early morning hours when outside temperatures are cooler and people are asleep, increase during the day as the outside temperature rises and people are active, then decrease again as darkness falls and people return to sleep.

The key item to notice in Figure 2.1 is how much variation in electrical load there can be in even a single 24 hour day. For this day on which the highest load for the year (10,000 MW) is reached (at hour 17 or 5 p.m.), the electrical load still drops significantly in the nighttime hours to approximately 5,150 MW (at hour 5 or 5 a.m.). Therefore, the highest load of the day is almost double the lowest load of the day.

Expanding this look to all 8,760 hours in a year gives another perspective of how electrical loads vary over the course of the year for our utility system. Figure 2.2 presents a graphic representation of these electrical loads for each of the 8,760 hours in a typical calendar year in a special version of a load curve that is typically referred to as "an annual load duration curve." In an annual load duration curve, the electrical loads are not shown in chronological order; i.e., the loads do not start with the 1 a.m. load on January 1, then the 2 a.m. load on January 1, etc. Instead, the highest load experienced during the year, regardless of what month, day, or hour of the day the highest load occurred on, is represented by the left-most value on the graph. This is the 10,000 MW value we just saw on the summer day graph in Figure 2.1.

The second highest load experienced during the year is then presented immediately to the right of the 10,000 MW value. This sequence is followed until the lowest load during the year (approximately 3,400 MW) is presented

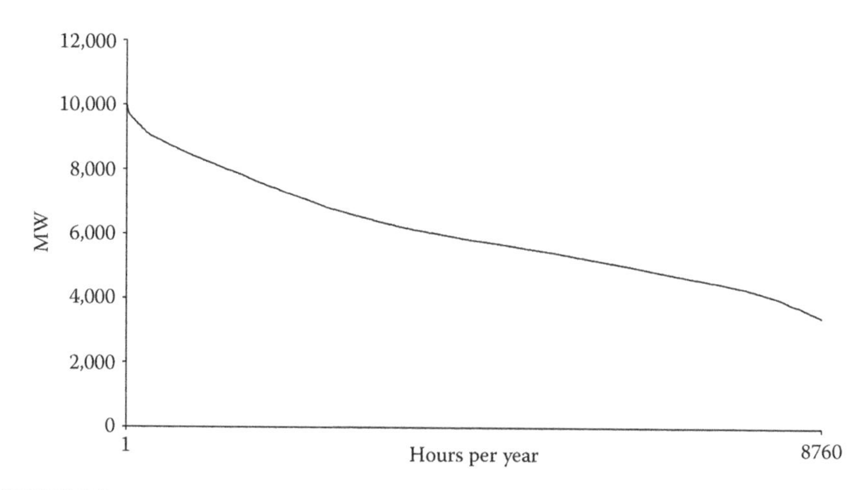

FIGURE 2.2
Representative annual load duration curve for a hypothetical utility with a peak load of 10,000 MW.

as the last value on the far right of the graph. Therefore, our utility system obviously has loads that are even lower than the 5,150 MW lowest daily load for our summer peak day that was shown previously.

As can be seen from comparing the previous load curve for the summer peak day with the annual load duration curve, the shape of the graph of the annual load duration curve is significantly different than the shape of the graph for the summer peak day*. From this curve, one can see there is even more variation in electrical loads over the course of a year (a range of 3,400–10,000 MW) than was for the summer peak day. In addition, there are many hours that can be labeled as relatively high-load hours and many other hours that can be labeled as relatively low-load hours. If one looks across the graphed line to where the approximate mid-point of the curve is, we see that this is roughly where the graphed line crosses the 6,000 MW level. What this means is that, for our hypothetical utility system, approximately half of the hours in the year will have an electrical demand of greater than 6,000 MW and approximately half of the hours in the year will have an electrical demand of less than 6,000 MW.

Therefore, in regard to the operation of the utility system, our utility will be serving a very wide range of electrical demand levels (3,400–10,000 MW)

* Figure 2.2 also uses FPL data as a starting point from which to create the annual load curve for our hypothetical utility. The 8,760 hourly load data from FPL was again reduced so that the highest hourly load was 10,000 MW and all other hourly loads were reduced proportionally.

TABLE 2.4

Assumed Capacity (MW) by Type of Generating Unit for the Hypothetical Utility System

Type of Generating Unit	MW	% of Total MW
Gas turbines	1,000	8.3%
Steam-oil/gas	3,500	29.2%
Combined cycle	3,000	25.0%
Steam–coal	3,500	29.2%
Nuclear	1,000	8.3%
Total =	12,000	100.0%

over the course of a year. Our utility will also be serving both a relatively high demand level (more than 6,000 MW) and a relatively low demand (less than 6,000 MW), about half of the hours in the year.

In order to see how this "profile" of hourly loads affects the operation of the hypothetical utility system's generating units, we will next "create" a set of generating units for the utility.

We start by assuming that our hypothetical utility system currently has one or more generating units of each of the five types of generating units previously discussed: (1) steam-oil/gas, (2) steam–coal, (3) gas turbine, (4) combined cycle, and (5) nuclear. The number of MW of generating capacity that we will assume are provided by each type of generating unit for our hypothetical utility system is based on a rough composite of a number of actual utility systems and not on any one specific utility system. The assumed number of MW supplied by type of unit is shown in Table 2.4.

The astute reader (you again) will ask the obvious question: "why does the utility have 12,000 MW of generating units if its highest load is only 10,000 MW?" The answer to the question has two parts. First, recall that regulated utilities are under an obligation from their respective regulatory agencies to provide electrical service to all customers at all times. Second, generating units, like automobiles and many other complex machines, both break unexpectedly (thus requiring time to be fixed) and need periodic planned maintenance. If a specific generating unit either unexpectedly breaks, or if it has been taken out of service for planned maintenance, it cannot operate and supply electricity during that time period.

Therefore, a utility system must have more generating units, or electric generating capacity (MW), than its highest load alone would seem to dictate. This additional generating capacity or "reserves" will be further discussed in Chapter 3.

Now Let's Operate Our Hypothetical Utility System*

Utility System Operation on the Summer Peak Day

Now that we have made an assumption regarding the amount of capacity that each type of generating unit can supply to our utility system, we can gain some insight into how the utility will actually operate those generating units. First, we recall that the representative fuel-based operating costs of the different types of generating units were previously presented in Table 2.3. This table showed that nuclear generating units were the least expensive units to operate at $6.60 per MWh. Therefore, the utility will operate nuclear units as often as possible. In utility parlance, the nuclear units will be "dispatched" first to meet the utility's electrical load before any other type of generating units are dispatched.

As previously shown in Table 2.4, we have assumed there is 1,000 MW of nuclear capacity on the utility system. By superimposing this amount of capacity on the summer peak day load duration curve previously presented in Figure 2.1, we get a picture of how much of the utility's load during the 24 hour period of the summer peak day can be supplied by nuclear generation. This picture is presented in Figure 2.3.

As shown in this figure, the shaded area at the bottom of the figure represents our utility system's nuclear generation capacity. It is clear that the

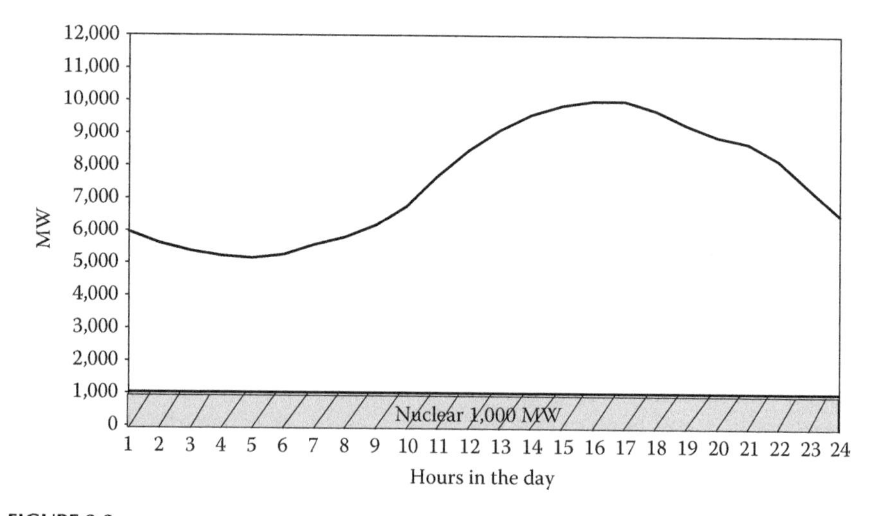

FIGURE 2.3
(See color insert.) The potential contribution from nuclear generation during the summer peak day.

* I realize that you may not have operated one of these babies before, but relax, you can handle it.

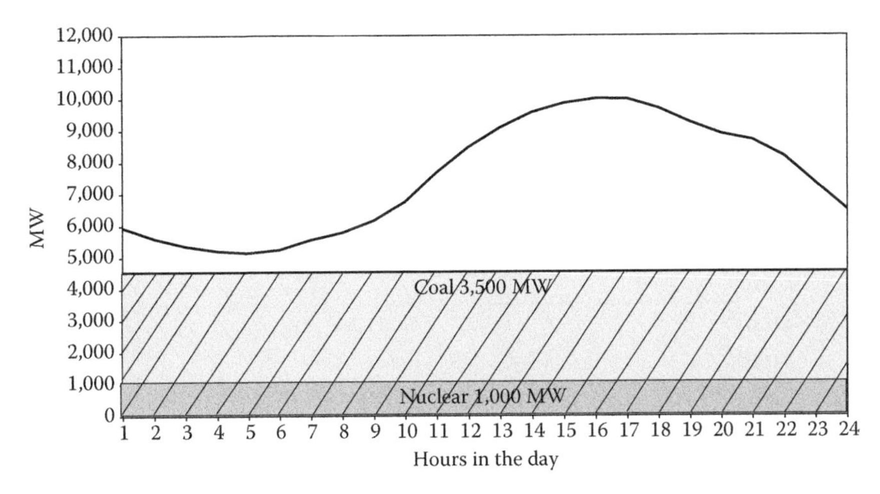

FIGURE 2.4
(See color insert.) The potential contribution from nuclear and coal generation during the summer peak day.

nuclear capacity does not "reach up" to the actual load level that is projected for any hour. Thus, nuclear capacity can supply only a relatively small portion of the electrical load even during the early morning hours (hours 1 through 8, or 1 a.m. to 8 a.m.) when the electrical load is at its lowest. (Recall that our utility system's lowest load on this peak summer day was approximately 5,150 MW and that this utility system's nuclear capacity can supply only 1,000 MW.)

Nevertheless, assuming that the nuclear generating capacity is available to operate (i.e., the nuclear units are not broken nor have the units been taken out of service for planned maintenance/refueling), the utility will operate the nuclear generation during all 24 hours because of the very low operating cost of nuclear units.

Our utility must now dispatch at least one more type of generating unit in order to meet the demand for electricity. Returning to Table 2.3, it is apparent that coal-fired steam units are the next most economical generation to operate at $20.00 per MWh. From Table 2.4, we see that our utility is assumed to have 3,500 MW of coal-fired generation. Figure 2.4 now shows how much of the utility's load during the 24 hour period is supplied by the combination of nuclear and coal.

As shown in this figure, the combination of our utility system's nuclear and coal generation (a total of 4,500 MW) represented by the total shaded area in the figure still doesn't meet even the lowest electrical load (5,150 MW) experienced during this peak day.

We will now "cut to the chase" and superimpose all of our utility system's capacity that can be supplied by each of the types of generation on this summer peak day load duration curve. This is presented in Figure 2.5.

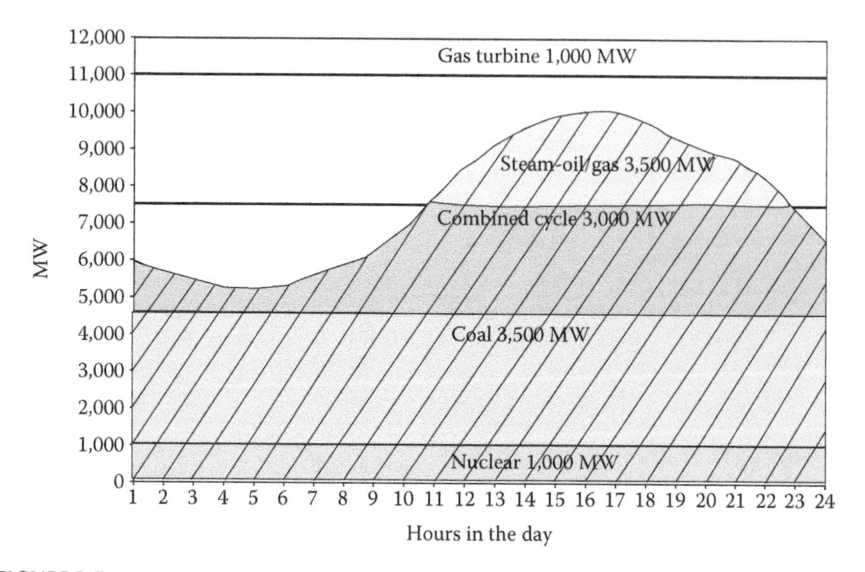

FIGURE 2.5
(See color insert.) The potential contribution from all types of generation during the summer peak day.

As shown by the third shaded area in this figure, the potential contribution from the next most economical type of generating units, combined cycle, can potentially be used in combination with nuclear and coal units to meet the load on the summer peak day from approximately midnight through 11 a.m. and then again during the hours from about 11 p.m. to midnight. As shown by the fourth and final shaded area, steam-oil/gas units would be needed in order for our utility to meet the remaining load during the hours from approximately hour 11 to hour 23 (11 a.m. to 11 p.m.).

The figure also seems to indicate that there is a "cushion" of exactly 1,000 MW from the top of the load curve (10,000 MW) to the bottom of the gas turbine band (11,000 MW). The logical question to ask is: "why aren't the gas turbine units used at all?"

The answer to the question is that the figure presents an idealized picture. What is being shown on the graph for the MW of each type of generating unit represents the maximum possible contribution that each type of generating unit can make.

The values shown on the figure assume that (i) no generating units of any type have broken down and are, therefore, not in service on this summer peak load day, and (ii) no generating units have been taken out of service for planned maintenance on that day.* But having generating units out of ser-

* Utilities typically strive, when possible, to schedule planned maintenance so that this work does not occur on peak load days. Thus one would not expect planned maintenance to diminish the generating capability of any type of generating unit on the summer peak day.

vice because the units have unexpectedly broken can easily happen on peak load days. Although a generating unit could break on any day, it is somewhat more likely to happen on a peak or other high load day. This is because a peak or high load day is frequently preceded by a number of high load days, weeks, or even months during which many of the utility system's generating units are operated at very high levels for long hours. This increases the chances that generating units may break.

If units of any specific type were to break during this peak load day, the amount of MW that is shown for this type of generating unit would be reduced, thus shrinking the height of the "band" shown across the figure for that type of generating unit.

For a moment, let's assume that on our utility's summer peak day, none of the nuclear, coal, or combined cycle units are out of service (i.e., they are not out for planned maintenance, nor have they broken). Therefore, the full capacity of these types of generating units (1,000 MW for nuclear units, 3,500 MW for coal units, and 3,000 MW for combined cycle units) are fully operational during the day. But let's also assume that a number of the steam-oil/gas units break during the day, reducing the available capacity from these units from their full capacity of 3,500 down to 2,000 MW. Therefore, 1,500 MW of steam-oil/gas capacity is unavailable that day. This case is depicted in Figure 2.6.

In this case, the former height of 3,500 MW for the steam-oil/gas units now shrinks to a height of 2,000 MW. In turn, the unchanged height of 1,000 MW for gas turbines now "drops down" in the figure so that some gas turbine capacity is used to meet our utility's very high loads during the mid-to-late afternoon hours.

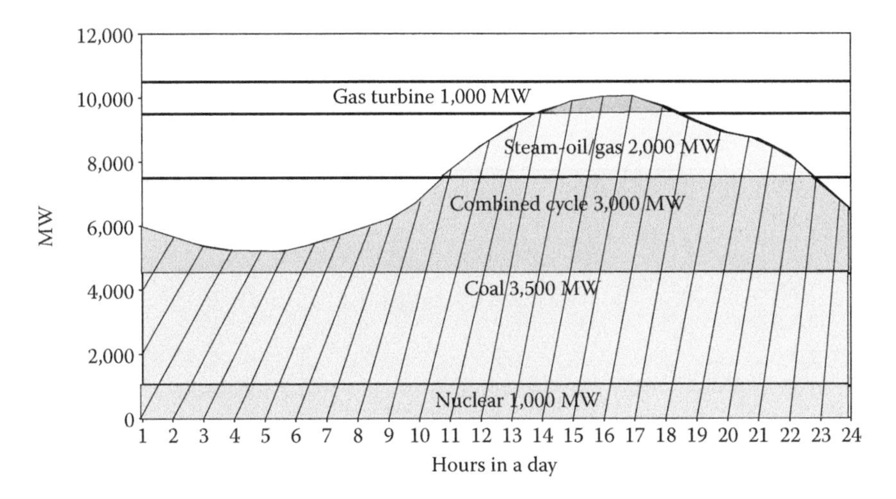

FIGURE 2.6
(See color insert.) The potential contribution from all types of generation during the summer peak day (assuming a reduction of 1,500 MW of steam-oil/gas capacity).

This example in which our utility system needs to operate gas turbines due to the unavailability (whether due to breakage or planned maintenance) of units in the other types of generating units that are more economical to operate than gas turbines, is actually quite representative of when and why utilities operate these expensive-to-operate gas turbine units.

Utility System Operation over the Course of a Year

Based on the insight we have gained from the discussion just completed regarding how our utility's different types of generating units will operate on the summer peak day, we can extend that look to how these types of generating units will be operated over the course of a year. Returning to the annual load duration curve presented earlier in Figure 2.2, we now make two changes to that figure.

First, we superimpose the maximum capacity of each type of generating unit (as we just did for the summer peak day load curve). Second, we shade the area that is actually under the annual load duration curve line in each of the bands representing the amount of capacity offered by each type of generating unit. The shaded area for each type of generating unit that is under the annual load duration curve line shows how much of each generating unit type is being operated at any given hour and helps demonstrate how many hours each type of generation will actually operate over the course of a year for our utility system. The result in presented in Figure 2.7.

This picture gives us the advantage of a much broader perspective (8,760 hours instead of 24 hours) from which to view how the generating

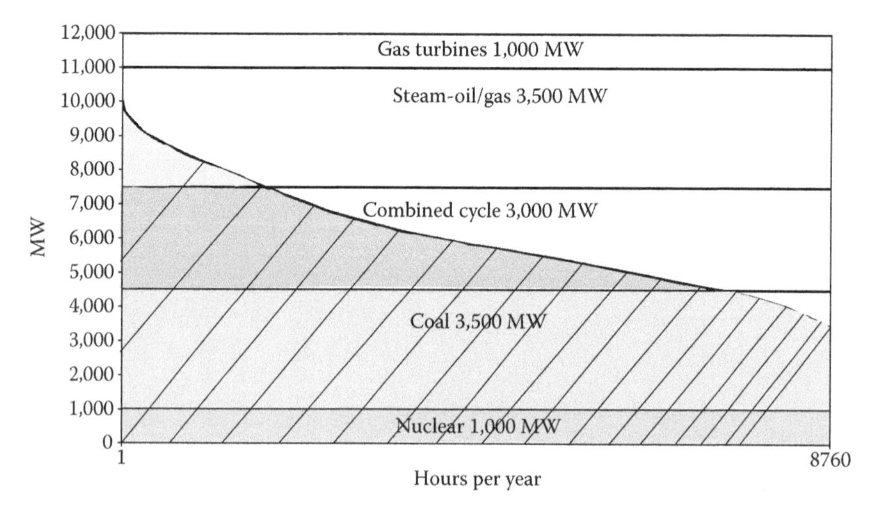

FIGURE 2.7
(See color insert.) The potential contribution from all types of generation during the course of a year.

units of our utility will be operated. Reminding ourselves that this picture is an idealized one (no unit breakage or planned maintenance is represented), we can use this broader perspective to draw certain conclusions about how our utility system's different types of generating units will be operated:

1. Due to their lowest operating costs, all of the nuclear generating capacity will be used all of the time these units are available to operate. (This is shown by the entire nuclear generating capacity band being shaded for all hours.)
2. The next most economical-to-operate type of generation, steam–coal units, will have a very high percentage of their capacity being used in all hours of the year. In addition, all of the coal-fired capacity will be used except for the very lowest load hours of the year. (This shows up as the small sliver in the coal generation band at the far right side that is unshaded.)
3. Combined cycle unit generation will be operated many of the hours of the year. The higher the load gets, the greater the amount of the combined cycle generation that will be used. As seen by the relatively large unshaded area to the right of the load curve that is within the combined cycle band, only a portion of the combined cycle generation will be operated during many of the hours in the year. There are also some hours in the year in which no combined cycle generation is used.
4. The steam-oil/gas capacity will play a much smaller role in the utility's operation over the course of a year. The small shaded area for steam-oil/gas generation on the upper left-hand side of the graph shows that these steam-oil/gas units will be used only in those hours when the electrical loads become quite high.
5. Finally, although the gas turbine capacity is not shaded in this graph, these units will have a role to play as discussed earlier for the summer peak day load curve. As significant amounts of other types of generating units become unavailable, the gas turbine units may be used, but only in a relatively few hours during the year. These hours typically represent the very high loads in the extreme upper left of the graph.

These general observations for our hypothetical utility system will be valid for many, if not most, electric utility systems. In regard to the importance of this low-cost-to-high-cost "dispatch order" of when the types of generating units are operated, a simple example of the differences in fuel costs that can occur between the operation of certain types of generating units is helpful.

In this example, let's assume that our hypothetical utility's 1,000 MW of nuclear generating capacity is out of service for 24 hours and that our utility supplies the 24,000 MWh of energy (=1,000 MW × 24 hours) that the nuclear capacity would have supplied with some of its steam-oil/gas units. We further assume that this "make up" energy will be supplied by these units burning only natural gas. What is the additional fuel cost from such an occurrence?

Using the assumed heat rate (11,000 BTU/kWh) and fuel cost ($0.60/mmBTU) for the nuclear capacity, we see that the cost of supplying 24,000 MWh, or 24,000,000 kWh, from the nuclear capacity would have cost $158,400 (=24,000,000 kWh × 11,000 BTU/kWh × 1 mmBTU/1,000,000 BTU × $0.60/mmBTU). However, because the nuclear capacity is out of service for this 24 hour period, the 24,000 MWh that will now be supplied by some of the steam units burning natural gas will cost $1,440,000 (=24,000,000 kWh × 10,000 BTU/kWh × 1 mmBTU/1,000,000 BTU × $6.00/mmBTU).

Therefore, our utility system will incur an additional $1,281,600 of additional fuel cost (=$1,440,000−$158,400) for this one 24 hour period. This simple example points out the reason for, and the importance of, operating a utility's generating units in a dispatch order of low-cost-to-high-cost because the utility's customers will typically be charged the full fuel costs of supplying electricity.*

Utilities also refer to the dispatch order of generating units in terms other than "low cost" and "high cost." A common naming convention is used instead that inherently recognizes the operating cost of the generating units by referring to how much of the time during the year a generating unit will operate. For example, nuclear units are typically referred to as "baseload" units because they continually operate to serve load that is always present for the utility system; i.e., "base" loads, due to the low operating cost of these units. For most utilities, coal-fired units are also considered as baseload units. At the other end of the spectrum, the gas turbines are referred to as "peaking" units because they typically operate only when electrical loads are at or near the utility's peak load due to their high operating costs.

In between these two types of generating units are the combined cycle units and the steam-oil/gas units. These types of units are often referred to as "intermediate" units because they typically operate at loads that fall between base loads and peak loads. (On certain utility systems, very efficient combined cycle units are also operated as baseload units.)

The terms "baseload," "peaking," and "intermediate" are meaningful because their use helps to quickly give a sense of how many hours of the year these units will operate or what "roles" these units will play in meeting load during the year. The roles that baseload units and peaking units will play in meeting customers' demand for electricity are simple. They function at opposite ends of the operation spectrum. At one end of the range,

* The example also points out the large fuel-savings benefits that can be obtained from nuclear generating units. Such units are expensive to build, but can generate enormous fuel savings.

baseload units will be operated during virtually every hour of the year in which these units are available to operate. At the opposite end of the range, peaking units will only operate a very few hours of the year in which electrical demand is very high and/or a large amount of other generating capacity is unavailable.

However, the role the intermediate units will play is actually much more interesting in regard to utility system operation. (And, as we shall see later, these units play a very important role in analyses that are conducted in order for the utility to plan for its future resource additions.) The role of the intermediate units for our utility system (i.e., combined cycle and steam-oil/gas units) can be seen graphically by returning to Figure 2.7 and putting a finger on the load duration curve itself.

Then, as one "follows" the curve up or down the load level (i.e., to the left or to the right on the load curve line), one would readily see that during almost the entire length of the curve (except for the extreme bottom right-hand portion of the curve), the intermediate units are the means by which the utility meets load that is continually varying from one hour to the next. This is carried out by using more or less capacity from the intermediate units.

Because the intermediate units for our utility system burn oil and gas, the volume of oil and gas that is burned varies from hour to hour to meet electrical demand that is continually changing. From virtually any point on the load duration curve, if we go up the curve (toward the upper left-hand corner) to the next higher load point, more oil and/or gas are being burned in either combined cycle, or steam-oil/gas, units. Likewise, if we go down the curve (toward the lower right-hand corner) to the next lower load point, less oil and/or gas is being burned in these same units.

In utility parlance, oil and gas can be referred to as our hypothetical utility system's primary "marginal" fuels because the amount of oil and gas used varies from hour to hour as the load goes up or down across almost all of the hours in the year. (The sole exception is during the relatively few hours of low load shown on the extreme right-hand side of the curve when coal serves as the marginal fuel). This is not the case for nuclear fuel, or for coal during most of the hours in the year, because the nuclear and coal-fired units that use these fuels operate as many hours as they are available to operate because of their low operating costs.*

As we shall see later in the book, the addition of new resources to our hypothetical utility system will have the most impact on our utility's oil and gas usage. Stated another way, the addition of these new resources will have the greatest impact on the number of hours that our utility's oil- and gas-fired existing intermediate and peaking generating units will be operated.

---

* Another utility with a larger amount of coal-fired generation would likely have coal as a marginal fuel for more hours in the year than shown for our utility system. For our discussion, the key point is that fossil fuels—oil, gas, or coal—will typically be a utility's marginal fuels. For our hypothetical utility, oil and gas will be the predominant marginal fuels.

This will be true regardless of whether the new resources to be added are Supply options (new generating units) or DSM options. These effects will be seen in three different areas: (i) the utility's system fuel usage, (ii) the utility's system air emissions, and (iii) the economics of selecting one new resource option versus another resource option.

But we are again getting ahead of ourselves. As I said, we will return to these topics in later chapters. Now it is time to back up and see what we have learned about the basic operation of our utility system, then peek around the corner to see what is coming next.

So What Have We Learned and Where Do We Go Next?

We have actually covered a lot of ground so far. We have discussed three facets of electric utility systems: (i) the electrical load, (ii) the type of generating units that are likely to exist for a utility system, and (iii) how those generating units will be operated in order to meet the electrical load. These three facets are fundamental to understanding both how utility systems work and, as we shall see later, how a utility decides what new resource options should be added to the system.

We have also illustrated these three facets through the use of a hypothetical utility. From the previous discussions that made use of our hypothetical utility and its electrical load, we can summarize what we have learned in the following five points:

1. Electrical load varies significantly for an electric utility both from hour to hour on a given day (as shown in the load curve for the summer peak day) and over the course of a year (as shown in the annual load duration curve).
2. There are five basic types of generating units (steam-oil/gas, steam–coal, combined cycle, gas turbine, and nuclear) that typically make up the bulk of the electrical generating capacity for a utility.
3. The operating costs for the five types of generating units vary significantly and utilities operate these units to the extent possible (subject to equipment maintenance, contractual, environmental, etc. constraints) so that the generating units with the lowest operating costs are used the most, and the units with the highest operating costs are used the least.
4. As a consequence, nuclear units (with the lowest operating costs) will operate as much as possible followed closely by coal-fired units. Conversely, gas turbines (with the highest operating costs) will

operate as little as possible. This leaves steam-oil/gas units, and gas-fired combined cycle units, for our utility as the primary units whose operation is dictated by the amount of load the utility must serve at any given hour.

5. Because the operation of both steam-oil/gas units, and the combined cycle units that burn natural gas, play the role of "ramping up" or "ramping down" to meet ever changing load levels over almost all of the hours in a year, the amount of oil and gas that is used is also constantly changing as the load changes. For this reason, oil and gas are referred to as the primary marginal fuels for our hypothetical utility system. (As previously mentioned in an earlier footnote, other utilities may have coal as a marginal fuel for more hours in the year than is assumed for our hypothetical utility system.)

These summary points provide a good overview picture of what our hypothetical utility system looks like in regard to its electrical load and generating units, plus how it will operate its generating units to meet the electrical load. Stated another way, we know how our utility system will meet its electrical load, at least for the year we have just examined.

In this chapter, we created a hypothetical utility system as a tool to assist us in subsequent chapters. In these chapters, we will use our utility system to help discuss and illustrate various electric utility concepts, issues, and analytical approaches. These discussions will be assisted by actually performing calculations using our hypothetical utility system.

We should also again note that what is really important in the discussions that follow are the concepts, issues, and analytical approaches that will be introduced and discussed, not the results of the actual calculations. This is because the calculation results will almost certainly differ if we had constructed our hypothetical utility's system of generating system and load patterns in a different way and/or used significantly different heat rate and fuel cost assumptions for our hypothetical utility system.

This leads me to introduce the first of my "Fundamental Principles of Electric Utility Resource Planning":

> **Fundamental Principle #1 of Electric Utility Resource Planning: "All Electric Utilities are Different"**
>
> **Each electric utility is different in regard to (at least) its electrical load characteristics and its existing generating units. Therefore, when faced with a particular problem or issue such as "which resource option is the best selection?," the correct answer for one electric utility may not be the correct answer for another electric utility.**

Of my Fundamental Principles that are introduced in this book, I believe this may be the most important to keep in mind, particularly for those who seek to influence the future direction of the electric utility industry through legislation and/or regulation.

In Chapter 3, we place our hypothetical utility system in a situation in which it needs to make a decision regarding the resources with which it serves its customers. We then begin an overview discussion of how resource planning is actually done by introducing three basic questions that utility resource planning must always answer.

3

Overview of Utility Resource Planning

In this chapter, an overview of the various aspects of utility resource planning is presented. The objective of this chapter is to simply introduce basic concepts that will be used in the remainder of the book. These concepts will be discussed in more detail in subsequent chapters.

One More Assumption Regarding Our Hypothetical Utility System

When discussing resource planning for an electric utility, it is helpful to use a specific example of why the utility may need to make a resource decision. There are a variety of reasons why a particular utility may need to make a decision regarding one or more resource options. A partial list of these reasons would include the following possibilities:

- A growing number of customers, and the accompanying growth in electrical demand and usage, that the utility must serve;
- Recognition that one (or more) existing electric generating unit is nearing the end of its useful life, necessitating replacement of this resource in order to maintain system reliability;
- Increasing costs of fuel for a utility system with older, less fuel efficient, existing generating units that result in increasing energy costs for the utility's customers; and,
- The introduction of new environmental regulations that will result in unacceptably high environmental compliance costs for the utility's customers unless new or more efficient resources are added to the utility system.

For purposes of the discussions that follow, we make the assumption that the reason our hypothetical utility system needs to make a resource decision is that the number of customers, and their accompanying electrical demand and usage, which the utility must serve is now forecasted to increase. This assumption has been chosen for two reasons. First, it allows a somewhat simpler discussion of the concept of "reliability analyses" (that will be discussed shortly) than might be possible if another reason for having to make a resource decision were chosen for use in our discussion. Second, virtually all utilities have faced increasing growth in customers and electric load in the past. Therefore, the use of this assumption is helpful in understanding how an electric utility "grew" into its present (or recent) system of electric generating units and other resources.

However, it is important to note that the resource planning concepts and approaches discussed in the remainder of this book for our hypothetical utility system that is assumed to face increasing growth are also generally applicable when a utility faces other reasons for having to make a resource decision. For example, regardless of the reason for why a utility must make a resource decision, the utility resource planner is faced with three basic questions.

Three Questions Utility Resource Planning Must Always Answer

As just stated, our utility system now forecasts that it is facing an increasing number of customers in the coming years. Assuming all else equal, then the electrical load that the utility must serve will also increase. Does the utility need to take any actions in order to meet this increased load? If so, when must these actions be taken? In utility parlance (you're starting to feel like a utility insider, aren't you?), the utility is trying to determine if it has a "resource need"; i.e., a need to add new resources in the future to meet this increased load.

In looking at this situation, a utility planner is faced with the following three questions which must always be answered:

1. When does the utility need to add new resources?
2. What is the magnitude (MW) of the new resources that are needed?
3. What is the best resource option with which to meet this need?

The first and second questions are not directly concerned with economics.* Instead, the focus is solely on the timing and magnitude of when new resources need to be added. Only if the answers to the first and second

* As we shall soon discuss, the first two questions relate to whether the utility system meets one or more pre-established "reliability" criteria. Economic considerations may enter the picture when these reliability criteria are established. However, once these criteria are established, the first two questions are typically answered by performing calculations that do not involve economic considerations.

questions—when are new resources needed and how much is needed—indicate that a significant amount of new resources are needed in a near enough time frame so that the utility must make decisions soon, does the resource planner need to move on to the third question. The answer to the third question is heavily focused on making the best choice for all of the utility's customers from both an economic perspective and a non-economic perspective.

We will provide an overview of how a utility approaches finding answers to these three questions in this chapter. Then, in subsequent chapters, we will provide examples of how these questions might actually be answered for our hypothetical utility system. We will start by examining the first and second questions. And, as you might guess, these two questions ("when does a utility need to add new resources?" and "what is the magnitude of the needed resources?") are related.

In the case of a utility facing increasing electrical demand, as is the case for our utility system, utilities usually address these two questions jointly and refer to the analyses designed to answer these two questions as "reliability analyses." In conducting reliability analyses, utilities use certain standards or criteria to gauge how reliable their utility system is projected to be. These reliability standards/criteria are frequently mandated by the utility's regulatory authority. As we shall see, because our utility system's electrical load is projected to increase, these criteria will be examined in light of the forecasted load growth to see if/when new resources need to be added to ensure that the utility continues to meet its reliability criteria.

Reliability Analysis: When Does a Utility Need to Add New Resources and What Is the Magnitude of Those Needed Resources?

Utilities typically make use of at least two perspectives in reliability analyses to ensure that a comprehensive picture of the utility's future resource needs is captured. Our discussion will focus on two perspectives commonly used at the time this book is written. Both of these perspectives are concerned with projections of the utility system in future years.

The first of these perspectives typically focuses on a specific hour for two days in the year: the hour on the one summer day in which the electrical load during the summer is projected to be the highest, and the hour on the one winter day in which the electrical load during the winter is projected to be the highest. These two hours are called the summer-peak hour and winter-peak hour, respectively. Analyses that use this perspective are generally referred to as "deterministic" analyses. The most commonly used form of a deterministic analysis is probably "reserve margin" analyses.* This perspective has the advantage of being easy to explain, and the analyses are performed on a simple spreadsheet or even with a calculator.

The second perspective examines all of the days in the year to determine the probability that generating units on the utility system may fail in such a manner that the utility will not have enough generation available to meet electrical load at some point during the year. These analyses are referred to as "probabilistic" analyses. These analyses are more complicated than deterministic analyses and require the use of fairly sophisticated computer models.

Reserve Margin Perspective (Simple to Calculate)

As previously mentioned, a reserve margin perspective looks at the highest hourly load (the peak hourly load) that is projected to be experienced in the summer and in the winter. The intent is to determine if the utility can safely meet these two peak loads with the resources it is projected to have at that time.

The criterion used in this type of analyses is called the "reserve margin criterion," and it is expressed as a percentage. The percentage represents the amount of generation capacity (MW) that a utility has on its system that is in excess of the highest projected load that the utility is expected to serve in both the summer and winter seasons for a given year. Values used as the reserve margin criterion for utilities often fall in the 15%–20% range.† This criterion is set at a level designed to ensure that utilities will have more generation capacity than their highest projected load, so the utility can still serve the load even if some generating units break on these peak load days and/or the load is higher than projected.‡

A reserve margin criterion is a minimum threshold criterion. In other words, a utility with a 20% reserve margin criterion is deemed to need additional resources if their projected reserve margin value for a given year drops

* There are variations of reserve margin analysis that are also in common use. (One such variation is termed capacity margin analysis.) These variations are similar to reserve margin analysis in their basic concept and there is little difference in the actual calculations. For these reasons, and for the sake of simplicity, our discussion will focus solely on reserve margin analysis.

† This range of values for the reserve margin criterion has been established over the years based both on actual utility operating experience and economic analyses. The economic analyses examined the inconvenience/cost to customers from interruptions in electric service if the criterion is set too low and the costs to customers from building more power plants if the criterion is set too high. Utility experience and the results of economic analyses have often indicated that a range of 15%–20% is sufficient. A variety of utility-specific considerations are taken into account when determining the actual value to use.

‡ In regard to this first consideration, you may recall that we earlier looked at something similar with our hypothetical utility system in Figure 2.6. Our utility system has 12,000 MW of generating capability and a system peak load of 10,000 MW. We saw that our utility could still meet the 10,000 MW peak load if 1,500 MW of its steam—oil/gas units were broken that day.

TABLE 3.1

Example of a Basic Reserve Margin Calculation

(1)	(2)	(3)	(4)
		= (1) – (2)	= (3)/(2)
Total Generating Capacity (MW)	**Peak Electrical Demand (MW)**	**Reserves (MW)**	**Reserve Margin (%)**
12,000	10,000	2,000	20.0

appreciably below 20%. However, if their projected reserve margin value for a given year is 20% or greater, the utility is viewed to be reliable from a reserve margin perspective and there is no need for additional resources to be added by that given year.

Using our hypothetical utility system, a basic reserve margin calculation is presented in Table 3.1.

(And you thought I was kidding when I said this calculation was simple.)

In this calculation, we see that our hypothetical utility system has a projected reserve margin of 20% for this particular year. If our hypothetical utility system has a reserve margin criterion of 20%, the utility's projected reserve margin value for this year of 20% meets that criterion. If a lower reserve margin criterion of 15% were used by the utility and/or its regulatory authority, the utility's projected 20% reserve margin value for this year would not only meet, but would exceed the 15% criterion. In either case, our hypothetical utility system is deemed to be reliable from a reserve margin perspective.

We now turn our attention to the second perspective commonly taken in utility reliability analyses.

Probabilistic Perspective (Not So Simple to Calculate)

This perspective requires analyses that are not nearly as simple as reserve margin calculations. Sophisticated computer models are typically used to perform these analyses.

The basic approach is to first obtain a variety of projections for each day of the year(s) that is being analyzed. This information includes, but is not limited to, the following: (i) the projected highest (peak) load for that day, (ii) the amount of generation that will be out of service due to planned maintenance on that day, and (iii) the projected likelihood that each individual generating unit will break on that day in which the unit is available to operate (i.e., any day that the unit is not already out of service due to planned maintenance).

The likelihood that a generating unit will break is typically labeled as a "forced outage rate" (FOR) for the generating unit. This value is expressed as a percentage. For example, a 2% FOR for a given generating unit (roughly speaking) means that, after accounting for the annual hours needed for planned maintenance, the unit has a 2% chance of breaking on any day it is expected to be available to operate. All else equal, the lower the FOR, the more reliable the generator is. Therefore, a unit with a 2% FOR is more reliable than a unit with a 3% FOR, one with a 3% FOR is more reliable than one with a 4% FOR, etc.*

Using this information, the probability that the utility will not have enough generation capacity available (after accounting for planned maintenance and the likelihood of breakage as expressed by each unit's FOR) to fully serve the electrical load for each day in the year is calculated. Then the probabilities of not being able to serve the load for each day in the year are summed to derive an annual value. This calculation is often referred to as a Loss-of-Load-Probability (LOLP) calculation.†

The criterion by which the results of LOLP calculations are judged is usually an annual probability value of 0.10 day per year that a utility may not be able to fully provide all of the electricity that customers demand. Loosely speaking, the LOLP criterion of 0.10 day per year means there is a 10% probability that the utility will not be able to meet the electrical demand at some point during the year. (The LOLP criterion is sometimes also discussed in terms of an equivalent 10-year perspective. In this case, the same LOLP criterion is referred to in terms of a probability of 1 day per 10 years. Both ways of expressing the LOLP criterion are typically used to refer to an equal level of reliability for the utility system.)

In contrast to the reserve margin criterion, the LOLP criterion is a maximum threshold criterion. In other words, as long as the LOLP value calculated for a given year is *less than* 0.10 day per year, the utility is deemed to be reliable from a probabilistic perspective. However, if the calculated LOLP value is higher than 0.10 (e.g., 0.15 or 0.25) day per year, the utility is no longer deemed to be reliable for that year from a probabilistic perspective and new resources need to be added.

As previously mentioned, a full LOLP calculation is a complicated one that is performed on sophisticated computer models. Therefore, a full LOLP analysis is not a simple calculation as is the case with the reserve margin reliability criterion. However, it is possible to provide a simple example of how the LOLP calculation process essentially works.

In our example, let's assume that we have a very small utility with three generating units (creatively labeled as Unit 1, Unit 2, and Unit 3). We also assume that each generating unit is 50 MW in size and each has a 4% forced outage rate (FOR). We also assume that on the particular day we shall be examining, the utility has a peak load of 100 MW and that none of the three generating units is scheduled to be on planned maintenance that day.

---

* Note that if a generating unit's FOR were to begin to approach double digits, a utility's thoughts may turn to converting that generating unit into an artificial reef.

† Just as there are variations of reserve margin analysis in common use, there are also variations of the probabilistic reliability analysis approach that are referred to by names other than LOLP. These variations are similar to LOLP analysis in their basic concept and there is little difference in the actual calculations. For these reasons, and for the sake of simplicity, our discussion focuses solely on LOLP analysis.

The question is how likely is it that the utility will be able to serve the peak load of 100 MW that day with its three generating units?

We begin by taking a look at the possible "situations" the utility could find itself in that day in regard to the operational status of the three generating units. There are eight possible situations:

1. Units 1, 2, and 3 are all operational (i.e., all 150 MW are operational);
2. Only Units 1 and 2 are operational (100 MW are operational);
3. Only Units 1 and 3 are operational (100 MW are operational);
4. Only Units 2 and 3 are operational (100 MW are operational);
5. Only Unit 1 is operational (50 MW are operational);
6. Only Unit 2 is operational (50 MW are operational);
7. Only Unit 3 is operational (50 MW are operational); and,
8. None of the three units are operational (0 MW are operational).

At first glance, it is clear that if the utility has a peak load that day of 100 MW, then it will be able to meet that load in four of the eight situations listed earlier because situations (1) through (4) would result in the utility having either 150 or 100 MW of generation operational. But this does not tell us how likely it is that the utility will be able to meet the 100 MW load that day. In order to determine this, we need to make use of the FOR values for each of the three generating units.

Our assumption of a 4% FOR value for each unit essentially means that, for any specific unit, there is a likelihood of 4% that the unit will be broken at any point in time (such as the one day in question). Conversely, it also means that there is a likelihood of 96% that the unit will be operational. We use this information to see what the likelihood is of situation (1) listed earlier occurring in which all three generating units are operational. This is calculated by multiplying the likelihood of Unit 1 being operational, times the likelihood of Unit 2 being operational, times the likelihood of Unit 3 being operational. In other words, the calculation becomes: $0.96 \times 0.96 \times 0.96 = 0.88474$. This means that there is a likelihood of 88.474% that all three units, or all 150 MW, will be operational on that day.

We next determine how likely it is that two of the three units will be operational, and the remaining unit will be broken, as is the case in situations (2), (3), and (4) mentioned earlier. For any one of these three situations, the calculation consists of the multiplication of the likelihood of one unit being operational times the likelihood that another unit is operational times the likelihood that the third unit is broken. This calculation is: $0.96 \times 0.96 \times 0.04 = 0.03686$ or 3.686%. Because there are three such possible situations, this value must be multiplied by a factor of 3 which results in a value of 0.11059 or a likelihood of 11.059% that only two units, or exactly 100 MW, will be operational on that day.

By looking at the likelihood that *at least* 100 MW will be operational on the day, we see that there is a likelihood of 88.474% of 150 MW being operational, and a likelihood of 11.059% of exactly 100 MW being operational. Therefore, the likelihood of the utility being able to meet its 100 MW load on this particular day is: 88.474% + 11.059% = 99.533%. Conversely, there is a likelihood of approximately 0.467% (100% − 99.533% = 0.467%) of the utility not being able to meet its 100 MW load on this particular day; i.e., that one of the situations (5) through (8) will occur on that day. From an LOLP perspective, the important value is the projected likelihood of 0.467% of the utility *not* being able to meet its 100 MW load on this particular day, or an LOLP value for this one day of 0.00467.

A full LOLP calculation essentially repeats this calculation for all 365 days in the year. These 365 calculations account for the highest load expected on each day and for whether one (or more) generating units is scheduled for planned maintenance on each day. The probability values of the utility not being able to meet its projected load for each day are then summed together to obtain an annual probability of not being able to meet its projected load at some point in the year. This sum is then compared to the LOLP criterion which is typically 0.1 (or 0.10000) day per year, as previously mentioned.

In our simple example, the LOLP value for that one particular day of 0.00467 would not, by itself, result in the utility failing the 0.10000 LOLP criterion for the year in question because this daily LOLP value of 0.00467 is a lower value than the 0.10000 LOLP annual criterion value. However, the daily LOLP values for the remaining 364 days would need to be accounted for before the utility would have completed its annual LOLP projection.

As this simple example has shown, a full LOLP calculation for a utility system is a complicated calculation. This is especially true when one considers that utility systems typically: (i) have many more than three generating units (recall that our hypothetical utility system has 10,000 MW of generating compared to the 150 MW used in this simple example), (ii) will have a different peak load value virtually every day, and (iii) will need to have generating units out for planned maintenance on certain days. Therefore, actual calculations of LOLP for our hypothetical utility system would be difficult to present concisely for the purpose of this book. For this reason, the rest of this book will utilize the simpler-to-use reserve margin calculation when discussing a reliability analysis for our hypothetical utility system. But we are not quite ready to leave our discussion of the probabilistic perspective to reliability analyses. This is because we now have two reliability analysis perspectives, deterministic and probabilistic. This leads to a logical question.

Which Reliability Perspective Is More Important?

We have two reliability analysis perspectives that a utility can, and often does, use. A logical question is whether one perspective is more important than the other when a utility is performing its reliability analyses?

Let's assume that a utility uses both perspectives in its reliability analyses of future years. Therefore, the utility develops projections of both reserve margin and LOLP for each year in its reliability analysis. For our hypothetical utility with its growing customer base and increasing electrical load, one of these perspectives will eventually show that its criterion (perhaps a reserve margin criterion of 20% or an LOLP criterion of 0.10 day per year) is not met, or is "violated," beginning in some future year. Regardless of which perspective is violated for that year, the utility will need to add resources in that year of a magnitude sufficient to either increase its generating capacity, and/or lower its projected peak demand, so that the criterion for this reliability perspective is no longer violated.

Let's assume that the utility's projections show that its reserve margin criterion of 20% will be violated in a particular year in the future, but that the LOLP criterion will be met for that same year. In this case, the reserve margin perspective can be said to be "driving" the utility's need for resources. Conversely, if the LOLP criterion was projected to be violated in that year, but the reserve margin criterion was projected to be met, then the LOLP perspective would be said to be driving the utility's need for resources.

In general terms, a utility with a lot of reserve generating capacity on its system, but whose generating units have relatively high FORs, is likely to find that reliability analysis based on the LOLP perspective is driving its need for resources. On the other hand, a utility whose generating units have relatively low FORs is more likely to be driven by the reserve margin criterion. Therefore, it is not unusual for either perspective to drive the need for additional resources for a particular utility depending upon the characteristics of the utility system.

Consequently, there is no universally correct answer to the question of which of the two reliability analysis perspectives is the more important perspective. Either of the two perspectives could be the one that drives a given utility's need for new resources.

Although it is hard to definitively answer which perspective is more important, it is not hard to answer which perspective is more commonly quoted. The reserve margin perspective is most often quoted by utilities when their reliability analyses are discussed. There are a couple of reasons for this.

First, as mentioned before, reserve margin analyses are much simpler to perform than LOLP analyses and are simpler to explain to utility regulators and the public.*

Second, even if LOLP analyses are driving a utility's need for additional resources, a utility can use the results of its LOLP analyses to reset its reserve margin criterion at an appropriate higher level. For example, a utility might change its reserve margin criterion from 15% to 18% if it is consistently found that the LOLP criterion was being violated with a 15% reserve margin criterion, but would not be violated if it were to use a higher 18% reserve

* I am tempted to add "certain utility executives" to this list, but that would be wrong.

margin criterion. In this way, its LOLP-based need for new resources may be "resolved" by the new, higher reserve margin criterion. This allows the utility to focus its reliability analyses on the simpler-to-calculate, and, perhaps more importantly, simpler-to-explain, reserve margin perspective.

Finally, many utilities have undertaken significant efforts over the years to improve the reliability of their generating units, thus driving the FORs of their electric generating units lower and lower. This effort has often resulted in lower LOLP projections, thus making it less likely that LOLP will be driving the utility's need for new resources.

For these reasons, the reserve margin perspective is the analytical approach most often discussed in regard to utility reliability analyses that are used to answer the first and second of the three basic questions a utility resource planner must answer. Therefore, this book will use only reserve margin calculations in subsequent chapters that look at the future resource needs of our hypothetical utility system.

With this overview of how utilities attempt to answer the first and second questions, we now turn our attention to a general discussion of the third question.

Resource Option Evaluation and Selection: What Is the Best Resource Option to Select for a Given Utility?

Assuming that a utility has completed its reliability analyses, and that the results of the reliability analyses show that the utility needs to add a certain amount of new resources by a certain year, the obvious next question is "what is the best resource option to add?" This is where the "fun" really begins in utility planning.* What the utility decides is the best answer to this third question is often the topic of regulatory approval hearings.† For those readers who have never had the pleasure of attending or participating in such events, these hearings typically take place over several days (at least). The hearings come after months of writing testimony and answering written and oral questions under oath. The hearing itself involves oral testimony in which one verbally answers even more questions. These hearings frequently involve input from various organizations who advocate for, and/or against, certain types of resource options. This is what we call "fun," especially if you are a witness in these hearings.

* This assumes a loose (and possibly perverse) definition of "fun."

† The first and second questions: "when does the utility need to add new resources?" and "what is the magnitude (MW) of the new resources that are needed?" are less frequently the topics of regulatory hearings. This is primarily because the reliability standards/criteria that the utility uses to answer those questions are often mandated by the utility's regulatory authority.

But we have a bit more to discuss before you're ready to dress up nicely and take the witness stand as a participant in these hearings. In this section, we will provide an overview of this aspect of utility planning and decision-making. We will do so by briefly discussing four topics that are crucial to an understanding of the third question itself and how it may be answered. These four topics are:

1. The two basic types of resource options;
2. The concept of integrated resource planning (IRP);
3. Economic evaluations; and,
4. Non-economic evaluations.

Two Basic Types of Resource Options: Supply and Demand Side Management Options

In an earlier chapter, we briefly introduced the fact that there are two basic types of resource options which utilities can select to meet a future need for additional resources. The first type of resource option includes all resource options that generate or supply electricity. These options will be referred to as "Supply" options. The list of Supply options include the previously discussed five types of generating units now commonly found on utility systems across the country: steam units that burn oil or gas, steam units that burn coal, gas/combustion turbines, combined cycle units, and nuclear units. The technology for some of these types of units (such as combined cycle units) has steadily advanced to the point where the currently available generating units of that type are significantly more efficient than the existing older units currently found on a utility system.

At the time this book is being written, the list of potential Supply options also includes a number of other types of generating units that are either at, or nearing, technical maturity for large-scale deployment on utility systems. A partial list of these, in alphabetical order, includes

1. Advanced technology coal steam units
2. Biomass facilities
3. Integrated (coal) gasification combined cycle (IGCC) units
4. Photovoltaic (PV) facilities
5. Solar thermal facilities
6. Waste-to-energy facilities
7. Wind turbines

In total, there are usually more than a dozen types of Supply options that may be potentially applicable for a given utility. Each of these types of Supply options shares one common trait: they generate electricity that can, at least theoretically, be used to meet a utility's need for additional resources. From that common point, these options differ significantly in regard to size (MW), cost, fuel use, air emissions, etc. In subsequent chapters, we will select a few of these options for further examination as we work through an analysis of how our hypothetical utility system might meet its growing need for electricity with a new Supply option.*

The second basic type of resource option with which a utility may meet its need for additional resources includes all resource options with which the utility can lower its customers' demand for electricity, particularly at the system's summer peak hour or winter peak hour (because it is the peak hour demand for electricity that is used in reserve margin calculations as previously seen in Table 3.1).

There are at least three paths by which utility customers' demand for electricity can be directly lowered:

1. Federal and/or state government-mandated appliance and lighting efficiency standards and building codes;
2. Voluntary actions taken by customers that may involve forms of electricity use not directly addressed by government-mandated standards/codes or by utility programs; and,
3. Programs offered by the utility to its customers that are designed to help participating customers reduce their electrical demand and usage.

The projected effects of government-mandated appliance and lighting efficiency standards and building codes, plus projections of voluntary customer efforts to conserve (often in reaction to higher electric rates and bills), are typically recognized up front by electric utilities in their resource planning efforts. This is frequently done by incorporating these projected effects into the utility's forecasts of future electric demand (typically referred to as the utility's "load forecast"†). At the time this book is written, these projected impacts are becoming an increasingly large factor in utilities' load forecasts, lowering the projected load below what it otherwise would have been.

Having already attempted to incorporate the projected effects of these two paths to reduced energy use in their load forecasts, utilities then turn to the

* Supply options can take the form of either utility-owned generating units or purchased power contracts. In the latter, another entity owns the generating unit and sells capacity (MW) and/or energy (MWh) to the utility under contractual terms. For simplicity's sake only, our discussions throughout the book will assume that the new Supply options being considered would be utility-owned.

† The subject of "load forecasting" for electric utilities is a complex one that would take many pages (if not a separate book) to adequately describe. Consequently, we will simply use a load forecast in our discussion of resource planning for our hypothetical utility system and not attempt to discuss how this load forecast was developed.

third path: utility programs. These utility programs will be referred to as demand side management (DSM) programs/options. There are two basic types of DSM programs: load management programs and energy conservation programs.

The two basic types of DSM programs that have traditionally been offered by utilities have the following general characteristics:

- Both types of DSM programs are designed to reduce the participating customers' electric load (kW) at the utility's summer and winter peak hours and to lower the total amount of electricity (kWh) the participating customers will use over the course of a year.
- Load management programs generally take one of two forms: (i) direct utility control of customers' appliances/equipment (i.e., the utility can remotely turn off, or otherwise regulate, the equipment's use of electricity during high load periods); or (ii) special electric rate structures (that feature higher prices during peak load hours, and lower prices during low load hours) that encourage a reduction of electricity usage during, and/or a shift of electricity usage away from, the utility's peak hours. In either case, these load management programs often result in relatively small impacts on the total amount of energy the customer consumes over a year compared to energy conservation programs.*
- Energy conservation programs typically do not feature direct control of the customer's load by the utility or offer special electric rate structures. Instead, incentives are typically offered by the utility to encourage customers to purchase a more efficient appliance, install higher levels of insulation, etc. Compared to load management programs, energy conservation programs generally result in relatively larger impacts on the total amount of energy a participating customer consumes over a year, but may result in lower peak hour demand (kW) savings than is the case with load management programs.†

Just as there are numerous Supply options a utility will have to choose from, there is also a very long list of DSM options that may be applicable to a utility based on its location, climate, its customers' electric usage patterns, etc. In fact, the list of potential DSM options may include hundreds of possibilities.

Even from this brief overview of the two basic types of resource options (Supply options and DSM options) a utility has to choose from in order to meet its resource needs, it is readily apparent that these two basic types of

* Load management type DSM programs are also referred to as "demand response" programs.

† Similarly, energy conservation type DSM programs are also referred to as "energy efficiency" programs.

resource options are fundamentally different. One type of option is designed to supply electricity and one type of option is designed to lower the demand for, and usage of, electricity. So how should a utility proceed to compare these two very dissimilar types of resource options in order to determine which option(s) is best for its customers? I'm glad you asked.

Integrated Resource Planning (IRP)

The concept of IRP has been around since at least the late 1980s. The basic concept behind an "integrated" resource planning approach is to ensure that both types of resource options, Supply and DSM, are evaluated on what I will term a "level playing field." This term will be used at various places throughout the remainder of this book so it is useful to explain what it means. This term is used in this book to mean that analyses are performed in a manner that shows no preference or bias toward either type of resource option. Utilizing an unbiased IRP approach for analyses of resource options helps ensure that a wide variety of resource options is examined, thus increasing the likelihood that the best possible choice of resource options will be selected for a utility's customers. The logic inherent in using such an analytical approach is readily apparent and by the early 1990s, utilities across the country regularly performed analytical approaches that were designed to implement the IRP concept.

However, also about that time, a number of states and utilities began a move toward an unregulated utility environment (and its underlying promise of lower electric rates that would, hopefully, be brought about by greater competition). As a result, these states and utilities largely abandoned the IRP concept because these utilities had been "split" apart into separate companies. One company would generate electricity; another company or organization would be in charge of transmitting electricity from the generators to specific geographic regions; then another company would distribute electricity to individual customers within the regions.

In such cases, there was less need to perform any type of planning except for the specific function (generation, transmission, or distribution) each company was responsible for. Furthermore, the two different types of resource options, Supply and DSM, would now typically be handled by two different companies. Typically, Supply options would be handled by a generation company, and DSM options would be handled by a distribution company or other entity. As a consequence, the IRP approach to evaluating the two types of resource options was usually abandoned by those states that had decided to pursue the path of an unregulated utility structure.

However, as previously mentioned, this bold "experiment" of an unregulated utility environment has had mixed results to-date. As this book is written, some of these states and utilities have moved back toward a regulated utility environment in which one utility company not only handles the three main utility functions (generation, transmission, and distribution), but also handles both Supply and DSM options. With that move has come a resurgence of interest in the IRP concept.

At this point, it is useful to provide my definition of IRP:

> **Integrated resource planning (IRP) is an analytical approach in which both types of resource options, Supply and DSM, are analyzed on a level playing field. For each resource option, an IRP analysis accounts for all known cost impacts on the utility system that are passed on to its customers through the utility's electric rates. In addition, noneconomic impacts to the utility system from the resource options are also evaluated. In this way, IRP analyses result in a comprehensive competition among resource options.**

The concept that resource options must *compete* with each other is explicit in this definition of IRP. In fact, this concept is a key principle of IRP analysis. Only in this way can a utility truly identify the best option(s) for its customers. It is necessary to include all of the cost impacts to the utility system that will result from the selection of a resource option because if one does not account for all cost impacts for a particular resource option, the competition can no longer be unbiased.

It is my belief that the IRP approach is the best way to analyze all resource options for electric utilities. That stated, it is instructive to point out that utilities in a number of states that profess to use an IRP approach actually use a modified version of my definition of IRP. This situation can occur for a variety of reasons, but occurs most often because of state regulations that mandate a certain amount of a particular type(s) of resource options to be included in a utility's resource plan. Examples of types of resource options that are currently most frequently mandated at certain levels are DSM and/or renewable energy options. In cases where specific levels of selected resource options such as these are mandated, the utility is no longer utilizing a true IRP approach according to the IRP definition just given.

However, a utility may then apply a modified IRP approach in which it "works around" these mandated levels of selected resource options to determine the optimum selection of Supply and DSM options to meet the *remaining* resource needs of the utility not addressed by the mandated resources. In such cases, the utility is seeking to utilize as much of an IRP approach as is possible given the mandates it faces. Another way to state this is that a restricted IRP approach is used that accounts for certain constraints (i.e., the mandates).

We will return to aspects of my IRP definition at various times in the book as we discuss various types of evaluations or analytical approaches. In Chapter 8, we will also discuss the issue of mandates/constraints that prevent a true IRP approach from being used and why a decision to step away from an IRP analytical approach can create problems for utilities and, most importantly, their customers.

With this basic definition of IRP analyses in hand, we now turn our attention to the first of two basic types of evaluations a utility undertakes in its IRP work: economic evaluations.

Economic Evaluations

Yes, we have now gotten to the point at which we talk about what these resource options cost. And when discussing resource options for electric utilities, we are talking about very large sums of money.

We start the discussion of economic evaluations by returning briefly to the IRP definition just provided. In the definition, an IRP analysis is described as an analysis that: "accounts for all known cost impacts on the utility system that are passed on to its customers through the utility's electric rates."

Inherent in that description is the concept that any resource option will have a variety of cost impacts on the utility system as a whole. Obviously, the actual cost to build or add a particular resource option will need to be accounted for. However, there are a number of other cost impacts. For example, the addition of a new generating unit will result in changes in how the utility dispatches its current generating units. This change in the dispatch order of a utility's existing system of generating units will result in changes in system fuel costs and system environmental compliance costs. Similarly, the addition of a DSM option will, by design, result in changes to the electrical load that the utility must serve. This changing electrical load shape, or pattern of electrical use, will also result in changes in how the utility dispatches its current generating units, as well as to changes in system fuel costs and system environmental compliance costs. Economic analyses of resource options must account for all such changes (and related cost impacts) in the operation of the utility system as a whole.

Between the many Supply and DSM options that a utility can choose from, the utility may be faced with the daunting task of having to analyze hundreds of potential options. For that reason, a utility may utilize a two-step approach in its economic evaluation.

First, a *preliminary* economic evaluation may be used to eliminate some of the options. (Note that this preliminary economic evaluation is not a necessary step in an IRP approach, but it may be a useful step to take.) This step is often called a preliminary economic "screening" of options in which the less economically competitive options are "screened out." This preliminary economic screening work can sometimes be performed using a spreadsheet approach. However, because of the inherent differences between Supply and DSM options, and the very different impacts each type of option may have on a utility system, the preliminary economic screening approaches for each of the two types of resource options typically differs as we shall discuss shortly.

Second, those options that survive the preliminary economic screening then compete with each other in what we will term a *final* economic evaluation to identify the option or options that are truly the best economic choice for the specific utility system.* In order to perform this work correctly, sophisticated computer models are used to evaluate Supply and DSM options in a way that accounts for all system cost impacts. This ensures that a level playing field exists for the analysis of resource options.

Continuing with our "overview" theme for this chapter, we will briefly discuss these preliminary and final economic evaluation approaches. In subsequent chapters, we will flesh out these economic evaluation approaches by utilizing our hypothetical utility system in the analyses of specific resource options.

Preliminary Economic Screening Evaluation of Supply Options

Supply options can vary in size from less than 1MW to more than 1,000MW. Intuitively, it makes sense that a Supply option of 1,000MW may have very different impacts on a utility system and its costs than a 1MW option. This difference in size is not a problem when evaluating these options in the final economic evaluation, because the final evaluation will utilize a sophisticated computer model that can accurately accommodate these different sizes in the evaluation. But if the utility is considering a large number of Supply options, it is often very time consuming to run these computer models for so many options. It would be helpful if some of these options could be screened out up front using a simpler analytical approach.

The question is how to address such huge differences in size without incurring the cost and staff person-hours necessary to use the sophisticated computer models? The most commonly used approach is called a "screening curve"† approach that is performed on a spreadsheet. The screening curve approach can be described as follows for an analysis of a generating unit:

1. The perspective taken in this approach is that of placing each Supply option to be compared "out in a field" by itself; i.e., the Supply options are not connected to the utility system. (As a consequence of this basic assumption that the Supply option is not connected to a utility

* "Final" economic evaluations are also referred to as "system" economic evaluations because such evaluations ensure that all of the system cost impacts are accounted for. These system cost impacts are often not fully accounted for in preliminary economic screening evaluations as will be discussed.

† The screening curve approach is also referred to as a "levelized cost of electricity (LCOE)" approach.

system, a number of impacts which the addition of the Supply option will actually have on the utility system are omitted in this analytical approach.)

2. The Supply option is assumed to operate for a number of hours per year.* Typically, a range of potential operating hours per year is examined.
3. All of the annual costs of building, maintaining, and operating the Supply option over each year of the option's projected life (typically 25–40 years) are then calculated.
4. These annual costs are first converted into what is termed "present value" costs. Then the present value costs are used to develop a levelized annual cost value.†
5. These levelized annual costs are then divided by the number of MWh of electricity the generating unit is assumed to produce annually. (The annual MWh value is simply the product of multiplying the MW of the unit by the number of hours per year the generating unit is assumed to operate.)
6. The result is a projected levelized cost to build and operate the generating unit and the result is typically expressed in terms of $/MWh or (equivalently) cents/kWh.
7. As previously mentioned, the calculation is often repeated using a range of potential annual operating hours (i.e., capacity factors). These results are often graphed. The shape of the graph is a curve, thus leading to the "screening curve" label commonly applied to this analytical approach.

Figure 3.1 shows the results for a typical screening curve analysis for two competing Supply options.

This figure shows that the projected levelized costs for Supply Option 1 are less than the projected levelized costs for Supply Option 2 throughout the range of capacity factors examined (from 50% to 80% of the hours in a year). Therefore, the cost of producing a MWh *solely from each of these generating units*, without any consideration of the impacts either new generating unit will have on the existing generating units on the utility system, will be lower with Supply Option 1 than from Supply Option 2.

* The number of hours a Supply option (generating unit) operates per year is referred to as its "capacity factor" which is expressed in terms of the percentage of hours in the year the generating unit runs. The number of hours the generating unit runs is divided by the 8,760 annual hours to derive its capacity factor. As we discussed in Chapter 2, the actual number of hours a generating unit will operate is determined by its operating cost; the lower the operating cost, the more hours the unit will run (thus the higher its capacity factor will be).

† Please see Appendix C for a discussion of present value costs and levelized costs.

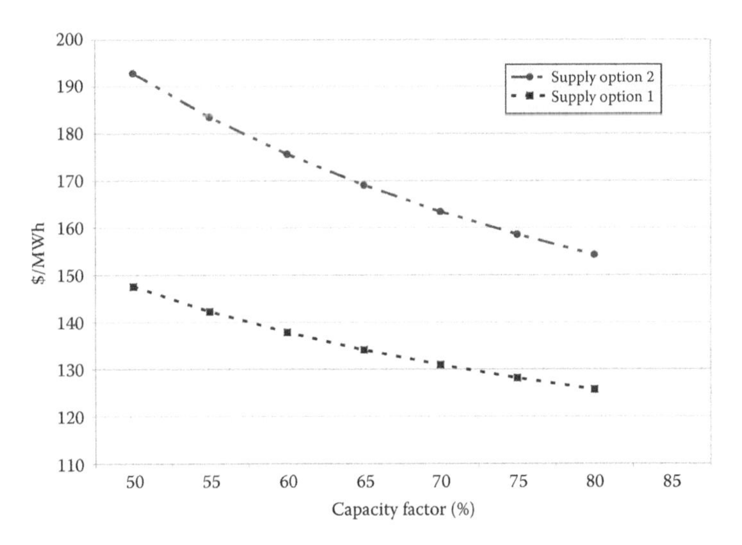

FIGURE 3.1
(See color insert.) Preliminary economic analysis: screening curve approach levelized $/MWh costs for two supply options.

The theory is that the lower the $/MWh value for a given Supply option, the more economic the option is. *However, the theory only holds true if the Supply options being compared are identical, or at least very similar, in regard to each of four key characteristics.* (For illustrative purposes at this point, we will mention that one of these key characteristics is the size (MW) of the resource options being considered. We will discuss in detail what each of these four key characteristics are later in the book.)

In those cases in which the Supply options being compared are identical, or very similar, in regard to all four key characteristics, a screening curve approach can be useful in identifying the more economical Supply option. However, the value of a screening curve analytical approach *becomes virtually meaningless* when any of the four key characteristics are dissimilar for the Supply options being compared.

This is because when any of the four key characteristics are dissimilar, the fact that a screening curve approach does *not* take into account a number of impacts of being connected to the utility system becomes of overriding importance. (Recall that the screening curve perspective is of a generating unit "alone in a field," unconnected to the utility system.) Therefore, a screening curve approach not only can, but will, give very inaccurate results in regard to which of these dissimilar Supply options is really the most economical choice for the utility.

Unfortunately, this fact is often overlooked and/or not understood. As a consequence, one may perform an analysis of vastly dissimilar resource options using a screening curve approach and believe that this analysis provides the answer of which resource option is more economical. Over the years, I have seen this mistake made numerous times by novice utility resource planners and by others who are trying to analyze resource options (or promote a selected resource option.) This mistake is usually a fatal one in terms of meeting the objective of accurately determining the most economic resource option for a utility system.

Therefore, we will now introduce my second Fundamental Principle of Electric Utility Resource Planning:

> **Fundamental Principle #2 of Electric Utility Resource Planning: "System Cost Impacts of Producing or Conserving Electricity are of Upmost Importance. Individual Resource Option Costs of Producing or Conserving Electricity are of Little or No Importance When Considered Separate from the Utility System as a Whole."**

The projected costs of producing/conserving electricity for any *individual* resource option by itself, often expressed in terms of cents/kWh or $/MWh, whether a Supply or DSM option, is of little/no consequence when performing economic evaluations whose objective is to select the most economic resource option for the utility system as a whole. When selecting a resource option for a particular utility, the objective is to identify the resource option that results in the lowest electric rates that are charged to customers. Only an analysis that accounts for all of the cost impacts that a resource option will have on the entire utility system can determine the most economic resource option.

We will return to a discussion of the screening curve approach, the four key characteristics of resource options, and the limitations to the use of screening curve analyses, in Chapter 5. In that chapter, we will flesh out this overview discussion by performing analyses that utilize our hypothetical utility system to examine Supply options. For now, simply keep in mind that a screening curve can be useful *only if each of four key characteristics of the Supply options being compared are identical or very similar.* If this is not the case, the results of a screening curve approach have little or no value. And, most importantly, a screening curve approach should *never* be used to make a final decision of which resource option to add to a utility system.

Preliminary Economic Screening Evaluation of DSM Options

The size of DSM options also offers a challenge in the evaluation of these options, but the challenge is a bit different than the type of "size challenge" previously discussed when evaluating two (or more) competing Supply options. With Supply options, the challenge was how to compare one generating unit that would supply 1 MW with a second generating unit that would supply 1,000 MW.

However, the challenge when evaluating competing DSM options is not due to a significant size difference from one DSM option to another. That is because the vast majority of DSM options will reduce load at a single participating customer's premises (home or business) by a relatively small amount. Consequently, there are seldom large differences in the amount of demand reduction per participating customer between two DSM options. For example, DSM options that are designed for residential and small commercial customers often are projected to have demand reduction values per participating customer of 1 kW or less. (Recall that 1,000 kW = 1 MW, so DSM's impact for one participating customer is typically much smaller than even the smallest Supply option.) Therefore, many DSM options will all have roughly the same demand reduction impact (approximately 1 kW) for a single participating customer.*

The challenge in evaluating DSM options is how to determine if signing up large numbers of customers to participate in any DSM option that reduces load by 1 kW (or less) per participating customer is a better economic decision than adding a new generating unit that may provide 1,000 MW of electrical generating capacity.

This problem is often addressed in final (or system) economic analyses by combining a number of DSM options that each are projected to have thousands of participating customers.† This approach results in a large amount of MW reduced by the DSM options competing against a comparable amount of MW of generating capacity from a Supply option. But how does one decide which DSM options are worth pursuing in an effort to sign up thousands of participating customers? In other words, how do we perform preliminary economic screening evaluations that can be used to screen all potential DSM options?

The answer is at once both similar to, and different from, the approach used in the screening curve analyses used to perform preliminary economic screening analyses of Supply options. The DSM preliminary evaluation approach is somewhat similar to a screening curve approach for Supply options in the sense that it also seeks to evaluate options on a comparable size basis. Recall that in screening curve analysis, each Supply option was compared on a $/MWh basis. In other words, what is the cost of producing a comparable amount of electricity (1 MWh)? This approach can be viewed as essentially evaluating Supply options from a 1 MW size perspective.‡

However, the screening curve analysis approach used for Supply options is based on comparing one Supply option versus another Supply option. The DSM preliminary evaluation screening approach that is most meaningful is

* DSM options for large commercial and industrial customers can have significantly larger demand reduction per participating customer values. However, as a general rule, individual DSM options typically vary much more in regard to the amount of energy (kWh) reduction they achieve for a single participating customer than they do in regard to the amount of demand (kW) reduction.

† In Chapter 6, we will simplify our discussion by assuming two DSM options, each with many thousands of participants.

‡ An equally valid way to view this approach is to say that the approach completely ignores the size of the Supply options being evaluated.

different in that respect.* This DSM approach does not compare one DSM option versus another DSM option, but instead compares one DSM option versus one Supply option.†

The actual DSM preliminary economic analysis approach combines these two facets by comparing the DSM option to a comparable sized generating unit. In practice, this means that one must "shrink" the generating unit down to a size comparable to the DSM option. This is typically done by assuming that the generating unit is 1 kW in size. One then calculates the costs for this 1 kW "mini" generating unit by dividing the total costs of the full-size generating unit by the total capacity (in terms of kW) to derive a $/kW cost. For example, assume that the cost to build the full-sized generating unit is $200 million and its capacity is 400 MW. The 400 MW equates to 400,000 kW.

Therefore, by dividing the $200,000,000 cost by 400,000 kW capacity, one derives a "pro-rata" $/kW cost to build the generating unit of $500/kW. One then assumes that a 1 kW-sized "mini" generating unit can be built for the same cost.‡ A 1 kW mini-sized generating unit with a cost of $500/kW can then be readily compared to similar sized DSM options.

There are three preliminary economic evaluation "tests" that are commonly applied when evaluating potential DSM options. One of these DSM "cost-effectiveness tests" (as they are commonly called), is the Participant Test. The Participant Test is designed to provide the perspective of a customer who might participate in the DSM option if it were offered by the utility. The other two commonly used tests, the Rate Impact Measure (RIM) Test and the Total Resource Cost (TRC) Test, are, in theory, designed to provide the perspective of whether the utility should offer the DSM option.§ (We will examine the three tests in detail in Chapter 6.)

The results of each of the three tests are presented in terms of a benefit-to-cost ratio. In regard to the Participant Test, the sum of the various benefits the participant receives is divided by the sum of the various costs the participant must pay. (The participating customer's benefits include annual savings on the electric bill, incentive payments received from the utility, and government tax credits, if applicable. The participating customer's costs include the up-front costs of buying and installing the more efficient appliance or

---

* Because this chapter is designed to introduce concepts, we will delay a discussion of why this preliminary economic screening analysis approach for DSM is most meaningful until later in Chapter 6 when more context for this issue has been provided.

† The Supply option selected for this comparison should be the type of generating unit the utility would actually add if it were going to address the need for new resources with a new generating unit. To choose any other type of generating unit would be relatively meaningless.

‡ In reality, it is not possible to do this. However, this simplifying assumption is useful in performing preliminary economic screening of individual DSM options.

§ These two tests, RIM and TRC, are in common use. There are other tests, including the Utility Cost test and the Societal test, that are also in use. These additional tests are essentially variations of the RIM and TRC tests. By understanding the RIM and TRC tests, one will also gain an understanding of the key components of these variations as well. Consequently, our discussion will focus solely on the RIM and TRC tests.

installing more insulation, etc., plus any on-going maintenance costs, if applicable.) The result of this division, using the present value of benefits and the present value of costs, produces a benefit-to-cost ratio.

If this benefit-to-cost ratio is 1.0, then benefits exactly equal costs, and the customer would theoretically be indifferent to participating in this DSM program if it were offered by the utility. If the ratio is greater than 1.0, then benefits are greater than costs, and, all else equal, it is in the customer's best economic interest to participate in the DSM program. Conversely, if the ratio is lower than 1.0, then benefits are less than costs and, all else equal, it is not in the customer's best economic interest to participate in the DSM program.

The RIM and TRC tests also use a benefit-to-cost ratio to present the results of each test. However, with these tests, the benefits represent the benefits to all of the utility's customers, participants, and non-participants alike, from implementing the DSM option and not building and operating the generating unit that otherwise would have been built. The costs represent the DSM-related cost impacts to all of the utility's customers from implementing the DSM option. And, just as with the Participant test, both the benefits and the costs are presented in present value dollars.

In general terms, the benefits from implementing DSM are the costs that are *avoided* by not building and operating the generating unit and by not serving as much energy to customers. (In other words, DSM benefits in these two tests are costs that are avoided.) The DSM costs are the costs incurred to implement and operate the DSM program.

Just as there were cautionary notes presented regarding the screening curve analytical approach used to perform preliminary economic screening evaluations of Supply options,* a cautionary note should also be made at this point. The cautionary note does not apply to the Participant Test. This test is widely recognized as a valid and necessary test to undertake when a utility is considering offering a DSM program. After all, if a customer won't benefit from the DSM program, why would/should a customer participate? Therefore the Participant Test is an essential test to undertake when evaluating DSM options.

The cautionary note is aimed at the other two tests that are designed to provide the perspective of whether the utility should offer the DSM option. A number of intelligent people reading this (and I am sure you are one of them) will be quick to ask "why does one need two tests that are designed to provide the same perspective?" The answer is that you don't need both, you only need one of these tests. But which of these two tests, RIM and TRC, should really be used?

---

* Some unfortunate attempts have been made to use a screening curve (cents/kWh) analysis approach to compare DSM options to Supply options. Unfortunately, the fact that DSM and Supply options are very dissimilar options violates the rule that this analytical approach can provide meaningful results *only* if the resource options being compared are identical, or very similar, in regard to each of four key characteristics. In other words, a screening curve approach cannot be used to produce meaningful results in comparing DSM and Supply options.

This has been the subject of debate at one time or another in virtually every state, and for many electric utilities, during the last several decades. In short, no universal consensus has yet been reached. This is unfortunate because one of these tests is clearly the logical choice in regard to the IRP concept of evaluating all resource options on a level playing field in which all cost impacts to a utility's customers are accounted for.

However, this conclusion is more easily explained using examples. Therefore, we will return to discuss these two tests in Chapter 6 when we flesh out the cost-effectiveness tests in the course of examining DSM options for our hypothetical utility system. For now, keep in mind that although these tests are commonly used for performing preliminary economic screening of DSM options, they (just as is the case with screening curve analysis results for Supply options) should *never* be used to make a final decision regarding which resource options should be added to a utility system. A more comprehensive analysis is needed to make a final decision regarding the resource options; i.e., a final (or system) economic evaluation.

Final (or System) Economic Evaluations

A final (or system) economic evaluation is one in which those resource options, Supply and DSM, that have survived the preliminary economic screening analyses (if preliminary economic analyses have been performed) are evaluated using analyses that account for all cost impacts to the utility system. This ensures that the competing resource options are evaluated on a level playing field to determine which options are the best choices for the utility system as a whole to meet the utility's future resource needs.

A utility's final (or system) economic evaluation is an IRP approach that includes several attributes. First, the evaluation should address all of the potential resource options that remain after any preliminary economic screening evaluation and which could realistically meet the utility's projected resource needs. As previously discussed, if preliminary economic screening evaluations are utilized, a number of Supply and/or DSM options may have been screened out in these preliminary evaluations.

Second, the evaluation should include all of the costs impacts that each resource option will have on the utility system. Third, the evaluation should ensure that the reliability criterion (i.e., the minimum reserve margin percentage and/or the maximum LOLP value) the utility utilizes are not violated by the selection of a particular resource option.

For these reasons, IRP final economic evaluations typically take the form of a comparison of different "resource plans" with each resource plan including one of the competing resource options. A resource plan is simply a projection of the resource options the utility will add in order to meet its projected resource needs over a number of years.

For example, suppose a utility is considering whether to add a new combined cycle generating unit to address a projected need for new resources in 5 years, or to add a comparable amount of DSM by the end of that same 5-year period. The utility would construct two resource plans that met their reliability criterion for each year in the analysis period. In one resource plan, the new combined cycle unit would be added. In the second resource plan, the needed amount of DSM would be added.

If the projected economic life of a combined cycle unit is assumed to be 25 years, the analysis would address the 5 years from the present to when the new combined cycle unit would be added, plus the 25 years of economic life for the unit from that point. In other words, the analysis would address a 30-year time period.

The first resource plan, Resource Plan A, would add the combined cycle unit in year 5. The second plan, Resource Plan B, would add the needed amount of DSM* so that by year 5, both resource plans would meet the utility's reliability criterion. Then additional resource options would be added to both resource plans in years 6 through 30 so that the reliability criterion continued to be met for both plans for all years in the evaluation.† This ensures that both of the competing resource plans will make the utility system equally reliable, thereby making the economic evaluation results more meaningful than would be the case if the two plans differed significantly in one or more years in regard to how reliable the utility system would be.

The idea is to eliminate any significant variations in the reliability of the two resource plans to ensure that the economic differences between the two resource plans are meaningful. In practice, utilities usually attempt to further eliminate significant variations between the two plans after the "decision year." (In this example, year 5 is the decision year because it is the year in which the utility first needs new resources.) This can be done by assuming that all of the new resources that are added to the resource plan in years (after the decision year) to maintain reliability are the same type of resource. For example, the utility might assume only combined cycle generating units would be added in these latter years. This approach eliminates other variables that could complicate the economic evaluation results and make it harder to be sure of which of the two resource options is the best choice for the decision year.

Once two resource plans are constructed that are of comparable reliability, the two plans are evaluated using sophisticated computer models that determine all utility system costs for each resource plan over the 30-year period.‡

* The term "needed amount of DSM" simply refers to the amount of DSM that allows the utility to meet its reliability criterion. This concept is discussed in more detail in Chapters 4 and 6.

† As will be discussed in Chapter 6, DSM options typically have shorter life "terms" than do new generating units. We will discuss how this difference between the two types of resource options be accounted for in developing resource plans involving DSM options in Chapter 6.

‡ At the time this book is written there are a number of computer models that are commonly used in IRP analyses. The number and names of such models are subject to change from time-to-time. For that reason, the names of applicable computer models will not be used in the book.

The economic impacts that will result from each resource plan are then compared in order to determine which resource plan is the best economic choice for the utility's customers. The competing resource option that is included in the most economic resource plan is thus identified as the most economic resource option for the utility's customers for the decision year.

The economic impacts of the resource plans can be compared on two economic bases: a total cost basis or an electric rate basis (in which the costs are divided by the total kWh of electricity used by customers to derive the cents/kWh electric rate that customers are charged). The most meaningful way to compare two resource plans is on an electric rate basis.

To see why this is the case, we first assume that we are comparing two resource plans that contain only Supply options (e.g., a new combined cycle unit and a new coal unit). The two resource plans are typically compared in regard to total costs that are referred to in terms of the "cumulative present value of revenue requirements (CPVRR)."* Suppose that Resource Plan X (the plan with the new combined cycle unit) was $100 million CPVRR less expensive than Resource Plan Y (the plan with the new coal unit). Resource Plan X is clearly the more economical choice from a cost basis. But is it also the economical choice from an electric rate basis?

The answer is "yes." To better understand this, think of an electric rate as a simple fraction. The numerator (i.e., the top value in the fraction) is the cost of the resource plan. The denominator (i.e., the bottom value) is the number of kWh the utility serves. If the denominator is identical in two fractions, then the fraction with the higher numerator will result in the larger value. For example, 3/4 is a larger value than 1/4.

When only Supply options are being compared in resource plans, the resource plans will serve the same number of kWh (i.e., the denominators are identical). Consequently, the resource plan with the lowest cost (i.e., the smallest numerator) will also be the resource plan with the lowest electric rate.

To flesh out this example using a simple example, suppose a (really small) utility would serve 1,000 kWh with either of two resource options. If the utility's total cost is projected to be $90 with one option, and $100 with another option. Assuming all else equal, from a total cost perspective, the option resulting in the $90 cost clearly is the economic choice. The same result occurs if we switch to an electric rate perspective because this option will result in an electric rate of $90/1,000 kWh = $0.09/kWh (or 9 cents/kWh) while the other option will result in an electric rate of $100/1,000 kWh = $0.10/kWh (or 10 cents/kWh).

Therefore, once it has been determined which resource plan (among the resource plans in which only Supply options are being evaluated) is more economical from the perspective of resource plan total costs, that same

* "Revenue Requirements" and CPVRR are simply "utility-speak" terms for the utility's total costs for each resource plan. See Appendix B for a definition of each term and see Appendix C for numerical examples of these terms.

resource plan will also be the more economical plan from an electric rate perspective as well.

Consequently, when evaluating resource plans that examine only Supply options, it is *not necessary* to take the additional step of examining the resource plans from an electric rate basis because we already know the outcome. (However, utilities may choose to perform a rate analysis in order to more clearly indicate the level by which a utility's electric rates will change.)

This is obviously not the case when comparing resource plans in which at least one of the resource plans contains a different amount of DSM. This is due to the fact that different amounts of DSM will result in differing amounts of energy (kWh) reductions occurring, thus changing the amount of kWh that the utility will serve. Or, in other words, the denominators for the two resource plans will no longer be identical as is the case when considering only Supply options.

To continue our earlier example, suppose that Resource Plan X (which features the combined cycle unit in the decision year) is now compared to Resource Plan Z, which features a comparable amount of DSM being added by the decision year. In this case, if one were to stop work after determining the total costs of each resource plan, might one be selecting a resource plan that resulted in higher electric rates (which would clearly not be the utility customers' choice for the best resource plan)? The answer is "yes," one might be doing just that unless one takes the next step of actually evaluating the projected electric rates for each resource plan.

Continuing our simple example, suppose that one were to stop work after determining that a DSM-based Resource Plan Z would result in a lower total cost than Resource Plan X (e.g., $89 for Resource Plan Z versus $90 for Resource Plan X). However, after taking into the account the change in the denominator caused by the DSM-induced reduction in kWh served by the utility (e.g., a reduction of 50 kWh), the electric rates for Resource Plan Z would be $89/(1,000 − 50 = 950 kWh) or $0.94/kWh (or 9.4 cents/kWh). The utility's customers would clearly prefer Resource Plan X because the electric rates they will be charged would be lower, 9 cents/kWh versus 9.4 cents/kWh.*

Therefore, the correct way to evaluate resource plans in which competing resource options are included is *to compare them from an electric rate perspective* in the final (or system) economic evaluation. As we have seen, it is generally acceptable to take a shortcut and look only at the resource plan total costs when only Supply options are being compared (because for Supply options, the lowest cost plan is automatically the plan with the lowest electric rates), but never when DSM options are being compared to Supply options.

We will return again to the topic of IRP final (or system) economic evaluations in Chapters 5 through 7 where we will flesh out our discussion of

* Just as we utilized a cumulative present value perspective to discuss utility costs over a number of years, we will also use a cumulative present value perspective of electric rates that will be introduced and discussed in Chapters 5, 6, and 7.

conducting a final economic evaluation of both Supply and DSM options utilizing our hypothetical utility system.

Before leaving this overview of the subject of final economic evaluations, another point needs to be made. The discerning reader (and you should definitely know who you are by now) will recall that we included cautionary notes when discussing the preliminary economic screening evaluations for both Supply and DSM options. The cautionary notes basically said that final resource decisions should *never* be based on the results of preliminary economic evaluations; one needs to perform a complete final (or system) economic evaluation which accounts for all impacts to the utility system. There is still one more cautionary note to make.

The resource plan and/or resource option that emerges from the final economic evaluation as the best economic choice may not always be the best *overall* choice. This can occur if there are important non-economic considerations by which the resource plans and/or resource options will also be judged. We will discuss a few of these non-economic considerations next.

Non-Economic Evaluations

We start by first discussing what we will refer to as "non-economic" considerations. The items that we will discuss as examples of non-economic considerations are often part of the final economic evaluations, but not always. We will briefly discuss three examples of such considerations.

I refer to these as non-economic considerations because they are usually referred to in non-economic terms such as "years," "tons," "mmBTU," or "percentages." In addition, the non-economic evaluation results for a given resource plan in regard to at least a few of these considerations are often compared to limits or standards that are themselves not discussed (at least directly) in terms of economics.

The three examples of considerations that we will discuss in this chapter are:

1. The length of time it takes before a resource plan and/or resource option becomes the economic choice;
2. The utility system's fuel usage for a resource plan; and,
3. The utility system's air emissions for a resource plan.

These three considerations are often used, to varying degrees, by electric utilities in their resource planning work.* We will discuss each of these considerations separately.

* There are other non-economic considerations that are used in utility resource planning. Two of these will be discussed in Chapter 8 in our discussion of "constraints" on resource planning. These two additional considerations are relatively new at the time this book is written and, therefore, are not yet in as widespread use as the three considerations discussed in this chapter. In addition, other non-economic considerations beyond those discussed in this book exist currently and new considerations are continually emerging.

Non-Economic Consideration Example (i): The Length of Time It Takes before a Resource Plan Becomes the Economic Choice

Suppose one has decided that a particular resource plan is the economic choice; i.e., it results in the lowest electric rate for the utility's customers. But is it the resource plan that most of the utility's current customers would actually choose today? If the resource plan provides the lowest electric rate over the time period addressed by the analyses, why wouldn't it be the choice of the utility's current customers?

To answer that, let's take a look at "when" the resource plan that is deemed to be the most economic actually becomes the most economic plan. Note that there are several ways in which one could look at a "time" question in regard to the economics of two (or more) resource plans. One could look either at costs (when examining only Supply options) or at electric rates (when comparing DSM and Supply options). And one could look at either nominal or present value costs or electric rates. Each different way of looking at this economic "timing" issue can provide a somewhat different perspective from which to view the resource plans. The value of each perspective may vary on a case-by-case basis.

For the sake of simplicity in our discussion, we will select and examine only one such perspective: cumulative present value costs. Take a look at Figures 3.2 and 3.3 in which two hypothetical resource plans, Resource Plan A and Resource Plan B, are used to evaluate two competing Supply options. These two resource plans are compared in regard to their CPVRR costs.*

In Figure 3.2, the resource plan that is judged to be the more economic plan of the two from a CPVRR cost perspective over the entire time period addressed in the evaluation is Resource Plan A. This is seen by looking at the ending (i.e., the last cost value on the right-hand side of the graph) CPVRR cost value. Resource Plan A has a CPVRR value of less than $4,000 million while Resource Plan B has a CPVRR value of greater than $4,000 million.

Now that the outcome is known, a relevant question is "how long does it take before Resource Plan A becomes the less expensive plan?" From looking at the graph, we see that Resource Plan A becomes the less expensive (or the more economic) plan starting in year 10 when the cost curves for the two plans "cross over" (i.e., Resource Plan A becomes the lower curve in the figure). Resource Plan A then remains the more economic plan for the remaining 20 years of the approximately 30-year analysis period. However, the "crossover" picture changes considerably if we assume different cost values as is presented in Figure 3.3.

* The graph used in the example would be essentially the same if we had chosen to compare the two resource plans in regard to the present value of electric rates. However, I have found it is easier for most people to readily understand a graph of cumulative present value costs from year to year than of cumulative present value electric rates from year to year. And, as previously discussed, when resource plans featuring only new Supply options are evaluated, the resource plan with the lowest costs will also have the lowest electric rates. (Note, however, that if the perspective had been nominal costs, instead of present value costs, the "cross over" point will usually be different.)

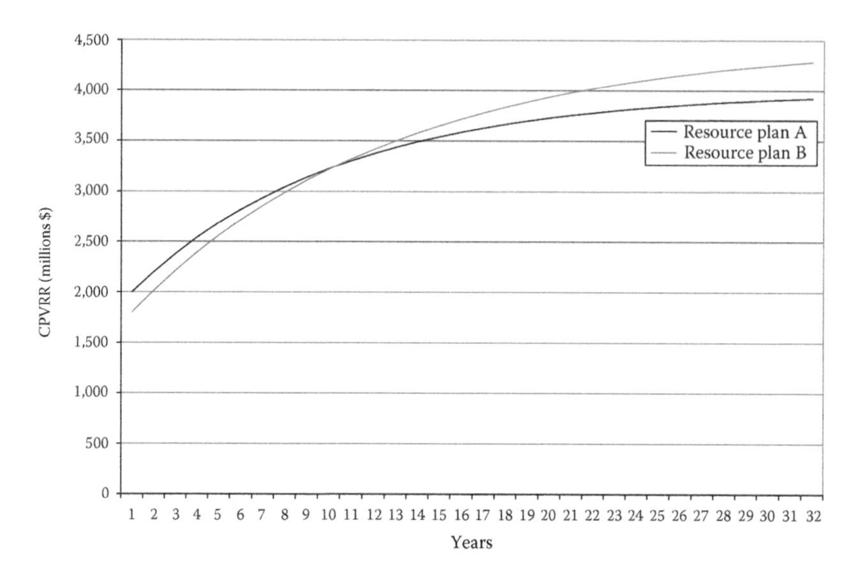

FIGURE 3.2
(See color insert.) "Cross over" graph of two hypothetical resource plans: cross over in 10 years.

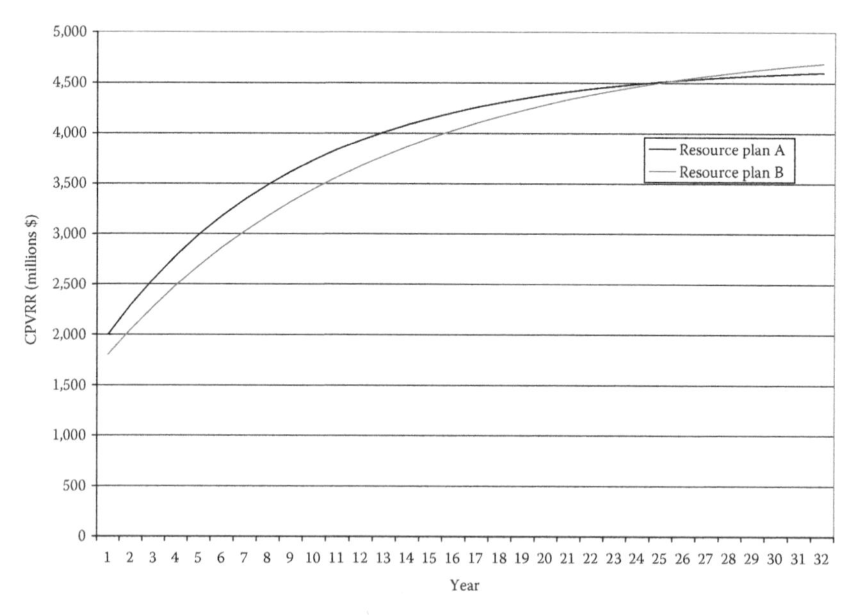

FIGURE 3.3
(See color insert.) "Cross over" graph of two hypothetical resource plans: cross over in 25 years.

In Figure 3.3, we have changed the costs for the two resource plans (as shown by the fact that the final CPVRR values for both plans now exceed $4,500 million). Resource Plan A remains the more economical resource plan, but it doesn't earn this designation until year 25 when the cost curves of the two plans cross over. Thus, customers must wait 25 years to see net cumulative present value benefits from Resource Plan A in this particular case.

Is Resource Plan A still the correct selection? There are no hard-and-fast rules regarding this, but it is unlikely that the utility's current customers would be thrilled to pay higher costs (on a present value basis) for 25 years, then begin to see present value net benefits that would largely be realized by the utility's customers 25 years from now; i.e., by the next generation of the population served by the utility.

For this reason, this particular non-economic consideration has been called a "generational equity" (or "intergenerational equity") issue in utility parlance. In "customer parlance," it might be called an issue of: "I have to wait *how many years* before I see net present value benefits?"*

It is relatively rare for a resource plan (or resource option, regardless of whether one is considering a Supply or DSM option) to immediately become the economic choice, and remain the economic choice for the full period of time addressed in the evaluation, when multiple resource options are being evaluated. In other words, there is usually some number of years before the "crossover" point in the projected economics of competing resource plans is reached. How long it takes for a resource plan to cross over is frequently of interest to the utility (and its regulators) in regard to the economics of the resource options in question.

However, the importance of the generational equity issue may be significantly reduced, or even eliminated, if a particular resource option in question has clear non-economic advantages in comparison to the competing resource options. Such non-economic advantages might include, but are not necessarily limited to, the ability of the particular resource option to significantly reduce the utility's dependence upon one or more types of fossil fuels and to significantly reduce the risk of future environmental regulations due to comprehensive reductions in air emissions from the utility system.

Before we leave this discussion of how long it takes before a resource plan becomes the more economic resource plan, let's introduce a different way to present the information shown by the two figures shown earlier. In certain cases, it can become difficult to easily distinguish the cross over point if presented in this type of graph. This can be especially true in cases in which more than two resource plans are being compared.

* Generational equity issues also arise with other utility projects that do not involve Supply and DSM options. For example, the high cost incurred today in burying overhead distribution lines when projected benefits may only be seen a number of years later can be seen as a generational equity issue. Non-utility projects such as the building of new canals, tunnels, etc., can also be viewed as involving generational equity issues.

For that reason, we will use a tabular format to present crossover information in the remainder of this book. Table 3.2 is an example of this tabular format. It presents the same basic information in regard to when Resource Plan A crosses over in comparison to Resource Plan B (i.e., in 10 years) as was shown in Figure 3.2. The table uses a designation of "1st" to denote the resource plan with the lower CPVRR cost through that year, and a designation of "2nd" to denote the resource plan with the higher CPVRR cost through that year.

TABLE 3.2

"Cross Over" Table for Two Hypothetical Resource Plans CPVRR Ranking of Resource Plans by Year

Year	Resource Plan A	Resource Plan B
1	2nd	1st
2	2nd	1st
3	2nd	1st
4	2nd	1st
5	2nd	1st
6	2nd	1st
7	2nd	1st
8	2nd	1st
9	2nd	1st
10	1st	2nd
11	1st	2nd
12	1st	2nd
13	1st	2nd
14	1st	2nd
15	1st	2nd
16	1st	2nd
17	1st	2nd
18	1st	2nd
19	1st	2nd
20	1st	2nd
21	1st	2nd
22	1st	2nd
23	1st	2nd
24	1st	2nd
25	1st	2nd
26	1st	2nd
27	1st	2nd
28	1st	2nd
29	1st	2nd
30	1st	2nd

1st = lowest CPVRR.

As shown in Table 3.2, Resource Plan A is shown to cross over in year 10 to become the lower CPVRR cost resource plan and it maintains that position for the remaining years of the analysis.

Because this tabular approach is easier to interpret when multiple resource plans are discussed, we will utilize this tabular format when discussing the economic crossover points for CPVRR costs and/or cumulative present value of electric rates in the remainder of the book.

Non-Economic Consideration Example (ii): The Utility System's Fuel Usage Due to a Resource Plan

The cost of fuel will already have been accounted for in all aspects of economic evaluations. However, for fuel usage, the non-economic consideration is often the question of how much a utility relies upon a particular type of fuel. In other words, it is an issue of fuel supply reliability or security, plus an issue of cost volatility (i.e., how rapidly a fuel's cost may increase in a relatively short time). Concerns by both utilities and regulators tend to increase when a utility becomes increasingly reliant on any one particular type of fuel.

Therefore, utilities strive to maintain and/or enhance their fuel "diversity" to avoid over-dependence on any one fuel type. Consequently, a resource plan that emerges from the final economic evaluation as the most economic choice, but which results in significantly more reliance upon a fuel that the utility is already heavily dependent on, may not be the best overall choice. In such a case, another resource plan that was not the best in terms of system economics, but is the resource plan that is the next-most economical and provides a desired level of fuel diversity, may emerge from the overall evaluation of both economics and non-economics as the resource plan of choice.

Non-Economic Consideration Example (iii): The Utility System's Air Emissions Due to a Resource Plan

This consideration is another that will have already been accounted for in the final economic evaluation. At the time of this writing, the types of air emissions that are most often included in IRP evaluations are: sulfur dioxide (SO_2), nitrogen oxides (NO_x), carbon dioxide (CO_2), mercury (Hg), and particulates.

The projected costs of complying with either current or projected environmental regulations that address each of these air emissions are accounted for in economic evaluations. This is typically done by including either the projected costs for pollution control equipment, or the projected costs of emission allowances. Nevertheless, resource plans may also be evaluated non-economically from the perspective of the amounts of projected system air emissions for each resource plan. This is one way in which one might gauge the viability of resource plans over many years in the future if environmental regulations were to be tightened over that time period.*

We now have an idea of the types of non-economic considerations that may be used to judge resource plans. But how do utilities address these non-economic considerations in IRP evaluations? One way is for the utility to use certain self-imposed constraints to all resource plans that are being evaluated.† In other words, no resource plan will be considered to be acceptable unless it meets these constraints. For example, the following self-imposed constraints could be used to address the three non-economic considerations we have just discussed:

1. A maximum length-of-time-to-crossover (e.g., x years or less) to a resource plan becoming the most economical resource plan.
2. The annual percentage of the amount of total energy supplied by a particular type of fuel cannot exceed preset annual limits.
3. The annual tonnage of a particular air emission cannot exceed preset annual limits.

In summary, the non-economic evaluation portion of IRP analyses could result in a resource plan that is the most economic choice *not* being selected as the best overall plan. In order to maximize efficiency and minimize time in their IRP evaluations, utilities may use preset constraints to address these non-economic considerations in their IRP final economic evaluation.

As this chapter concludes, we note that the chapter has presented an overview discussion of key concepts important in electric utility resource planning. We will now return to our hypothetical utility system and begin to apply these concepts.

* Another way to gauge the viability of resource plans in regard to the possibility of tightened environmental regulations is to use different projected cost values for environmental compliance. We will return to the subject of potential tightening of environmental regulations in Chapter 8 in our discussion of constraints on the utility resource planning process.

† In Chapter 8, we will discuss a number of other constraints on utility resource planning. These will include both constraints that are self-imposed by the utility and constraints that are not self-imposed.

4

Reliability Analyses for Our Utility System

At this point, let's recap where we are in our examination of electric utility systems. In Chapter 2, we discussed some fundamental concepts of how the electric load for utilities varies on both a daily and an annual basis. We also briefly examined five basic types of generating units that are commonly found on utility systems. We also discussed how a utility uses the operating cost of each type of generating unit in deciding which type of generating unit to operate at any given time to meet the continually varying electric demands of its customers.

In Chapter 3, we discussed the fact that utilities plan for the future by finding answers to three fundamental questions. We then discussed how reliability analyses, using both reserve margin and probabilistic perspectives, are utilized to answer two of these three questions: "When does the utility need to add new resources?" and "What is the magnitude (MW) of the new resources that are needed?" We then examined how an integrated resource planning (IRP) approach is the best way to determine the answer to the third question: "What is the best resource option with which to meet this future need for resources?"

We also briefly discussed how both economic and non-economic evaluations are often required before the final answer to the question of "which resource option is best overall for a specific utility?" can be determined.

In summary, the previous two chapters have provided an overview of how utilities operate their existing generating units (and why they do it this way), plus provided an overview of the IRP process that electric utility systems can (and should) use to plan for the future.

It is now time to move beyond the overview presentations of these concepts and examine these concepts in more detail. In so doing, we will see how these concepts actually work in practice. Starting with this chapter, and continuing for several more chapters, we shall begin to "flesh out" these concepts using our hypothetical utility system. So we start by returning to our hypothetical utility system and conducting a reliability analysis for our utility system to see when it needs new resources and what the magnitude of the future resource need is.

TABLE 4.1

Example of a Basic Reserve Margin Calculation

(1)	(2)	(3)	(4)
		= (1) − (2)	= (3)/(2)
Total Generating Capacity (MW)	Peak Electrical Demand (MW)	Reserves (MW)	Reserve Margin (%)
12,000	10,000	2,000	20.0

When Does Our Utility System Need New Resources?

When we last left our hypothetical utility system, we had conducted a reliability analysis that showed, for a given future year, a reserve margin projection that was presented in Table 3.1. We now provide this same projection again in Table 4.1.

As shown in the table, the projected reserve margin value for this particular year is 20%.

Let's now assume that the reserve margin perspective is the one driving the need for new resources for our utility system. In other words, the reserve margin criterion is violated in future years before the LOLP criterion is violated.

Let's also assume that our utility's reserve margin criterion is 20%. Clearly for the particular year addressed in Table 4.1, our utility system is deemed to be reliable because its projected reserve margin value is equal to the reserve margin criterion of 20%.

Now let's expand our view to look at a few more years. We do so in Table 4.2.

Instead of a reserve margin projection for just one year, we now have a projection for 6 years. The first thing to notice is that the projection for the one particular year that was presented in Table 4.1 is for the year now labeled as "Current Year + 4." In other words, this projection is for 4 years from now (i.e., 4 years from the "Current Year").*

In looking back to the Current Year, we see in Column (1) that our utility system actually has 400 MW less generating capacity than what it is projected to have in 4 years: 11,600 MW in the Current Year instead of the 12,000 MW of generating capacity in Current Year + 4 that we just discussed. Our utility's peak electrical demand is also lower in the Current Year by 400 MW: 9,600 MW instead of the 10,000 MW. This combination of lower generating capacity and lower peak demand result in a reserve margin value for the Current Year of 20.8% as shown in Column (4). Because this reserve margin

* This term "Current Year" is used throughout this book to denote the year our hypothetical utility is conducting its analysis. This approach avoids the use of a specific year (such as the year this book was written) which will, of course, no longer be the actual "current year" when someone reads this book years after it is written.

TABLE 4.2

Reserve Margin Analysis Projection for the Hypothetical Utility

	(1)	(2)	(3)	(4)
			= (1) − (2)	= (3)/(2)
Year	**Total Generating Capacity (MW)**	**Peak Electrical Demand (MW)**	**Reserves (MW)**	**Reserve Margin (%)**
Current Year	11,600	9,600	2,000	20.8
Current Year + 1	12,000	9,700	2,300	23.7
Current Year + 2	12,000	9,800	2,200	22.4
Current Year + 3	12,000	9,900	2,100	21.2
Current Year + 4	12,000	10,000	2,000	20.0
Current Year + 5	12,000	10,100	1,900	18.8

value is higher than our utility's 20% reserve margin criterion, our utility is deemed to be reliable in the Current Year.*

However, by examining the projected annual load values in Column (2) in this table, we see that our utility system is projecting growth in its peak demand of 100 MW per year for each of the next 5 years. Because of this projected growth, our utility's projected reserve margin one year in the future (Current Year + 1) would drop below the 20% criterion to 19.6% unless new resources are added. (Although this is not shown in the table, it is easily calculated: 11,600 MW of generation capacity minus 9,700 MW of peak load would result in 1,900 MW of reserves. Then 1,900 MW of reserves divided by 9,700 MW of peak load would yield a reserve margin of 19.6%.)

Although a utility might decide not to add new resources in that year because its reserve margin criterion of 20% is projected to be violated by a relatively small amount, the projection of continued growth in peak load each year would cause our utility's projected reserve margin value of 19.6% one year in the future to drop to 18.4% two years in the future. The reserve margin would then continue to drop each year thereafter. Consequently, our utility had previously decided to add 400 MW of generating capacity in Current Year + 1 to bring its total generating capacity to 12,000 MW. This increased amount of generation capacity is projected to ensure that the 20% reserve margin criterion would be met for 4 years (Current Year + 1 through Current Year + 4).

Yet the growth in peak load is projected to continue past this fourth year into the future, causing the projected reserve margin to drop significantly below the 20% criterion, to 18.8%, in the 5th year.† Our utility now knows the

* I'm relieved to hear this. Clearly, this indicates superior resource planning was carried out for our utility system in prior years.

† Just a reminder that our utility system could have also found itself in a similar situation of needing new resources in Current Year + 5, even without increased load growth, if, for example, an existing generating unit was scheduled to be retired in that year, thus decreasing the Total Generating Unit Capacity (MW) value in Column (1).

answer to the first question: "when does it need to add new resources?" The answer is 5 years from the Current Year (Current Year + 5).

We now turn our attention to how reserve margin analyses can be used to answer the second question: "what is the magnitude of the resources needed?" Perhaps unexpectedly, we shall see that the answer depends upon what type of resource the utility chooses to add.

What Is the Magnitude of the New Resources Needed by Our Utility System?

From Table 4.2 it is clear that, because the peak load is projected to grow at 100 MW per year, the utility needs to add new resources by Current Year + 5 because the projected reserve margin has dropped to 18.8%. But what is the magnitude (MW) of the new resources that are needed? Table 4.2 does not provide an answer to this question.

Also, as we discussed in Chapter 3, there are two basic types of resource options that the utility could add to meet a need for future resources: Supply options and DSM options. Our answer(s) to the question of "what is the magnitude of new resources needed?" must be responsive to both of these types of resource options that the utility could choose. This will ensure that the correct amount of either type of resource option has been chosen so that the utility's reserve margin criterion of 20% is met.

We shall start by looking at one of the two types of resource options the utility might choose: Supply options. Then we shall see how we would answer the question of the magnitude of resources needed in Current Year + 5 if the utility chooses to meet the need by adding new generating capacity. In order to do this, we turn our attention to Table 4.3.

Table 4.3 is an expanded version of Table 4.2 in which one new column, Column (5), has been added. In Column (5), we show how the generating capacity of the utility would have to change in order for a 20% reserve margin criterion to be met each year. The calculation formula to derive the values shown in Column (5) is provided in the table after the (5) in the column heading.

This formula is straightforward. The projected peak load for a given year (shown in Column (2)) is multiplied by 1.20 to determine how many MW of generating capacity would be needed to have total generating capacity that would be 20% higher than the projected peak load. This amount of generating capacity would allow the utility to meet its 20% reserve margin criterion. Then the expected annual total generating capacity of the utility for that year (shown in Column (1)) is subtracted from this calculated value to determine how far off the utility is in regard to the generating capacity needed to meet the 20% criterion.

TABLE 4.3

Reserve Margin Analysis Projection for the Hypothetical Utility: MW Needed if Only New Generation Is Added

	(1)	(2)	(3)	(4)	(5)
			= (1) − (2)	= (3)/(2)	= ((2) * 1.20) − (1)
Year	Total Generating Capacity (MW)	Peak Electrical Demand (MW)	Reserves (MW)	Reserve Margin (%)	Generation Only MW Needed to Meet Reserve Margin (MW)
Current Year	11,600	9,600	2,000	20.8	(80)
Current Year + 1	12,000	9,700	2,300	23.7	(360)
Current Year + 2	12,000	9,800	2,200	22.4	(240)
Current Year + 3	12,000	9,900	2,100	21.2	(120)
Current Year + 4	12,000	10,000	2,000	20.0	0
Current Year + 5	12,000	10,100	1,900	18.8	120

As shown in Column (5), this calculated value is negative (i.e., values presented in parentheses indicate negative values) for the Current Year and through 3 years in the future (Current Year through Current Year + 3). These negative values denote that the utility has more generating capacity than it needs to meet its 20% reserve margin criterion. This is not exactly news because, as shown in Column (4), the utility's projected reserve margin exceeds the 20% criterion in these same years.

Consequently, the utility does not need to add generating capacity in these years. Looking ahead one more year to Current Year + 4, we see that the value calculated for this year in Column (5) is 0 MW of needed new generating capacity. This is again expected because the utility's projected reserve margin for that year is exactly 20%.

However, for Current Year + 5, Column (5) presents a calculated value of 120 MW of needed new generation capacity to return the utility to a projected reserve margin level of 20%. From the formula presented for Column (5), we see that this value is derived by multiplying the projected load for that year, 10,100 MW, by 1.20. The result is a total of 12,120 MW of resources needed to meet the 20% reserve margin criterion. Then, after subtracting the current generation resources of 12,000 MW, the difference is 120 MW. The formula tells us that this is the magnitude of new resources that need to be added if the new resource is new generating capacity. But is 120 MW of new generation really the correct answer?

We can check that calculation by altering the table to allow the 120 MW of new generating capacity to be entered back into the reserve margin calculation. We do so in Table 4.4, which is a modified version of Table 4.3 that we just discussed. In order to focus better on this calculation "check," only information for the last 2 years that were shown in Table 4.3 is presented in Table 4.4.

TABLE 4.4

Reserve Margin Analysis Projection for the Hypothetical Utility: MW Needed if Only New Generation Is Added

	(1a)	(1b)	(1c)	(2)	(3)	(4)
			= (1a) + (1b)		= (1c) − (2)	= (3)/(2)
Year	Previously Projected Generating Capacity (MW)	Newly Projected Generating Capacity (MW)	Total Projected Generating Capacity (MW)	Peak Electrical Demand (MW)	Reserves (MW)	Reserve Margin (%)
Current Year + 4	12,000	0	12,000	10,000	2,000	20.0
Current Year + 5	12,000	120	12,120	10,100	2,020	20.0

In Table 4.4, Column (1) from the previous table has now been expanded into three columns. Column (1a) presents the same information as Column (1) from the previous table; i.e., our utility's previously projected generating capacity. Two new columns have also been added. Column (1b) shows the newly projected generating capacity that will be added to our utility system and Column (1c) shows the sum of the previously projected and newly projected generating capacity. Columns (2) through (4) are essentially unchanged from the previous table.

As expected, there is no change for Current Year + 4 because no new generating capacity is added. However, the calculation for Current Year + 5 shows that adding 120 MW of new generation in that year will indeed return our utility's projected reserve margin to 20%. This check confirms the original projection presented in Table 4.3: 120 MW of new generation is needed if the utility decides to meet its projected resource needs 5 years in the future with a Supply option.*

But is this projection of 120 MW of new resources also valid if the utility decides to meet its need for new resources with a DSM option? The answer is "no" as shown in Column (5) of Table 4.5. This table is a DSM-based version of the previously presented Table 4.3, which addressed the magnitude of resources needed if the new resource is new generating capacity. Table 4.5 is modified to perform the same calculation if the new resource is DSM.

This table shows that the projected value for the magnitude of new resources needed to meet a 20% reserve margin criterion is only 100 MW

* The astute reader might also have noticed that, starting with the 20% reserve margin value in Current Year + 4, the utility is adding 100 MW of load in the next year. Because the utility's reserve margin is 20%, the utility would need to add 1.20 × 100 MW of new load = 120 MW to "keep pace" with the new load in Current Year + 5. (However, the question of how much generating capacity is needed is rarely answered so easily.)

TABLE 4.5

Reserve Margin Analysis Projection for the Hypothetical Utility: MW Needed if Only New DSM Is Added

	(1)	(2)	(3)	(4)	(5)
			=(1) − (2)	=(3)/(2)	=(((2) * 1.20) − (1))/1.20
Year	**Total Generating Capacity (MW)**	**Peak Electrical Demand (MW)**	**Reserves (MW)**	**Reserve Margin (%)**	**DSM Only MW Needed to Meet Reserve Margin (MW)**
Current Year	11,600	9,600	2,000	20.8	(67)
Current Year+1	12,000	9,700	2,300	23.7	(300)
Current Year+2	12,000	9,800	2,200	22.4	(200)
Current Year+3	12,000	9,900	2,100	21.2	(100)
Current Year+4	12,000	10,000	2,000	20.0	0
Current Year+5	12,000	10,100	1,900	18.8	100

TABLE 4.6

Reserve Margin Analysis Projection for the Hypothetical Utility: MW Needed if Only New DSM Is Added

	(1)	(2a)	(2b)	(2c)	(3)	(4)
				=(2a) − (2b)	=(1) − (2c)	=(3)/(2c)
Year	**Total Generating Capacity (MW)**	**Previously Forecasted Electrical Demand (MW)**	**Newly Projected DSM (MW)**	**Firm Electrical Demand (MW)**	**Reserves (MW)**	**Reserve Margin (%)**
Current Year+4	12,000	10,000	0	10,000	2,000	20.0
Current Year+5	12,000	10,100	100	10,000	2,000	20.0

if the need is met by new DSM resources.* As before, let's check this by inserting the 100 MW of DSM back into the calculation. We do so in Table 4.6 that again focuses only on the last 2 years.

Similar to Table 4.4, one of the columns from the original table (Table 4.3) has again been expanded. However, in this case, Column (2) has been expanded. Column (2a) presents the previously forecasted peak demand and two new

* The astute reader may have already determined that the answer to this question is 100 MW. Table 4.1 showed that, in Current Year+4, our utility had exactly a 20% reserve margin. Then the load increased 100 MW in the next year. Therefore, the addition of new DSM that lowers the peak load 100 MW would return the utility to a 20% reserve margin. (However, the question of how much DSM is needed is rarely answered so easily.)

columns have been added to this table. Column (2b) shows the amount of projected new DSM that will be added to the utility system. Column (2c) then shows the result of a calculation of how the new DSM will lower the forecasted peak demand to provide a lower projection of electrical demand. This projection of load after DSM is accounted for is commonly called the "firm" demand for the utility. Columns (1), (3), and (4) are essentially unchanged from those presented earlier in Table 4.5.

As expected, there is no change for Current Year + 4 because no new DSM is added in that year. However, the calculation for Current Year + 5 shows that adding only 100 MW of new DSM in/by that year will indeed return the utility's projected reserve margin to 20%.* This check confirms the original projection presented in Table 4.5 that 100 MW of new DSM is needed if the utility decides to meet its projected resource needs 5 years in the future with new DSM only.

What Have We Learned and What Is Next?

We have now taken the first step in fleshing out our discussion of resource planning concepts by determining that our hypothetical utility system has a resource need 5 years in the future. We have also determined that the resource need is either 120 MW if the need is met solely by a new Supply option, or 100 MW if the need is met solely by a new DSM option.

The fact that there is a difference in the amount of new resources needed depending on the type of resource option that is selected, new Supply or new DSM, is a basic concept of utility resource planning. A reserve margin-driven utility (such as our hypothetical utility system) will always have a greater resource need if that need is met by a new Supply option than if the resource need is met by a new DSM option.† The difference between the two amounts is determined by a utility's reserve margin criterion (which is 20% for our hypothetical utility system, and which is representative of the reserve margin criterion used by many utilities).

In other words, if a utility's new resource needs can be met by 100 MW of new DSM, it will take 120 MW of a new Supply option to meet the utility's projected resource needs, assuming that the utility's reserve margin criterion is 20%. Looked at another way, if the utility's projected peak demand

* In reality, a utility would almost certainly begin adding DSM at least several years earlier instead of waiting until the year the full DSM resource is actually needed. The earlier implementation of DSM is driven both by economics and by the practical problems of signing up thousands of participating customers for the DSM program(s). Therefore, the picture presented here is a simplified one in which we were solely concerned with establishing the magnitude of the needed DSM resources by Current Year + 5.

† The situation may be less clear cut if LOLP is driving a utility's resource needs.

will increase in the future by 100 MW, and it currently exactly meets its 20% reserve margin criterion, then it will need 120 MW of new generation to maintain the 20% reserve margin level.

At this point, you may be asking yourself the following question: "Because the utility only needs 100 MW of a DSM option to meet its reliability criterion, versus 120 MW of a Supply option, doesn't the DSM option have a built-in advantage over the Supply option when it comes time to choose a resource option?"

Perhaps surprisingly, the answer is "yes" and "no." (I hate these types of answers, so let's explain.) From an economic perspective of just the initial cost of installing or acquiring each resource option, DSM does have an advantage. Assume for a moment that the costs of acquiring 1 MW of both the DSM and Supply options are equal. In such a case, and assuming all else equal, DSM has an economic advantage in regard to the cost of acquiring the resource because the utility only has to obtain 100 MW of the DSM resource versus 120 MW of the Supply resource.

In practice, the acquisition costs of DSM and Supply options will likely vary considerably. If the acquisition cost of DSM is lower than that of the competing Supply option, then the DSM advantage is increased due to the fact that fewer MW of DSM compared to Supply are needed. Conversely, if the acquisition cost of DSM is higher than that of the competing Supply option, this DSM disadvantage is lessened for the same reason. Therefore, the initial cost of acquiring a resource option provides an advantage to DSM because the utility doesn't need as much of that type of resource. This particular advantage shows up in economic evaluations of resource options.

However, because the utility doesn't need to acquire as much DSM as it would if it were to choose a Supply option, DSM will also have less of an impact on the operation of the utility system than would be the case if the same MW amount of DSM resource had been acquired as would be acquired if a Supply option had been chosen (120 MW in our example). In other words, the impacts to the utility system operation from acquiring only 100 MW of DSM are less than the impacts would be if 120 MW of DSM had been acquired. Two of the types of impacts to system operation will be: (i) changes to the utility system's fuel usage, and (ii) changes to the utility system's air emissions. The implementation of 100 MW of DSM will have less of an impact on system fuel use and system emissions than if 120 MW of DSM had been implemented.

Consequently, DSM's potential impact on either system fuel usage or system air emissions will be less than it would have been if the same amount of DSM were added to the utility system as would be added with Supply options. This can actually work to DSM's disadvantage when evaluating the utility's system fuel usage and/or system air emissions. These particular operational disadvantages show up in both economic and non-economic evaluations of resource options.

In summary, the fact that a smaller amount of a DSM option is needed, compared to the amount of a Supply option, to ensure the same reliability level for a utility system results in both advantages and disadvantages for DSM versus Supply options.

A utility could attempt to remove the system operation-based disadvantages inherent in implementing a smaller amount of DSM by implementing the exact same amount (MW) of DSM as would be needed if a Supply option had been chosen. However, by doing so, the economic advantage of having to acquire a smaller amount of DSM would be eliminated. Consequently, utilities typically do not attempt such an approach, and we will not take such an approach, as we continue to work with our utility system in evaluating Supply and DSM options in the next several chapters.

In Chapter 5, we turn our attention to some Supply options that our utility system will consider for meeting its resource needs 5 years in the future. We will then follow these Supply options through an economic evaluation to determine which one of these Supply options is the best economic choice for our utility. In Chapter 6, it will be DSM's turn. We will follow some DSM options through an economic evaluation to determine which DSM option is the best economic choice for the utility. Then, in Chapter 7, we will pull together the Supply and DSM options and evaluate them from both an economic and a non-economic perspective.

5

Resource Option Analyses for Our Utility System: Supply Options

In this chapter, we follow our utility system through economic analyses of four different Supply options it will consider as possible choices with which to meet its projected resource need 5 years in the future (in Current Year + 5). In practice, a utility would likely consider more than four Supply options. However, in order to simplify the discussion, while still illustrating the important points regarding economic analyses of Supply options, our utility system has graciously condensed its list to only four Supply options.

Types of Supply Options under Consideration

The four Supply options our utility system is considering include: two gas-fired combined cycle (CC) units (imaginatively labeled as CC Unit A and CC Unit B), one gas-fired combustion turbine (CT) unit, and a photovoltaic (PV) facility that uses sunlight to generate electricity. We will make the simplifying assumption that the operating life for each of these four Supply options is 25 years. The rest of the assumptions for the key inputs for the economic evaluation of these four Supply options are presented in Table 5.1.

For discussion purposes, the inputs presented in the table will be grouped into general categories that are directly or indirectly related to three types of costs: (i) capital costs, (ii) other fixed costs, and (iii) operating costs.

Capital Costs: Rows (1) through (3)

We start the discussion of inputs related to capital costs with the capacity (MW) shown on row (1) of the table. Most generating units come in pre-determined sizes. The CT size of 160 MW is representative of certain CT units available at the time this book is written. Because CC units use CT units as their basic "building block," CC units can be built in various sizes depending on how many CTs are used. CC units being built at the time this book is written typically use from one to four CT units, plus one or more heat recovery steam generators. The 500 MW capacity assumed for both of our CC options is representative of some of these CC units.

TABLE 5.1

Key Inputs for Economic Evaluation: Supply Options

Input	Units of Measurement	CC Unit A	CC Unit B	CT	PV
(1) Capacity	MW	500	500	160	120
(2) Capital cost	$ (millions)	$500	$520	$104	$600
(3) Capital cost	$/kW	$1,000	$1,040	$650	$5,000
(4) Fixed O&M	$/kW	$6.00	$6.00	$4.00	$5.00
(5) Capital replacement	$/kW	$10.00	$10.00	$8.00	NA
(6) Firm gas transportation	$/mmBTU	$2.25	$2.25	$2.25	NA
(7) Firm gas needed	mmBTU/day	100,000	100,000	10,000	NA
(8) Variable O&M	$/MWh	$2.00	$1.50	$0.20	NA
(9) Type of fuel	—	Natural gas	Natural gas	Natural gas	Sunlight
(10) Heat rate	BTU/kWh	6,600	6,700	10,400	NA
(11) Availability	% of hours per year	92%	92%	92%	20%
(12) Capacity factor	% of hours per year	80%	80%	5%	20%
(13) SO_2 emission rate	lbs/mmBTU	0.006	0.006	0.006	0.000
(14) NO_x emission rate	lbs/mmBTU	0.010	0.010	0.033	0.000
(15) CO_2 emission rate	lbs/mmBTU	119	119	118	0

Although the size of the two CC options, CC Unit A and CC Unit B, are identical, the capital costs for the two CC units are slightly different as shown in row (2) of the table.

Because CC and CT options come in pre-determined sizes, our utility system will be considering CC and CT options of a size considerably larger than its next projected resource need of 120 MW. However, because PV options consist of very small (approximately one to several kW) modules that are grouped together, a PV option can be built that more exactly matches a utility's projected resource need. For our purposes, we will assume that our utility will add 120 MW of PV so that its resource needs of 120 MW are exactly met if the PV option is selected.

Let me state here that, for purposes of our analyses, we are going to make an unrealistically optimistic assumption about the PV option. Generation from PV is dependent upon an intermittent energy source: sunlight. Cloud cover can reduce or virtually eliminate the output from PV for periods of time ranging from minutes to hours. Consequently, utilities generally assume that a PV installation can be counted on to consistently produce only a relatively small percentage of the "nameplate" rating of the PV installation at the utility system's peak hour. (In our example, the nameplate rating is 120 MW.) These percentages are determined by computer modeling of expected capacity and energy outputs and/or by actual operating experience at the specific site(s) where the PV modules will be installed. Therefore, the percentage represents an average amount of capacity that can be expected from the PV installation based on a number of years of either projected or actual data.

For example, let's arbitrarily assume that the percentage of the PV output that will reliably be delivered at a utility's peak hour from PV facilities located at various sites is 20%. What this means in the real world is that only 20% of the 120 MW of the PV installation would typically be accounted for in reserve margin calculations instead of the full 120 MW. Thus, in the real world, one might assume only 24 MW from PV for reliability analyses instead of the full 120 MW. This utility would have to add other resources to "make up" the difference between the 120 MW necessary to meet the resource need and the 24 MW from PV.*

In addition, the output of PV equipment can be expected to slightly degrade each year by approximately 1/2 of 1% per year for a number of years due to chemical effects in the PV modules from long-term exposure to the sun and moisture.

However, for purposes of our discussion, we are going to ignore both of these factors and assume that our PV installation is credited with 100% of its 120 MW nameplate rating for reserve margin calculation purposes, and that there is no annual degradation in PV output. Although both of these assumptions are unrealistic, and will result in our analyses being more favorable for PV than would actually be the case, it will simplify both the analyses and our discussion of the results of the analyses.†

Before leaving row (1), we note that the differences in the capacity (MW) offered by these four Supply options will play an important role in the economic evaluation. We shall see why this is so later in this chapter.

Rows (2) and (3) of Table 5.1 provide the capital cost for the options. By capital cost we mean the cost of building the unit. The assumed capital cost for each option is presented in the table in two ways. First, in row (2), the capital cost is presented in terms of the actual cost in millions of dollars.‡ For example, CC unit A is assumed to have a cost of $500 million.

Second, in row (3), the capital cost is presented in terms of cost per a common unit of capacity (i.e., cost per kW) for the option. This value is derived by dividing the capital cost (in millions of dollars) by the capacity (MW) in row (1). (The capacity MW value is first converted to a kW by multiplying

---

* The utility could choose to add five times more PV, i.e., to add approximately $5 \times 120\,MW = 600\,MW$ (nameplate rating) of PV at various sites to ensure that it reliably received 120 MW from PV at the utility's peak load hour. The utility could also add other non-PV resources instead.

† We shall return to the issue of the intermittent nature of certain renewable energy resource options (such as PV) in Chapter 8.

‡ These assumed capital costs account for the costs of the generating facility itself including interest charges during construction, the land upon which the facility will be built, and the transmission additions needed both to connect the generating facility to the rest of the utility system, and to upgrade other areas in the overall transmission system to accommodate handling the output of the new generating unit.

the MW value by 1,000 because 1,000 kW = 1 MW.) The result is expressed in terms of $/kW as shown in row (3). The reason for expressing the capital cost in terms of $/kW is that Supply options can vary so widely in terms of their capacity that it is difficult to gauge the relative cost of building units if one looks only at the cost in terms of millions of dollars. The use of $/kW values allows one to better gauge the relative costs of generating units on a common (1 kW) size basis.

As shown in Table 5.1 on row (2), the total cost for the two CC units are $500 and $520 million, respectively, while the CT unit is $104 million and the PV option is $600 million. (Just as we noted in Chapter 1, and in our introductory discussion of assumed fuel costs in Chapter 2, we again point out that the capital (and other) costs of these types of generating units that we are assuming are not important to this discussion of resource planning concepts and analytical approaches. Although the assumed generating unit cost values we are using are representative of costs at the time this book was written, at any point in the future these cost (and other) values will certainly be different because they are continually changing. However, the concepts and analytical approaches we are discussing will remain valid regardless of the assumption values used to illustrate these concepts and analytical approaches.)

From the information presented on row (2), the CT appears to be an inexpensive choice relative to the CC and PV units while PV appears to be an expensive choice relative to the CC and CT units. However, because the capacity (MW) of each of these four options varies so much; i.e., from 500 MW to 120 MW, it is difficult at first glance to gauge the relative costs of building a set amount of capacity with each option. As previously mentioned, converting the capital costs into $/kW values in row (3) allows one to easily see the cost on a "cost per unit" (/kW) basis. After performing this conversion, we see that the cost per kW value for the CC and CT options has narrowed significantly because the two CC units have costs of $1,000/kW and $1,040/kW, respectively, while the CT option's cost is $650/kW. Conversely, this "cost per unit" perspective shows that PV is truly a more expensive option to build because it has a cost of $5,000/kW compared to a cost range of $650/kW to $1,040/kW for the CC and CT options.

Therefore, our first impression of the economics of the four Supply options is that the renewable energy option, PV, is very expensive to construct compared to the fossil fueled units, CC and CT. Although this impression is correct, we must remember that our cost picture is far from complete.

Other Fixed Costs: Rows (4) through (7)

We turn next to what we will refer to as other "fixed costs" which include costs not related to building or directly operating the unit, but which are incurred essentially to keep the unit available for operation. Table 5.1 lists three such costs: Fixed operation and maintenance (O&M) on row (4), Capital Replacement on row (5), and Firm Gas information on rows (6) and (7).

Fixed O&M costs typically refer to costs incurred for salaries of the generating unit's operating personnel and for some regular maintenance activities. These costs are typically expressed in terms of $/kW. As shown on row (4), the assumed Fixed O&M costs for these four options vary little with costs ranging from $4/kW to $6/kW. Consequently, this cost category is likely to play a relatively minor role in the economic evaluation of these four Supply options.

The second fixed cost category is termed "capital replacement." We are using this term to refer to expenditures that are incurred to periodically replace major equipment that wears out over time as the generating unit is operated. (A non-utility analogy might be an automobile tire that wears out and must be replaced after some threshold number of miles driven has been reached.) These values are also typically expressed in terms of $/kW. The assumed capital replacement cost values shown on row (5) for the CC and CT units again vary little with values ranging from $8 to $10/kW. However, PV has no moving parts and, therefore, is typically perceived to have no such capital replacement costs. In this case, the PV option has a clear cost advantage in regard to this cost category.

This is also the case when discussing the third of the fixed cost categories, "firm gas." This term refers to the cost that is incurred to ensure that a natural gas-fired generating unit will have sufficient gas to burn when the unit is operating. Because of its gaseous nature, large quantities of natural gas are not typically stored next to a generating unit as readily as one can store large quantities of oil or coal. Natural gas must be delivered by pipeline, often by a company separate from the electric utility, and the electric utility will want to ensure that a sufficient amount of gas is readily available at all times the generating unit is likely to be operating.

Therefore, the electric utility frequently enters into a contract with the gas supplier to "reserve" a set amount of natural gas per day for use by the utility. In exchange, the utility pays a type of reservation charge, often referred to as a "firm gas transportation" charge, as shown on row (6). This charge is typically expressed in terms of dollars per million BTUs ($/mmBTU). These costs vary but are often, at the time this book is written, in the general vicinity of approximately $1.75 to $2.50/mmBTU. On row (6), we have assumed a firm gas transportation cost of $2.25/mmBTU for the CC and CT options. Because PV does not need natural gas (or any fuel other than sunlight), this cost is not applicable ("N.A.") for the PV option.

In regard to the set amount of gas the utility is reserving for the two different types of gas-fired generating units, we have assumed 100,000 mmBTU/day for the CC options, and 10,000 mmBTU/day for the CT option, as shown on row (7).* For the PV option, consideration of a needed amount of gas per day assumption is again not applicable.

When the assumed firm gas transportation charge of $2.25/mmBTU is combined with the assumed amount of gas that is reserved per day, and this is then extended to all 365 days/year, the overall annual firm gas cost is significant. Using the CC option as an example, our calculation for the annual firm gas transportation cost becomes:

$$\text{Annual firm gas transportation cost}$$
$$= \$2.25/\text{mmBTU} \times 100{,}000 \text{ mmBTU/day} \times 365 \text{ days/year}$$
$$= \$82.1 \text{ million/year for the CC options.}$$

The calculation for the CT option is identical except for the lower 10,000 mmBTU/day value and the result of the calculation for the CT option is approximately $8.2 million/year. Consequently, the PV option again has an economic advantage over both the CC and CT options in regard to this fixed cost category, and the CT option has an advantage over the CC options.

Operating Costs: Rows (8) through (15)

The remaining rows of Table 5.1 provide information related to the costs and other impacts of actually operating these Supply options on our utility system.† The first of these operating costs is shown on row (8), the variable O&M cost that is expressed in terms of $/MWh. These costs are directly tied to the number of hours the generating unit operates and may include expenses for water, lubricants, etc., but typically does not include the cost of the fuel.

As mentioned earlier, variable O&M costs typically are in the range of approximately $2.00/MWh, especially for types of Supply options that will operate at fairly high capacity factors. Accordingly, we have assumed that the two CC options have variable O&M costs of $2.00/MWh and $1.50/MWh, respectively, and the CT option's variable O&M cost is $0.20/MWh. Because the PV option has no moving parts and uses virtually no water except for cleaning the modules, we assume that a variable O&M cost is again not applicable (N.A.) for this option (i.e., it has a variable O&M cost of $0.00/MWh).

* The amount of incremental firm natural gas that a utility will actually require if it chooses to acquire a new natural gas-fired generating unit is completely utility-specific. The amount will be dependent upon a number of factors including the amount of firm natural gas the utility already has under contract, the amount of gas the new generating unit is projected to require, etc. In our discussion, we are assuming that our utility system has to acquire a significant amount of additional natural gas to operate the new gas-fired units, particularly for a new CC unit.

† Recall that the fuel-based operating costs were briefly discussed in Chapter 2.

The table next lists the type of fuel used by each option on row (9). Natural gas is used by the CC and CT options, while sunlight is used by the PV option. (The costs of natural gas were briefly discussed in Chapter 2. We will further discuss this assumption later in this chapter.)

The heat rate (or efficiency of converting fuel into electricity) of the generating unit is then listed on row (10) and is expressed in terms of BTU/kWh. The two CC options are assumed to have very low heat rates of 6,600 and 6,700 BTU/kWh, respectively, which indicate a high level of efficiency in converting fuel into electricity. By contrast, the CT option's heat rate of 10,400 BTU/kWh shows that the CT option is a much less efficient type of generating unit.

For purposes of this part of our discussion, the PV option's "heat rate" is not applicable. This is because the PV option's fuel (sunlight) is free and, therefore, the efficiency with which the PV option converts its fuel into electricity is simply not a factor in economic analyses in regard to heat rate. (However, PV's conversion efficiency is indirectly accounted for in other inputs for PV including the projected $/kW cost and capacity factor.)

The next two input categories shown in rows (11) and (12), respectively, are availability and capacity factor. Availability, shown on row (11), refers to how many hours of the year a generating unit is capable of operating after accounting for planned maintenance and projected unplanned maintenance (i.e., the unit's forced outage rate, or FOR).* The other input, capacity factor, shown on row (12), refers to how many hours of the year a generating unit is expected to operate on a particular utility system. As we have seen in Chapter 2, this value is driven by economics pertaining to the utility system in question. However, the less expensive the operating cost of a unit is, the more hours the unit will typically operate as we previously discussed.

On row (11), we have assumed availability factors of 92% for the CC and CT options based on an assumption that planned and unplanned maintenance will require approximately 8% of the hours in the year. Consequently, the CC and CT options are available to run the remaining 92% of the hours in a year.

Because PV options require relatively little planned maintenance (because there are no moving parts, at least with PV installations that do not include machinery that moves the PV modules throughout the day to track the sun's movements), the availability of PV is primarily limited by the number of expected daylight hours.† Consequently, when developing an availability

* Generating unit availability is a facet of the electric utility industry that is actually quite complicated. Therefore, the approach we're using in regard to discussing generating unit availability is a simplified one.

† Certain PV-related equipment, such as inverters that convert a PV's direct current (DC) output to alternating current (AC), can fail. Such equipment failure will also limit the output of PV facilities. However, the availability of sunlight is typically the overriding factor in estimating PV availability.

assumption for PV, 50% of the hours each year—all of the night time hours—are tossed out first. Then, one must account for a second factor: the reduction in sunlight during those daylight hours when cloud cover, rain, etc., reduce the amount of available sunlight. This second factor will vary considerably depending upon the geographic location of the PV installation. A PV installation in a very dry location will experience less of this reduction than will a PV installation in a more humid location.

Then, unless the PV module is designed to track the movement of the sun across the sky during the day, there is a third factor: the reduction in the amount or intensity of sunlight striking the PV modules at hours when the sun is not directly overhead. The output of PV modules is at a maximum when the sun strikes the modules directly; i.e., when the sun is straight overhead at midday. The PV output is reduced in earlier morning hours, and in later afternoon hours, as the sunlight strikes the PV modules at a less direct angle.

For purposes of this discussion, we have assumed that the effects of these three factors will result in our PV option having an availability of 20%, a value that is representative for PV in a number of locations at the time this book is written.

On row (12) we have assumed capacity factors of 80% for the CC options, 5% for the CT option, and 20% for the PV option. The low heat rate values (indicative of high efficiency) for the CC options will result in low operating costs, thus driving the CC options' high capacity factor value of 80%. Conversely, the high heat rate value (indicating low efficiency) for the CT option results in high operating costs, thus resulting in a low capacity factor value for the CT option of 5%. In regard to PV, because there is no fuel cost and we are assuming zero variable O&M costs, PV has no operating cost. Therefore, it will run as much as possible; as much as it is available to run. Therefore, its assumed capacity factor is identical to its assumed availability of 20%.

The last three rows of the table, rows (13), (14), and (15), provide inputs for the generating options' emission rates for three types of air emissions: sulfur dioxide (SO_2), nitrogen oxides (NO_x), and carbon dioxide (CO_2). These emission rates are typically expressed in terms of pounds of emission per million BTUs of fuel burned (lbs/mmBTU).

Row (13) shows the SO_2 emission rates for the four Supply options. For the CC and CT options, we have assumed an emission rate is 0.006 lb/mmBTU and for the PV option, this emission rate is zero because no fuel is burned. Row (14) shows the assumed NO_x emission rate is 0.010 lb/mmBTU for the CC options, 0.033 lb/mmBTU for the CT option, and zero for the PV option. Row (15) shows the assumed CO_2 emission rate is 119 lb/mmBTU for the CC options, 118 lb/mmBTU for the CT option, and zero for the PV option. The emission rate inputs shown in these rows are representative values for these types of new generating units at the time this book is written. (In the next section of this chapter, we will discuss "environmental compliance cost" values, i.e., projected costs associated with emissions, for each of these three types of air emissions.*)

This completes the introduction of the inputs for the four Supply options that will be used in our utility's economic evaluations. However, before we leave this section, let's look back and see what appears to be obvious, and perhaps more importantly, what is not obvious, from looking at these inputs.

First, from looking at the inputs, the following statements about these four Supply options appear to be obvious:

- In regard to the capital costs, the CC and CT options are much less expensive than the PV option, regardless of whether the capital costs are compared on a total cost basis or a $/kW basis. The CT option has a capital cost advantage versus the CC options.
- In regard to the other fixed costs, the converse is true; the PV option has lower costs than either the CC or CT options. The CT option again has an advantage in regard to these other fixed costs compared to the CC options.
- In regard to the operating costs, the PV option again has lower costs than either the CC or CT options, simply because it has no fuel costs and we are assuming it has no variable O&M costs. As for the CC and CT options, it is somewhat difficult to judge at a glance which of these options has an operating cost advantage. The CT option has lower variable O&M costs, a lower CO_2 emission rate, and will run much less of the time (a 5% capacity factor) than will the CC options (with an 80% capacity factor). Therefore, the fuel cost to operate the CT option during a year will be lower than with either of the CC options. Conversely, the CC options are much more efficient, and have lower NO_x emission rates, than the CT options. (There is no difference in the SO_2 emission rates between the CC and CT options.)

At this point, one might be tempted to make two statements. The first statement is that it appears obvious that the CT option has a capital and other fixed cost advantage over the CC options because: (i) the CT option is lower in both capital and other fixed costs and (ii) a comparison of the various operating costs versus the CC options are a mixed bag with the CT option being lower in regard to some of these costs and higher in regard to other costs.

The second statement one might be tempted to make is that it is not obvious whether the significant capital cost disadvantage of the PV option is

* At the time this book is written, there are both state and federal regulations for SO_2 and NO_x in the United States, and there is an established market for allowances for these two emissions. The same is not true at this time for CO_2. However, due to the expectation that U.S. federal regulations for CO_2 will eventually be established (as is currently the case in Europe), we will also be projecting compliance costs for CO_2 emissions in our discussion.

overcome by PV's clear advantages in regard to other fixed costs and operating costs.

However, one might also make a third statement. At this point it is not obvious how much effect the truly significant differences in both the capacity (MW), and the capacity factor (% of hours/year the options will operate), of the four Supply options will have in regard to determining which option is the best economic choice for our utility system.*

We will see how all of this actually plays out in the next three sections of this chapter. We will start by performing a preliminary economic screening evaluation of the four Supply options.

Preliminary Economic Screening Evaluation of the Supply Options

We shall next perform a preliminary economic screening evaluation of these four Supply options and we will use a screening curve analytical approach that was briefly discussed in Chapter 3. However, before we can conduct a screening curve analysis, there are a couple of inputs that are needed for this analysis that are not directly tied to these four specific Supply options.

These inputs are the projected costs of natural gas (the fuel that the CC and CT options utilize) and the projected environmental compliance costs for the three types of air emissions that our utility system is including in its analyses: SO_2, NO_x, and CO_2. Recalling that the screening curve analytical approach is a multi-year analysis, we will need projected cost values for fuel and environmental compliance for a number of years. Assuming that the four Supply options have an operating life (or, perhaps more accurately, an accounting "book life") of 25 years, and knowing that our utility's resource needs begin 5 years in the future, our projections for fuel cost and environmental compliance costs will need to address 30 years.†

These assumed cost projections do not attempt to account for projected timing of various market corrections, changes in economic conditions, proposed or potential environmental/energy legislation, and/or potential advancement in generating or emission control equipment. One reason for this is that our economic evaluation is not designed to look at an actual

* If one could accurately pick out the most economic resource option just glancing at the inputs that will be used in economic and non-economic analyses, then there would be no need for this book. (And I could have spent this time playing more golf.) Sadly, it is not that easy.

† We again offer a reminder that the assumed projected values for fuel costs and environmental costs would, in reality, certainly change many times in the years after this book is written. However, the assumed values are necessary tools used to illustrate resource planning concepts and analytical approaches. These concepts and analytical approaches will remain valid regardless of the assumed values used to illustrate them.

specific starting year (e.g., 2015 or 2065), but is presented in terms of a "Current Year" and 29 years into the future from that year. For this reason, it made no sense to attempt to factor in the timing of such potential occurrences of the types listed earlier. (In practice, because the actual starting year for the forecast is known, utility forecasters of load, fuel costs, environmental compliance costs, etc., often do attempt to incorporate the impacts of projected specific events.)

And, as previously mentioned, utilities attempt to account for uncertainty in various forecasts by utilizing multiple forecasts in their resource planning work. For simplicity's sake in discussing the concepts and analytical approaches of utility resource planning, we are using single forecasts for load, fuel costs, environmental compliance costs, etc.

The annual costs for natural gas and the annual compliance costs for SO_2, NO_x, and CO_2 emissions that we are assuming for the economic evaluations of the four Supply options are presented in Table 5.2. As previously mentioned, 30 years of projected costs, from "Current Year" through "Current + 29" year, are presented.

Armed with the Supply Option–specific cost assumptions previously provided in Table 5.1, and the assumptions for natural gas and environmental compliance costs presented in Table 5.2, we can now conduct a preliminary economic screening evaluation of the four Supply options using a screening curve approach.*

Figure 5.1 presents the results of the screening curve analysis for these four Supply options.

As we can see, a $/MWh value is presented for each capacity factor value in increments of 5%, i.e., at 5%, 10%, etc. (Note that no value for a capacity factor of 0% is shown because that would reflect only the capital and other fixed costs for the option; no operating hours—and therefore no operating costs—would be assumed.) Then, for each subsequent 5% increase in capacity factor, the option's operating costs are added to the capital and other fixed costs. Then that total cost is divided by the number of MWh that represents that option's capacity (500 MW for CC, 160 MW for CT, or 120 MW for PV) operating at the given capacity factor. For example, for the CT option, the $/MWh value shown at the 5% capacity factor mark represents the total costs divided by 160 MW × 8,760 h/year × 5% (or the total costs divided by 70,080 MWh).

As we can see from Figure 5.1, there is a wide range of values driven partly by the capital costs of the Supply options, but driven mostly by the wide disparity in the assumed maximum capacity factors for each Supply option:

* Yes, I know that I said earlier that a screening curve approach cannot provide meaningful information when the resource options being compared are dissimilar in regard to four key characteristics. We are ignoring the differences in the Supply options for the moment solely for the purpose of allowing you to get more comfortable with how screening curve results "look." Fear not, we will return to discuss the significant limitations of screening curve analyses in short order.

TABLE 5.2

Assumed Costs for Natural Gas and Environmental Compliance

Year	Natural Gas Price ($/mmBTU)	Environmental Compliance Costs		
		SO_2 ($/ton)	NO_x ($/ton)	CO_2 ($/ton)
Current	$6.00	$1,100	$1,800	$5
Current + 1	$6.12	$1,155	$1,836	$8
Current + 2	$6.24	$1,213	$1,873	$11
Current + 3	$6.37	$1,273	$1,910	$14
Current + 4	$6.49	$1,337	$1,948	$17
Current + 5	$6.62	$1,404	$1,987	$20
Current + 6	$6.76	$1,474	$2,027	$23
Current + 7	$6.89	$1,548	$2,068	$26
Current + 8	$7.03	$1,625	$2,109	$29
Current + 9	$7.17	$1,706	$2,151	$32
Current + 10	$7.31	$1,792	$2,194	$35
Current + 11	$7.46	$1,881	$2,238	$38
Current + 12	$7.61	$1,975	$2,283	$41
Current + 13	$7.76	$2,074	$2,328	$44
Current + 14	$7.92	$2,178	$2,375	$47
Current + 15	$8.08	$2,287	$2,423	$50
Current + 16	$8.24	$2,401	$2,471	$53
Current + 17	$8.40	$2,521	$2,520	$56
Current + 18	$8.57	$2,647	$2,571	$59
Current + 19	$8.74	$2,780	$2,622	$62
Current + 20	$8.92	$2,919	$2,675	$65
Current + 21	$9.09	$3,065	$2,728	$68
Current + 22	$9.28	$3,218	$2,783	$71
Current + 23	$9.46	$3,379	$2,838	$74
Current + 24	$9.65	$3,548	$2,895	$77
Current + 25	$9.84	$3,725	$2,953	$80
Current + 26	$10.04	$3,911	$3,012	$83
Current + 27	$10.24	$4,107	$3,072	$86
Current + 28	$10.45	$4,312	$3,134	$89
Current + 29	$10.66	$4,528	$3,197	$92

5% for CT, 20% for PV, and 80% for CC. As a glance at this particular figure shows, the wide range of values typically presented in a screening graph can make quite difficult to visually distinguish between the values for the different resource options. In order to clear up this picture a bit, we will take a closer look at the options, but this look will examine only pairs of options with similar maximum capacity factors. This will allow us to narrow the focus in regard to the capacity factors addressed in the figure.

We first look at the two options, CT and PV, which have the lower maximum capacity factors. This view is provided in Figure 5.2.

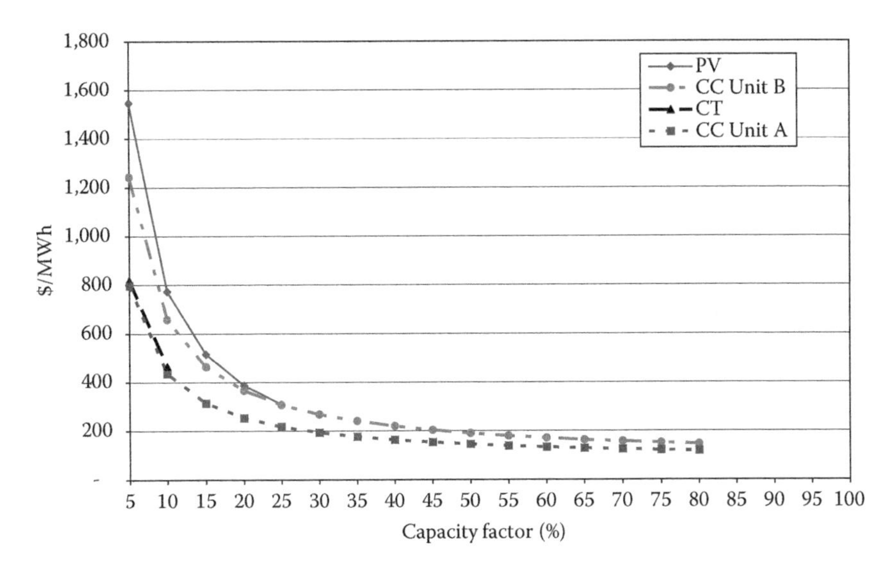

FIGURE 5.1
(See color insert.) Preliminary economic screening analysis: Screening approach levelized $/MWh costs for all four supply options.

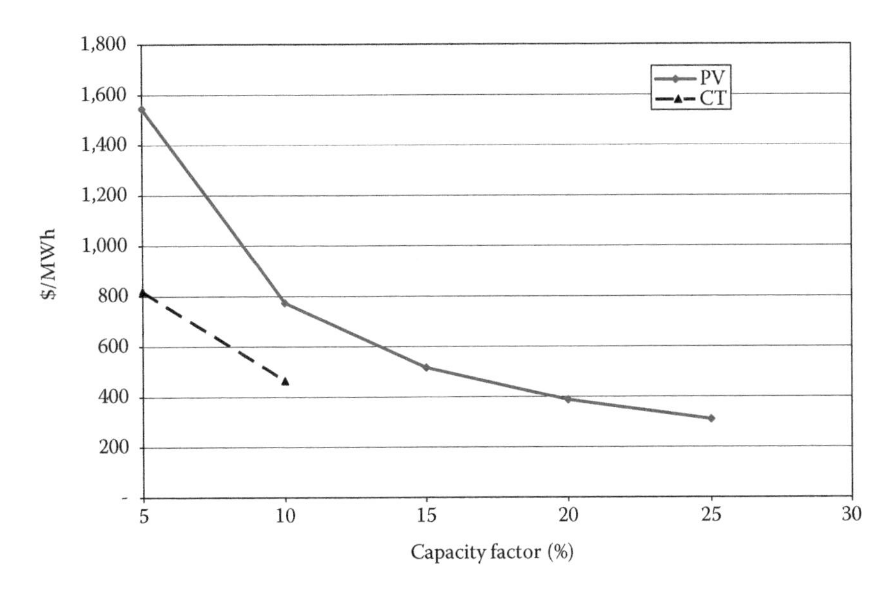

FIGURE 5.2
(See color insert.) Preliminary economic screening analysis: Screening approach levelized $/MWh costs for CT and PV options.

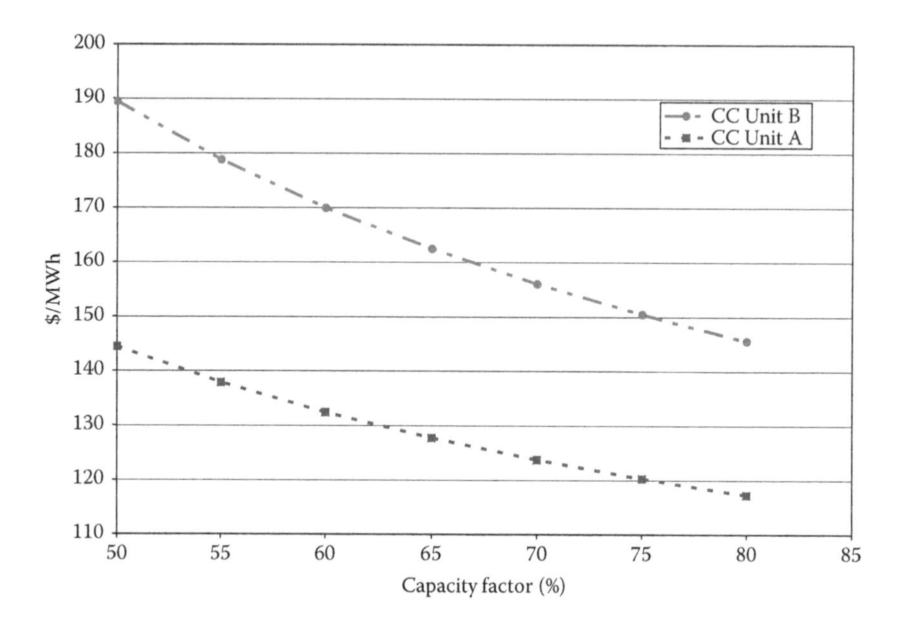

FIGURE 5.3
(See color insert.) Preliminary economic screening analysis: Screening approach levelized $/MWh costs for CC Unit A and CC Unit B.

In this figure, we narrow our focus to the range of capacity factors that the two resource options will operate in: 5% to 20%.* This narrower focus allows us to examine these two lower capacity factor options more closely. From this figure it is clear that the $/MWh cost for the PV option is definitely higher than the $/MWh cost for the CT option at a 5% capacity factor level. However, at the 20% capacity factor level where the PV option is expected to operate, the PV option's $/MWh cost becomes lower than the CT option's cost at a 5% (or 10%) capacity factor.

In Figure 5.3, we now examine the projected $/MWh costs for the two CC options. We also narrow the focus for the two CC options as well.†

In this figure, our focus is narrowed to a range of capacity factors (50% to 80%) that is closer to how these CC units will actually operate on our utility system (80%). This view shows that CC Option A is clearly the less expensive of the two CC options throughout the range of capacity factors.

So what do we do with this screening curve information? Recall the discussion in Chapter 3 in which we said that a basic rule of using a screening curve approach is that the resource options being analyzed must be identical,

* For illustrative purposes, we have extended the CT option's capacity factor maximum value to 10% to avoid having only one point of data in the figure for that option. We have performed the same capacity factor extension for the PV option as well (so it would not feel slighted).

† You may note that you have seen this figure before, only with slightly different labeling of the two resource options, in Figure 3.1 in Chapter 3.

or at least very similar, in each of four key characteristics? This seems like a good time to state what those key characteristics are.

The four key characteristics of resource options that must be identical, or very similar, for a screening curve approach to yield meaningful results for a comparison of those resource options are:

1. Capacity (MW);
2. Annual capacity factor;
3. The percentage of the option's capacity (MW) that can be considered as firm capacity at the utility's system peak hours; and,
4. The projected life of the option.

We will soon see why these characteristics are so important in regard to screening curve analyses as we work through the remainder of this chapter.

Armed now with this knowledge of what the four key characteristics are, we return first to the screening curve analysis of CC Option A and CC Option B. Both of these CC units have identical capacity (500 MW) and annual capacity factor (80%) assumptions, both CC units are assumed to have 100% of their capacity as firm capacity at the utility system's peak hour, and both CC units have identical projected life values (25 years).

In other words, our rule that all four key characteristics must be identical, or very similar, for a screening curve to provide meaningful results is met in regard to the two CC options. Therefore, we can safely eliminate CC Option B using the screening curve results because its $/MWh cost is definitely higher than is CC Option A's $/MWh cost throughout the capacity factor range. So we can safely say "goodbye" to CC Option B.

However, we note that two of the four key characteristics for CT and PV are neither identical, nor even very similar, to the CC options (or to each other). In regard to both the CT and PV options, their capacities of 160 MW and 120 MW, respectively, are very dissimilar to the 500 MW capacity of the CC units. There are also very large differences in the capacity factors of the three types of Supply options: 5% for CT, 20% for PV, and 80% for CC. Therefore, even though the assumptions we have made for the remaining two key characteristics (the percentage of capacity that can be considered as "firm" capacity at the peak hour,* and the projected life of the option†) for the CT and PV options are identical to those for the CC option, the basic rule for utilizing a screening curve approach is violated in regard to these three Supply options.

Therefore, a screening curve approach cannot be used to eliminate any of the three remaining Supply options: CC Unit A, the CT option, or the PV

* Recall that we have made the unrealistically optimistic assumption for PV that 100% of its output will be consistently available at our utility's peak hour. If we had used a more realistic assumption for PV, it would have also "failed" the basic rule regarding the use of a screening test in regard to this third key characteristic.

† Recall that we are assuming that all of the Supply options have a 25-year life.

option. A screening curve approach is simply not adequate to provide meaningful information regarding such dissimilar resource options.

Consequently, the preliminary economic screening evaluation using a screening curve approach has eliminated only one Supply option, CC Option B. The remaining Supply options; CC Unit A, the CT option, and the PV option, are carried forward to the final (or system) economic evaluation. We prepare for that evaluation in the next section. (And, near the end of this preparation, we shall see why dissimilarities in even one of the four key characteristics of resource options can severely limit the usefulness of a screening curve analytical approach.)

Creating the Competing "Supply Only" Resource Plans

As discussed in Chapter 3, an integrated resource planning (IRP) approach looks at long-term resource plans in order to ensure that all of the impacts that a proposed resource addition will have on the utility system over an extended period of time are captured in the analyses. We are now at a point where we will create three separate resource plans. Each resource plan will utilize one of the three remaining Supply options. And, for purposes of distinguishing these three resource plans from other resource plans that we will develop later when examining DSM options, we will refer to these three resource plans collectively as "Supply Only" resource plans.

In creating these Supply Only resource plans, we must first extend the projection of the utility's reserve margin beyond the 6-year period (Current Year through Current Year + 5) we examined in Table 4.3. This is necessary because the addition of any one of the three Supply options in Current Year + 5 will only meet the utility's projected needs for a few years at best if we assume (and we do) that the utility's load forecast continues to grow at 100 MW/year.

Our utility's reserve margin projection, assuming no resource additions after Current Year + 1, can be extended to encompass the assumed 25-year life of our three remaining Supply options. That projection is shown in Table 5.3.

Note that the information contained in Table 5.3 for the Current Year through Current Year + 5 is identical to that shown previously in Table 4.3. Then information for 24 more years (through Current Year + 29) is presented. We see that the projection of continued load growth of 100 MW/year (as seen in Column (2)) continues to decrease the utility's projected reserve margin each year (as seen in Column (4)). In fact, the projected reserve margin values even turn negative starting in Current Year + 25.

Consequently, it is clear that the addition of any of the three new Supply options in Current Year + 5 will only carry our utility so far. Other resource options will need to be added at various times in later years to ensure that our utility's reserve margin criterion continues to be met. But when must

TABLE 5.3

Long-Term Projection of Reserve Margin for the Hypothetical Utility: Assuming No Resource Additions after "Current Year + 1"

	(1)	(2)	(3)	(4)	(5)
			= (1) − (2)	= (3)/(2)	= ((2)*1.20) − (1)
Year	Total Generating Capacity (MW)	Peak Electrical Demand (MW)	Reserves (MW)	Reserve Margin (%)	Generation Only MW Needed to Meet Reserve Margin (MW)
Current Year	11,600	9,600	2,000	20.8	(80)
Current Year + 1	12,000	9,700	2,300	23.7	(360)
Current Year + 2	12,000	9,800	2,200	22.4	(240)
Current Year + 3	12,000	9,900	2,100	21.2	(120)
Current Year + 4	12,000	10,000	2,000	20.0	0
Current Year + 5	12,000	10,100	1,900	18.8	120
Current Year + 6	12,000	10,200	1,800	17.6	240
Current Year + 7	12,000	10,300	1,700	16.5	360
Current Year + 8	12,000	10,400	1,600	15.4	480
Current Year + 9	12,000	10,500	1,500	14.3	600
Current Year + 10	12,000	10,600	1,400	13.2	720
Current Year + 11	12,000	10,700	1,300	12.1	840
Current Year + 12	12,000	10,800	1,200	11.1	960
Current Year + 13	12,000	10,900	1,100	10.1	1,080
Current Year + 14	12,000	11,000	1,000	9.1	1,200
Current Year + 15	12,000	11,100	900	8.1	1,320
Current Year + 16	12,000	11,200	800	7.1	1,440
Current Year + 17	12,000	11,300	700	6.2	1,560
Current Year + 18	12,000	11,400	600	5.3	1,680
Current Year + 19	12,000	11,500	500	4.3	1,800
Current Year + 20	12,000	11,600	400	3.4	1,920
Current Year + 21	12,000	11,700	300	2.6	2,040
Current Year + 22	12,000	11,800	200	1.7	2,160
Current Year + 23	12,000	11,900	100	0.8	2,280
Current Year + 24	12,000	12,000	0	0.0	2,400
Current Year + 25	12,000	12,100	(100)	−0.8	2,520
Current Year + 26	12,000	12,200	(200)	−1.6	2,640
Current Year + 27	12,000	12,300	(300)	−2.4	2,760
Current Year + 28	12,000	12,400	(400)	−3.2	2,880
Current Year + 29	12,000	12,500	(500)	−4.0	3,000

these other resource additions be added and what do we assume these new resource additions will be?

For purposes of this discussion we will refer to these later resource additions as "filler units" because their purpose is to "fill in" as needed in later years when our utility's reserve margin criterion would not otherwise be met. However, our utility is not making any definitive decision at this time regarding what type of resource addition would be made in these future years because it does not have to make that decision now. The only definitive resource option decision our utility has to make now is for Current Year + 5.

For that reason, we will refer to Current Year + 5 as the "decision year." And, as discussed in Chapter 3, the addition of the filler units will ensure that each resource plan meets the utility's reliability criterion in all years, thus ensuring that the comparison of competing resource plans is meaningful.*

We have three Supply options that we are discussing; CC Unit A, CT, and PV. (Recall that we will examine DSM options in the next chapter.) Rather than introduce a fourth Supply option to play the role of the filler units, we will simply select one of these three Supply options to also play the role of the filler unit. For illustrative purposes, we will select the 500 MW CC Unit A to play the role of filler unit.

We are now ready to take a look at how the introduction of any of the three remaining Supply options in Current Year + 5 will lead to the creation of a resource plan. This is done in two steps. First, we look at how the introduction of one of these three Supply options in Current Year + 5 impacts the utility's projected reserve margin. Second, using that information, we add a 500 MW filler unit in the appropriate years so that the reserve margin criterion is met in all years.

We will start with the introduction of CC Unit A in Current Year + 5 and see how the utility's projected reserve margin is impacted in Table 5.4.

Let's first discuss the two changes we have made to the table itself. First, Column (1) in the previous Table 5.3 has been expanded to encompass Columns (1a) through (1f). The values in Column (1a) are identical to what appeared in Column (1) in the previous table. The values in Columns (1b) through (1e) represent the cumulative number of new Supply options of the various types that are assumed to be added: the number of CC options in Column (1b), the number of CT options in Column (1c), the number of PV options in Column (1d), and the number of filler units in Column (1e). Finally, Column (1f) then shows the total generating capacity for the utility for each year.

Second, the capacity (MW) for each of these types of Supply options is shown in a small table that appears below Columns (1b) through (1d). The number of Supply options of each type is then multiplied by the MW for

* The questions of how many years in the future one continues to add filler units, and what type(s) of filler unit is used, are often discussed by utilities as "end effects" in resource planning. The real issue is to attempt to ensure that these choices are not unduly influencing the choice of the resource option being selected for the near-term decision year. Therefore, the number of years for which filler units are added, and/or the types of filler units, may be varied by the utility in additional analyses of resource plans in order to determine how much influence these assumptions may have.

each Supply option type shown in this small table. For example, by looking at the row for Current Year + 5 and Column (1b), we see that 1 CC unit, CC Unit A, is added in that year. This number of CC units, 1 unit, is multiplied by the 500 MW for each CC unit from the small table to yield a gain of 500 MW in that year. Likewise, no (zero) CT, PV, or filler units are added in that year so 0 MW are added in that year from these types of Supply options.

Column (1f) then adds the previously projected generating capacity in Column (1a) to the additional MW from each of the types of Supply options to derive the total projected generating capacity for that year. This total is presented in Column (1f). For Current Year + 5, the calculation is 12,000 MW + (1 × 500 MW) + (0 × 160 MW) + (0 × 120 MW) + (0 × 500 MW) = 12,500 MW. (This formula is shown on the table starting at the bottom of Column (1f).)

By examining the reserve margin values in Column (4) we see that the addition of CC Unit A in Current Year + 5 allows the utility to not only meet its reserve margin criterion for that year, but also for 3 more years. The first year in which the reserve margin criterion is not met is now Current Year + 9. That year is highlighted by shading the Current Year + 9 row in the table. Consequently, no filler unit addition is needed until Current Year + 9. In subsequent years, additional filler units will also be needed. We see how many filler units are needed, and when they are needed, in Table 5.5.

By examining Column (1e) of this table, we see that a cumulative total of five filler units have been added through the 30-year analysis period. These filler units are added, one each, in: Current Year + 9, Current Year + 13, Current Year + 18, Current Year + 22, and Current Year + 26.*

We now have a resource plan that adds CC Unit A in Current Year + 5, then adds 5 filler units over the course of the next 25 years. We will (very creatively) name this resource plan Supply Only Resource Plan 1 (CC).

We will now construct similar resource plans based on adding either the CT option or the PV option in Current Year + 5, then adding the appropriate number of filler units as needed in subsequent years. Table 5.6 shows the projected reserve margin if the CT option is added.

As shown in Table 5.6, the addition of the 160 MW CT option in Current Year + 5 results in the utility still having a resource need in the next year (Current Year + 6) because the projected reserve margin is 19.2% for that year. That row is again shaded in the table. Table 5.7 shows the subsequent addition of filler units that are needed starting in Current Year + 6.

* The alert reader may ask why no filler unit is added in certain years such as Current Year + 17 in which the projected reserve margin is 19.5%. In practice, utilities may allow the reserve margin to fall slightly below their reliability criterion for a particular year. Our hypothetical utility has taken that approach and will not add a filler unit unless the projected reserve margin drops below 19.50%, i.e., below a value that will round up to the reserve margin criterion of 20%.

TABLE 5.4

Long-Term Projection of Reserve Margin for the Hypothetical Utility: With New CC Unit Added in "Current Year+5"

	(1a)	(1b)	(1c)	(1d)	(1e)	(1f)	(2)	(3)	(4)	(5)
						=(See Formula Below)		=(1f) − (2)	=(3)/(2)	=((2)*1.20) − (1f)
		Cumulative No. of New Unit Additions								
Year	Previously Projected Generating Capacity (MW)	CC Unit A (No. Units)	CT (No. Units)	PV (No. Units)	Filler Units (No. Units)	Total Projected Generating Capacity (MW)	Peak Electrical Demand (MW)	Reserves (MW)	Reserve Margin (%)	Generation Only MW Needed to Meet Reserve Margin (MW)
Current Year	11,600	0	0	0	0	11,600	9,600	2,000	20.8	(80)
Current Year+1	12,000	0	0	0	0	12,000	9,700	2,300	23.7	(360)
Current Year+2	12,000	0	0	0	0	12,000	9,800	2,200	22.4	(240)
Current Year+3	12,000	0	0	0	0	12,000	9,900	2,100	21.2	(120)
Current Year+4	12,000	0	0	0	0	12,000	10,000	2,000	20.0	0
Current Year+5	12,000	1	0	0	0	12,500	10,100	2,400	23.8	(380)
Current Year+6	12,000	1	0	0	0	12,500	10,200	2,300	22.5	(260)
Current Year+7	12,000	1	0	0	0	12,500	10,300	2,200	21.4	(140)
Current Year+8	12,000	1	0	0	0	12,500	10,400	2,100	20.2	(20)
Current Year+9	12,000	1	0	0	0	12,500	10,500	2,000	19.0	100
Current Year+10	12,000	1	0	0	0	12,500	10,600	1,900	17.9	220
Current Year+11	12,000	1	0	0	0	12,500	10,700	1,800	16.8	340
Current Year+12	12,000	1	0	0	0	12,500	10,800	1,700	15.7	460
Current Year+13	12,000	1	0	0	0	12,500	10,900	1,600	14.7	580
Current Year+14	12,000	1	0	0	0	12,500	11,000	1,500	13.6	700
Current Year+15	12,000	1	0	0	0	12,500	11,100	1,400	12.6	820

Current Year + 16	12,000	1	0	0	0	12,500	11,200	1,300	11.6	940
Current Year + 17	12,000	1	0	0	0	12,500	11,300	1,200	10.6	1,060
Current Year + 18	12,000	1	0	0	0	12,500	11,400	1,100	9.6	1,180
Current Year + 19	12,000	1	0	0	0	12,500	11,500	1,000	8.7	1,300
Current Year + 20	12,000	1	0	0	0	12,500	11,600	900	7.8	1,420
Current Year + 21	12,000	1	0	0	0	12,500	11,700	800	6.8	1,540
Current Year + 22	12,000	1	0	0	0	12,500	11,800	700	5.9	1,660
Current Year + 23	12,000	1	0	0	0	12,500	11,900	600	5.0	1,780
Current Year + 24	12,000	1	0	0	0	12,500	12,000	500	4.2	1,900
Current Year + 25	12,000	1	0	0	0	12,500	12,100	400	3.3	2,020
Current Year + 26	12,000	1	0	0	0	12,500	12,200	300	2.5	2,140
Current Year + 27	12,000	1	0	0	0	12,500	12,300	200	1.6	2,260
Current Year + 28	12,000	1	0	0	0	12,500	12,400	100	0.8	2,380
Current Year + 29	12,000	1	0	0	0	12,500	12,500	0	0.0	2,500

CC Unit A =	500	MW	Formula: (1f) = (1a) + ((1b) × 500) + ((1c) × 160) + ((1d) × 120) + ((1e) × 500)
CT =	160	MW	
PV =	120	MW	
Filler Units =	500	MW	

TABLE 5.5

Supply Only Resource Plan 1 with New CC Unit Added in "Current Year + 5" Plus Filler Units

	(1a)	(1b)	(1c)	(1d)	(1e)	(1f)	(2)	(3)	(4)	(5)
						= (See Formula Below)		= (1f) − (2)	= (3)/(2)	= ((2)*1.20) − (1f)
		Cumulative No. of New Unit Additions								
Year	Previously Projected Generating Capacity (MW)	CC Unit A (No. Units)	CT (No. Units)	PV (No. Units)	Filler Units (No. Units)	Total Projected Generating Capacity (MW)	Peak Electrical Demand (MW)	Reserves (MW)	Reserve Margin (%)	Generation Only MW Needed to Meet Reserve Margin (MW)
Current Year	11,600	0	0	0	0	11,600	9,600	2,000	20.8	(80)
Current Year + 1	12,000	0	0	0	0	12,000	9,700	2,300	23.7	(360)
Current Year + 2	12,000	0	0	0	0	12,000	9,800	2,200	22.4	(240)
Current Year + 3	12,000	0	0	0	0	12,000	9,900	2,100	21.2	(120)
Current Year + 4	12,000	0	0	0	0	12,000	10,000	2,000	20.0	0
Current Year + 5	12,000	1	0	0	0	12,500	10,100	2,400	23.8	(380)
Current Year + 6	12,000	1	0	0	0	12,500	10,200	2,300	22.5	(260)
Current Year + 7	12,000	1	0	0	0	12,500	10,300	2,200	21.4	(140)
Current Year + 8	12,000	1	0	0	0	12,500	10,400	2,100	20.2	(20)
Current Year + 9	12,000	1	0	0	1	13,000	10,500	2,500	23.8	(400)
Current Year + 10	12,000	1	0	0	1	13,000	10,600	2,400	22.6	(280)
Current Year + 11	12,000	1	0	0	1	13,000	10,700	2,300	21.5	(160)
Current Year + 12	12,000	1	0	0	1	13,000	10,800	2,200	20.4	(40)
Current Year + 13	12,000	1	0	0	2	13,500	10,900	2,600	23.9	(420)
Current Year + 14	12,000	1	0	0	2	13,500	11,000	2,500	22.7	(300)
Current Year + 15	12,000	1	0	0	2	13,500	11,100	2,400	21.6	(180)
Current Year + 16	12,000	1	0	0	2	13,500	11,200	2,300	20.5	(60)

Current Year + 17	12,000	1	0	0	2	13,500	11,300	2,200	19.5	60
Current Year + 18	12,000	1	0	0	3	14,000	11,400	2,600	22.8	(320)
Current Year + 19	12,000	1	0	0	3	14,000	11,500	2,500	21.7	(200)
Current Year + 20	12,000	1	0	0	3	14,000	11,600	2,400	20.7	(80)
Current Year + 21	12,000	1	0	0	3	14,000	11,700	2,300	19.7	40
Current Year + 22	12,000	1	0	0	4	14,500	11,800	2,700	22.9	(340)
Current Year + 23	12,000	1	0	0	4	14,500	11,900	2,600	21.8	(220)
Current Year + 24	12,000	1	0	0	4	14,500	12,000	2,500	20.8	(100)
Current Year + 25	12,000	1	0	0	4	14,500	12,100	2,400	19.8	20
Current Year + 26	12,000	1	0	0	5	15,000	12,200	2,800	23.0	(360)
Current Year + 27	12,000	1	0	0	5	15,000	12,300	2,700	22.0	(240)
Current Year + 28	12,000	1	0	0	5	15,000	12,400	2,600	21.0	(120)
Current Year + 29	12,000	1	0	0	5	15,000	12,500	2,500	20.0	0

CC Unit A =	500	MW
CT =	160	MW
PV =	120	MW
Filler Units =	500	MW

Formula: (1f) = (1a) + ((1b) × 500) + ((1c) × 160) + ((1d) × 120) + ((1e) × 500)

TABLE 5.6

Long-Term Projection of Reserve Margin for the Hypothetical Utility: With New CT Unit Added in "Current Year+5"

	(1a)	(1b)	(1c)	(1d)	(1e)	(1f)	(2)	(3)	(4)	(5)
						=(See Formula Below)		=(1f) – (2)	=(3)/(2)	=((2)*1.20) – (1f)
		Cumulative No. of New Unit Additions								
Year	Previously Projected Generating Capacity (MW)	CC Unit A (No. Units)	CT (No. Units)	PV (No. Units)	Filler Units (No. Units)	Total Projected Generating Capacity (MW)	Peak Electrical Demand (MW)	Reserves (MW)	Reserve Margin (%)	Generation Only MW Needed to Meet Reserve Margin (MW)
Current Year	11,600	0	0	0	0	11,600	9,600	2,000	20.8	(80)
Current Year+1	12,000	0	0	0	0	12,000	9,700	2,300	23.7	(360)
Current Year+2	12,000	0	0	0	0	12,000	9,800	2,200	22.4	(240)
Current Year+3	12,000	0	0	0	0	12,000	9,900	2,100	21.2	(120)
Current Year+4	12,000	0	0	0	0	12,000	10,000	2,000	20.0	0
Current Year+5	12,000	0	1	0	0	12,160	10,100	2,060	20.4	(40)
Current Year+6	12,000	0	1	0	0	12,160	10,200	1,960	19.2	80
Current Year+7	12,000	0	1	0	0	12,160	10,300	1,860	18.1	200
Current Year+8	12,000	0	1	0	0	12,160	10,400	1,760	16.9	320
Current Year+9	12,000	0	1	0	0	12,160	10,500	1,660	15.8	440
Current Year+10	12,000	0	1	0	0	12,160	10,600	1,560	14.7	560
Current Year+11	12,000	0	1	0	0	12,160	10,700	1,460	13.6	680
Current Year+12	12,000	0	1	0	0	12,160	10,800	1,360	12.6	800
Current Year+13	12,000	0	1	0	0	12,160	10,900	1,260	11.6	920
Current Year+14	12,000	0	1	0	0	12,160	11,000	1,160	10.5	1,040
Current Year+15	12,000	0	1	0	0	12,160	11,100	1,060	9.5	1,160

Current Year + 16	12,000	0	1	0	0	12,160	11,200	960	8.6	1,280
Current Year + 17	12,000	0	1	0	0	12,160	11,300	860	7.6	1,400
Current Year + 18	12,000	0	1	0	0	12,160	11,400	760	6.7	1,520
Current Year + 19	12,000	0	1	0	0	12,160	11,500	660	5.7	1,640
Current Year + 20	12,000	0	1	0	0	12,160	11,600	560	4.8	1,760
Current Year + 21	12,000	0	1	0	0	12,160	11,700	460	3.9	1,880
Current Year + 22	12,000	0	1	0	0	12,160	11,800	360	3.1	2,000
Current Year + 23	12,000	0	1	0	0	12,160	11,900	260	2.2	2,120
Current Year + 24	12,000	0	1	0	0	12,160	12,000	160	1.3	2,240
Current Year + 25	12,000	0	1	0	0	12,160	12,100	60	0.5	2,360
Current Year + 26	12,000	0	1	0	0	12,160	12,200	(40)	–0.3	2,480
Current Year + 27	12,000	0	1	0	0	12,160	12,300	(140)	–1.1	2,600
Current Year + 28	12,000	0	1	0	0	12,160	12,400	(240)	–1.9	2,720
Current Year + 29	12,000	0	1	0	0	12,160	12,500	(340)	–2.7	2,840

CC Unit A =	500	MW
CT =	160	MW
PV =	120	MW
Filler Units =	500	MW

Formula: (1f) = (1a) + ((1b) × 500) + ((1c) × 160) + ((1d) × 120) + ((1e) × 500)

TABLE 5.7

Supply Only Resources Plan 2 with New CT Unit Added in "Current Year + 5" Plus Filler Units

	(1a)	(1b)	(1c)	(1d)	(1e)	(1f)	(2)	(3)	(4)	(5)
						= (See Formula Below)		= (1f) – (2)	= (3)/(2)	= ((2)*1.20) – (1f)
		Cumulative No. of New Unit Additions								
Year	Previously Projected Generating Capacity (MW)	CC Unit A (No. Units)	CT (No. Units)	PV (No. Units)	Filler Units (No. Units)	Total Projected Generating Capacity (MW)	Peak Electrical Demand (MW)	Reserves (MW)	Reserve Margin (%)	Generation Only MW Needed to Meet Reserve Margin (MW)
Current Year	11,600	0	0	0	0	11,600	9,600	2,000	20.8	(80)
Current Year + 1	12,000	0	0	0	0	12,000	9,700	2,300	23.7	(360)
Current Year + 2	12,000	0	0	0	0	12,000	9,800	2,200	22.4	(240)
Current Year + 3	12,000	0	0	0	0	12,000	9,900	2,100	21.2	(120)
Current Year + 4	12,000	0	0	0	0	12,000	10,000	2,000	20.0	0
Current Year + 5	12,000	0	1	0	0	12,160	10,100	2,060	20.4	(40)
Current Year + 6	12,000	0	1	0	1	12,660	10,200	2,460	24.1	(420)
Current Year + 7	12,000	0	1	0	1	12,660	10,300	2,360	22.9	(300)
Current Year + 8	12,000	0	1	0	1	12,660	10,400	2,260	21.7	(180)
Current Year + 9	12,000	0	1	0	1	12,660	10,500	2,160	20.6	(60)
Current Year + 10	12,000	0	1	0	2	13,160	10,600	2,560	24.2	(440)
Current Year + 11	12,000	0	1	0	2	13,160	10,700	2,460	23.0	(320)
Current Year + 12	12,000	0	1	0	2	13,160	10,800	2,360	21.9	(200)
Current Year + 13	12,000	0	1	0	2	13,160	10,900	2,260	20.7	(80)
Current Year + 14	12,000	0	1	0	2	13,160	11,000	2,160	19.6	40
Current Year + 15	12,000	0	1	0	3	13,660	11,100	2,560	23.1	(340)

Current Year + 16	12,000	0	1	0	3	13,660	11,200	2,460	22.0	(220)
Current Year + 17	12,000	0	1	0	3	13,660	11,300	2,360	20.9	(100)
Current Year + 18	12,000	0	1	0	3	13,660	11,400	2,260	19.8	20
Current Year + 19	12,000	0	1	0	4	14,160	11,500	2,660	23.1	(360)
Current Year + 20	12,000	0	1	0	4	14,160	11,600	2,560	22.1	(240)
Current Year + 21	12,000	0	1	0	4	14,160	11,700	2,460	21.0	(120)
Current Year + 22	12,000	0	1	0	4	14,160	11,800	2,360	20.0	0
Current Year + 23	12,000	0	1	0	5	14,660	11,900	2,760	23.2	(380)
Current Year + 24	12,000	0	1	0	5	14,660	12,000	2,660	22.2	(260)
Current Year + 25	12,000	0	1	0	5	14,660	12,100	2,560	21.2	(140)
Current Year + 26	12,000	0	1	0	5	14,660	12,200	2,460	20.2	(20)
Current Year + 27	12,000	0	1	0	6	15,160	12,300	2,860	23.3	(400)
Current Year + 28	12,000	0	1	0	6	15,160	12,400	2,760	22.3	(280)
Current Year + 29	12,000	0	1	0	6	15,160	12,500	2,660	21.3	(160)

CC Unit A=	500	MW
CT =	160	MW
PV =	120	MW
Filler Units =	500	MW

Formula: $(1f) = (1a) + ((1b) \times 500) + ((1c) \times 160) + ((1d) \times 120) + ((1e) \times 500)$

By examining Column (1e) of this table we see that a cumulative total of six filler units have been added through the 30-year analysis period. These filler units are added, one each, in: Current Year + 6, Current Year + 10, Current Year + 15, Current Year + 19, Current Year + 23, and Current Year + 27.

Therefore, the addition of the CT option results both in (i) a greater number of filler units being added overall and (ii) a change in the timing of when filler units are added (i.e., in what years they are added). This second Supply option resource plan, based on the addition of the CT option, is definitely different than the first Supply option resource plan which was based on the CC option. Continuing to be led by our highly creative instincts, we will refer to this second resource plan as Supply Only Resource Plan 2 (CT).

We now turn to the PV option and Table 5.8 shows the projected reserve margin if the PV option is added in Current Year + 5.

Similar to the projected reserve margin picture when the CT option is added, we see that our utility will also not meet its reserve margin criterion in Current Year + 6. That row is again shaded in the table. Table 5.9 shows how the filler unit additions must occur if the PV option is selected.

We once again examine Column (1e) for this table and we see that six filler units have again been added through the 30-year analysis period. These filler units are added, one each, in the same years as filler units were added for the previous resource plan featuring the CT option with one exception; we now add a filler unit 2 years earlier in Current Year + 13 instead of in Current Year + 15. Therefore, although the addition of the PV option results in the same number of filler units being added as with the CT option, there is still a change in the timing of one of those filler units. Therefore, this third resource plan differs from the previous two resource plans. And, as you surely have guessed by now, we will refer to this third resource plan as Supply Only Resource Plan 3 (PV).

If we were to put these three resource plans side-by-side, we could more clearly see the differences between them. Therefore, we do so with Table 5.10.

Table 5.10 shows the significant differences between the three Supply Only resource plans, particularly in regard to the timing and number of filler units being added in each resource plan. By looking at this table, one can see that in the years after Current Year + 5, there are only 5 years (such as Current Year + 9) in which the cumulative number of filler units is identical for the three resource plans.

These differences in the timing of filler units, plus the difference in the total number of filler units added in each resource plan, *is solely due to the difference in the amount of capacity that is added in the decision year of Current Year + 5*. As these three resource plans demonstrate, even a relatively small difference in capacity between two resource options (such as a difference of 40 MW (=160 MW capacity of the CT option versus 120 MW capacity of the PV option) can result in a shift in the timing of future resource additions). And, as you might expect, this change in the timing of future resource additions in a given year will change the utility's total cost projection for that year and each year thereafter. (We shall actually see this in the next section.)

This effect is magnified greatly when the difference in the capacity being added is more significant. This can be seen by comparing Supply Only Resource Plan 1 (CC) with either Supply Option Resource Plan 2 (CT) or Supply Only Resource Plan 3 (PV).

A discerning reader may see a lightbulb go off right about now. Recall that one of the four key characteristics that two competing resource options must be identical, or very similar, in regard to in order to allow a screening curve approach to give meaningful results is the capacity (MW) of the two resource options. We have just seen one reason why that is true. Even a relatively small difference of 40 MW between two Supply options, CT and PV, is enough to change the timing of future resource additions. This change in the timing of when future resource options are added will directly impact the costs for the utility from that year forward.* This change in the timing of future resource option additions, and the associated costs with this change, are simply not captured in a screening curve analysis approach.

If our utility had (incorrectly) eliminated one of these two resource options after completing only the screening curve-based preliminary economic evaluation, it would have missed accounting for this one impact to the resource plans in future years. The impact of this seemingly small difference in capacity between the PV and CT options clearly shows up when comparing long-term resource plans based on these options. In turn, the impact of this difference in the resource plans can only be fully addressed in an economic evaluation of the resource plans themselves. We do so next.

Final (or System) Economic Evaluation of Supply Options

Overview

Now that we have created a separate resource plan for each of the three remaining Supply options, the final (or system) economic evaluation can be carried out. This evaluation of each resource plan consists of two steps. The first step provides the perspective of the utility system costs with each resource plan. The second step provides the perspective of the electric rates that our utility's customers will be charged to recover all of our utility's costs.

In the first step, the cumulative present value of annual revenue requirements (CPVRR) for our utility system with each resource plan is calculated.

* As we shall later see, there are other reasons why dissimilarities between resource options for even this one key characteristic of capacity (MW) will result in significant impacts to the costs of resource plans that are not captured in a screening curve approach.

TABLE 5.8

Long-Term Projection of Reserve Margin for the Hypothetical Utility: With New PV Added in "Current Year + 5"

	(1a)	(1b)	(1c)	(1d)	(1e)	(1f)	(2)	(3)	(4)	(5)
						= (See Formula Below)		= (1f) – (2)	= (3)/(2)	= ((2)*1.20) – (1f)
		Cumulative No. of New Unit Additions								
Year	Previously Projected Generating Capacity (MW)	CC Unit A (No. Units)	CT (No. Units)	PV (No. Units)	Filler Units (No. Units)	Total Projected Generating Capacity (MW)	Peak Electrical Demand (MW)	Reserves (MW)	Reserve Margin (%)	Generation Only MW Needed to Meet Reserve Margin (MW)
Current Year	11,600	0	0	0	0	11,600	9,600	2,000	20.8	(80)
Current Year + 1	12,000	0	0	0	0	12,000	9,700	2,300	23.7	(360)
Current Year + 2	12,000	0	0	0	0	12,000	9,800	2,200	22.4	(240)
Current Year + 3	12,000	0	0	0	0	12,000	9,900	2,100	21.2	(120)
Current Year + 4	12,000	0	0	0	0	12,000	10,000	2,000	20.0	0
Current Year + 5	12,000	0	0	1	0	12,120	10,100	2,020	20.0	0
Current Year + 6	12,000	0	0	1	0	12,120	10,200	1,920	18.8	120
Current Year + 7	12,000	0	0	1	0	12,120	10,300	1,820	17.7	240
Current Year + 8	12,000	0	0	1	0	12,120	10,400	1,720	16.5	360
Current Year + 9	12,000	0	0	1	0	12,120	10,500	1,620	15.4	480
Current Year + 10	12,000	0	0	1	0	12,120	10,600	1,520	14.3	600
Current Year + 11	12,000	0	0	1	0	12,120	10,700	1,420	13.3	720
Current Year + 12	12,000	0	0	1	0	12,120	10,800	1,320	12.2	840
Current Year + 13	12,000	0	0	1	0	12,120	10,900	1,220	11.2	960
Current Year + 14	12,000	0	0	1	0	12,120	11,000	1,120	10.2	1,080
Current Year + 15	12,000	0	0	1	0	12,120	11,100	1,020	9.2	1,200
Current Year + 16	12,000	0	0	1	0	12,120	11,200	920	8.2	1,320

Current Year + 17	12,000	0	0	1	0	12,120	11,300	820	7.3	1,440
Current Year + 18	12,000	0	0	1	0	12,120	11,400	720	6.3	1,560
Current Year + 19	12,000	0	0	1	0	12,120	11,500	620	5.4	1,680
Current Year + 20	12,000	0	0	1	0	12,120	11,600	520	4.5	1,800
Current Year + 21	12,000	0	0	1	0	12,120	11,700	420	3.6	1,920
Current Year + 22	12,000	0	0	1	0	12,120	11,800	320	2.7	2,040
Current Year + 23	12,000	0	0	1	0	12,120	11,900	220	1.8	2,160
Current Year + 24	12,000	0	0	1	0	12,120	12,000	120	1.0	2,280
Current Year + 25	12,000	0	0	1	0	12,120	12,100	20	0.2	2,400
Current Year + 26	12,000	0	0	1	0	12,120	12,200	(80)	−0.7	2,520
Current Year + 27	12,000	0	0	1	0	12,120	12,300	(180)	−1.5	2,640
Current Year + 28	12,000	0	0	1	0	12,120	12,400	(280)	−2.3	2,760
Current Year + 29	12,000	0	0	1	0	12,120	12,500	(380)	−3.0	2,880

CC Unit A =	500	MW
CT =	160	MW
PV =	120	MW
Filler Units =	500	MW

Formula: (1f) = (1a) + ((1b) × 500) + ((1c) × 160) + ((1d) × 120) + ((1e) × 500)

TABLE 5.9

Supply Only Resource Plan 3 with New PV Added in "Current Year + 5" Plus Filler Units

	(1a)	(1b)	(1c)	(1d)	(1e)	(1f)	(2)	(3)	(4)	(5)
						= (See Formula Below)		= (1f) – (2)	= (3)/(2)	= ((2)*1.20) – (1f)
		Cumulative No. of New Unit Additions								
Year	Previously Projected Generating Capacity (MW)	CC Unit A (No. Units)	CT (No. Units)	PV (No. Units)	Filler Units (No. Units)	Total Projected Generating Capacity (MW)	Peak Electrical Demand (MW)	Reserves (MW)	Reserve Margin (%)	Generation Only MW Needed to Meet Reserve Margin (MW)
Current Year	11,600	0	0	0	0	11,600	9,600	2,000	20.8	(80)
Current Year + 1	12,000	0	0	0	0	12,000	9,700	2,300	23.7	(360)
Current Year + 2	12,000	0	0	0	0	12,000	9,800	2,200	22.4	(240)
Current Year + 3	12,000	0	0	0	0	12,000	9,900	2,100	21.2	(120)
Current Year + 4	12,000	0	0	0	0	12,000	10,000	2,000	20.0	0
Current Year + 5	12,000	0	0	1	0	12,120	10,100	2,020	20.0	0
Current Year + 6	12,000	0	0	1	1	12,620	10,200	2,420	23.7	(380)
Current Year + 7	12,000	0	0	1	1	12,620	10,300	2,320	22.5	(260)
Current Year + 8	12,000	0	0	1	1	12,620	10,400	2,220	21.3	(140)
Current Year + 9	12,000	0	0	1	1	12,620	10,500	2,120	20.2	(20)
Current Year + 10	12,000	0	0	1	2	13,120	10,600	2,520	23.8	(400)
Current Year + 11	12,000	0	0	1	2	13,120	10,700	2,420	22.6	(280)
Current Year + 12	12,000	0	0	1	2	13,120	10,800	2,320	21.5	(160)
Current Year + 13	12,000	0	0	1	2	13,120	10,900	2,220	20.4	(40)
Current Year + 14	12,000	0	0	1	3	13,620	11,000	2,620	23.8	(420)
Current Year + 15	12,000	0	0	1	3	13,620	11,100	2,520	22.7	(300)

Current Year + 16	12,000	0	0	1	3	13,620	11,200	2,420	21.6	(180)
Current Year + 17	12,000	0	0	1	3	13,620	11,300	2,320	20.5	(60)
Current Year + 18	12,000	0	0	1	3	13,620	11,400	2,220	19.5	60
Current Year + 19	12,000	0	0	1	4	14,120	11,500	2,620	22.8	(320)
Current Year + 20	12,000	0	0	1	4	14,120	11,600	2,520	21.7	(200)
Current Year + 21	12,000	0	0	1	4	14,120	11,700	2,420	20.7	(80)
Current Year + 22	12,000	0	0	1	4	14,120	11,800	2,320	19.7	40
Current Year + 23	12,000	0	0	1	5	14,620	11,900	2,720	22.9	(340)
Current Year + 24	12,000	0	0	1	5	14,620	12,000	2,620	21.8	(220)
Current Year + 25	12,000	0	0	1	5	14,620	12,100	2,520	20.8	(100)
Current Year + 26	12,000	0	0	1	5	14,620	12,200	2,420	19.8	20
Current Year + 27	12,000	0	0	1	6	15,120	12,300	2,820	22.9	(360)
Current Year + 28	12,000	0	0	1	6	15,120	12,400	2,720	21.9	(240)
Current Year + 29	12,000	0	0	1	6	15,120	12,500	2,620	21.0	(120)

CC Unit A =	500	MW
CT =	160	MW
PV =	120	MW
Filler Units =	500	MW

Formula: (1f) = (1a) + ((1b) × 500) + ((1c) × 160) + ((1d) × 120) + ((1e) × 500)

TABLE 5.10

Overview of Three Supply Only Resource Plans

Year	Supply Only Resource Plan 1 (CC)	Supply Only Resource Plan 2 (CT)	Supply Only Resource Plan 3 (PV)
Current Year	—	—	—
Current Year + 1	—	—	—
Current Year + 2	—	—	—
Current Year + 3	—	—	—
Current Year + 4	—	—	—
Current Year + 5	CC Unit A (500 MW)	CT (160 MW)	PV (120 MW)
	Cumulative No. of Filler Units		
Current Year + 6	0	1	1
Current Year + 7	0	1	1
Current Year + 8	0	1	1
Current Year + 9	1	1	1
Current Year + 10	1	2	2
Current Year + 11	1	2	2
Current Year + 12	1	2	2
Current Year + 13	2	2	2
Current Year + 14	2	2	3
Current Year + 15	2	2	3
Current Year + 16	2	3	3
Current Year + 17	2	3	3
Current Year + 18	3	3	3
Current Year + 19	3	4	4
Current Year + 20	3	4	4
Current Year + 21	3	4	4
Current Year + 22	4	4	4
Current Year + 23	4	5	5
Current Year + 24	4	5	5
Current Year + 25	4	5	5
Current Year + 26	5	5	5
Current Year + 27	5	6	6
Current Year + 28	5	6	6
Current Year + 29	5	6	6

The CPVRR value represents the total costs for each resource plan.* This calculation accounts for all of our utility system's costs that are driven by the Supply option selected for the decision year, plus all of the subsequent filler units for that resource plan. At a minimum, the following types of costs are accounted for in final (or system) economic evaluations:

* Please see Appendix C for a mini-lesson regarding various economic terms, such as CPVRR and revenue requirements, that we will be discussing in this chapter, and in subsequent chapters.

- The capital, fixed O&M, variable O&M, and capital replacement costs for the selected Supply option and the subsequent filler units;
- The fixed costs associated with fuel delivery (such as firm gas transportation costs for all new Supply options that are fueled by natural gas);
- The administrative and incentive costs for any DSM options that are selected, plus the savings from any transmission and distribution costs that are avoided or deferred by the addition of a DSM option that reduces future load growth*; and,
- The fuel costs and environmental compliance costs for the entire utility system, including costs for both the new Supply options and the utility's existing generating units.

All of these individual annual costs (or annual "revenue requirements") are then summed to derive the total annual revenue requirements for a specific resource plan. These total annual revenue requirements for each year in the analysis period are then "present valued" and each year's present value revenue requirement value is summed. The result is a single CPVRR number that represents the present value cost of the utility's revenue requirements for the entire analysis period that are driven by a specific resource plan.

The second step consists of first combining the resource plan-specific annual costs with other annual costs which our utility will have that are not affected by the addition of a resource plan. For example, a list of these other costs will include, but not be limited to, the following: the remaining (after depreciation) capital costs of existing utility generating units, the remaining capital costs of transmission and distribution lines, buildings, salaries for existing staff, etc. Note that, although these other costs of the utility system are not affected by the selection of any of the resource options being evaluated, they are important in conducting final economic evaluations of resource options. This is because these other costs are factors in projecting what the utility's electric rates will be.

The sum of those system costs that are directly tied to resource options, and the other costs described earlier, represents the total annual cost of our utility that is passed on to its customers through electric rates. The annual total costs are then divided by the total number of annual sales (kWh) of the utility to derive an annual system average electric rate that is usually expressed in terms of cents per kilowatt-hour (cents/kWh).

Just as the total present valued cost of the resource plan-specific costs is represented by a single CPVRR value, a single value can also be calculated to

* Note that the DSM-related cost impacts are listed here only for purposes of describing the types of costs that are included in an IRP final (or system) economic evaluation. In this chapter, we are not yet evaluating additional DSM resource options. Consequently, in this chapter, the DSM-related costs are zero. DSM options are addressed in the next chapter.

represent the electric rate perspective. This is most often done by converting the annual system average electric rate values to a present valued electric rate value for each year, then summing these present valued annual electric rates. (We will refer to this sum as the "original" present valued sum of electric rates.)

Then a single, constant electric rate value is assumed for each year that, when present valued for each year and summed, results in the identical sum as the original present valued sum of electric rates. This single electric rate that is held constant in this calculation is referred to as the "levelized" system average electric rate.*

In summary, each resource plan can be described by two economic values: the CPVRR value that represents all of our utility system's costs that are driven of the resource plan itself, and the levelized system average electric rate value that represents the electric rates that will be charged to our utility's customers.

We now turn our attention to calculating these CPVRR and levelized system average electric rate values for each of the three Supply Only resource plans for our utility system.

Total Cost Perspective (CPVRR) for the Supply Only Resource Plans

We begin with the total cost perspective for the three Supply Only resource plans. We will first calculate the CPVRR cost for the Supply Only Resource Plan 1 (CC). The total CPVRR cost for this resource plan is determined by a number of cost components. Rather than attempt to show all of these individual cost components on a single spreadsheet, the individual cost components will first be grouped into three groups of costs: "Fixed" Costs, "DSM" Costs, and "Variable" Costs. Each of these three groups of costs will be presented and discussed separately.

The Fixed Costs for Supply Only Resource Plan 1 (CC) are presented in Table 5.11.

The Fixed Costs component represents the following costs for the CC unit and the subsequent filler units: generation capital costs in Column (2), fixed O&M costs in Column (3), capital replacement costs in Column (4), and firm gas transportation costs in Column (5). The sum of these Fixed Costs is shown in Column (6). The cost values presented in the rows for all years are nominal costs. Within each column these costs are then present valued and summed with the result shown on the "Total CPVRR" line at the bottom of the table. The CPVRR Fixed Cost total value for this resource plan is $3,582 million CPVRR as shown at the bottom of Column (6).

---

* I realize that both the total cost perspective (CPVRR) and the electric rate perspective (levelized system average electric rate) are likely a bit (or more) confusing at this point. Hang in there! Both perspectives should be much clearer as we work through an example of how these values are developed.

TABLE 5.11

(See color insert.) Fixed Costs Calculation for Supply Only Resource Plan 1 (CC)

Year	(1) Annual Discount Factor 8.000%	(2) Generation Capital (Nominal $, Millions)	(3) Generation Fixed O&M (Nominal $, Millions)	(4) Generation Capital Replacement (Nominal $, Millions)	(5) Firm Gas Transportation (Nominal $, Millions)	(6) = Sum of Cols. (2) through (5) Total Generation Fixed Costs (Nominal $, Millions)
Current Year	1.000	0	0	0	0	**0**
1	0.926	0	0	0	0	**0**
2	0.857	0	0	0	0	**0**
3	0.794	0	0	0	0	**0**
4	0.735	0	0	0	0	**0**
5	0.681	100	3	5	82	**190**
6	0.630	96	3	5	82	**186**
7	0.583	92	3	5	82	**182**
8	0.540	88	3	5	82	**179**
9	0.500	192	6	11	164	**374**
10	0.463	184	7	11	164	**366**
11	0.429	176	7	11	164	**358**
12	0.397	167	7	11	164	**350**
13	0.368	276	11	18	246	**551**
14	0.340	263	11	18	246	**538**
15	0.315	250	11	18	246	**526**
16	0.292	237	11	19	246	**513**
17	0.270	224	11	19	246	**501**
18	0.250	340	16	26	329	**710**
19	0.232	322	16	26	329	**693**
20	0.215	304	16	27	329	**676**
21	0.199	286	16	27	329	**658**
22	0.184	408	21	35	411	**874**
23	0.170	384	21	36	411	**852**
24	0.158	360	22	36	411	**829**
25	0.146	336	22	37	411	**806**
26	0.135	464	27	45	493	**1,030**
27	0.125	434	28	46	493	**1,001**
28	0.116	404	28	47	493	**973**
29	0.107	374	29	48	493	**944**
Total CPVRR =		**$1,685**	**$79**	**$131**	**$1,687**	**$3,582**

The second group of costs that will be discussed is that of DSM Costs associated with new DSM resource options that our utility has chosen. However, as we have discussed, no incremental DSM resources have been added for this Supply Only resource plan. Consequently, there are no incremental DSM costs for this resource plan. However, because the tabular cost format that is being introduced in this chapter will also be used later in Chapter 6 when we discuss resource plans with DSM resource options, we shall present the zero incremental DSM costs in order to introduce the complete tabular format.

Thus, the values for the DSM Costs grouping are presented in Table 5.12 which also "carries over" the sum of the Fixed Costs (Column (6)).

As mentioned earlier, because this is a Supply Only resource plan, there are no new DSM resources added. Consequently, the various DSM costs categories in Columns (7) through (10) have zero costs. Therefore, the CPVRR cost for the DSM Costs is zero as indicated at the bottom of Column (10). And, after accounting for both Fixed and DSM costs, the CPVRR cost for Supply Only Resource Plan 1 (CC) with CC Unit A is \$3,582 million (Fixed) + \$0 million (DSM) = \$3,582 million. However, we have not yet accounted for Variable Costs associated with this resource plan.

We now account for the Variable Costs in Table 5.13. In this table, we carry over the total Fixed Costs (Column (6)) and the total DSM Costs (Column (10)).

The Variable Costs grouping represents the following three types of costs: generation variable O&M costs in Column (11), total utility system net fuel costs in Column (12), and total utility system environmental compliance costs in Column (13).* The CPVRR Variable Cost value for this resource plan is \$44,246 million as presented at the bottom of Column (14).

As suggested by the values in this table, the Variable Cost value for a resource plan (\$44,246 million CPVRR for Supply Only Resource Plan 1(CC)) is typically much larger than the Fixed Cost value (\$3,582 million CPVRR for this same resource plan) for several reasons. First, the two largest (by far) Variable Cost components, system fuel costs and system environmental compliance costs, address costs for *all* of a utility's generating units, existing and new, while the Fixed Costs typically address only the new generating units that are added to the system. There are typically many more existing units on a utility system than the number of new units that will be added in any analysis period. For this reason, a utility's annual fuel and environmental compliance costs in any given year will typically be much greater than new Fixed Costs for the same year.

---

* In this table, the variable O&M costs in Column (11) show zero values until Current Year + 5 when the first new Supply Unit begins operation. Thus we are accounting for variable O&M costs for new generating units only. Although existing generating units do have variable O&M costs, some utilities account for these costs as fixed costs in their budgeting process. This can make it difficult to accurately calculate a projection of variable O&M costs for existing generating units. For that reason we do not attempt this projection, but we will account for variable O&M costs for existing generating units later when we discuss how levelized system average electric rates are calculated.

TABLE 5.12

(See color insert.) DSM Costs Calculation for Supply Only Resource Plan 1 (CC)

	(1)	(6)	(7)	(8)	(9)	(10)
		= Sum of Cols. (2) through (5)				= Col (7) + Col (8) – Col (9)
Year	Annual Discount Factor 8.000%	Total Generation Fixed Costs (Nominal $, Millions)	DSM Administrative Costs (Nominal $, Millions)	DSM Incentive Payments (Nominal $, Millions)	T&D Costs Avoided by DSM (Nominal $, Millions)	DSM Net Costs (Nominal $, Millions)
Current Year	1.000	**0**	0	0	0	**0**
1	0.926	**0**	0	0	0	**0**
2	0.857	**0**	0	0	0	**0**
3	0.794	**0**	0	0	0	**0**
4	0.735	**0**	0	0	0	**0**
5	0.681	**190**	0	0	0	**0**
6	0.630	**186**	0	0	0	**0**
7	0.583	**182**	0	0	0	**0**
8	0.540	**179**	0	0	0	**0**
9	0.500	**374**	0	0	0	**0**
10	0.463	**366**	0	0	0	**0**
11	0.429	**358**	0	0	0	**0**
12	0.397	**350**	0	0	0	**0**
13	0.368	**551**	0	0	0	**0**
14	0.340	**538**	0	0	0	**0**
15	0.315	**526**	0	0	0	**0**
16	0.292	**513**	0	0	0	**0**
17	0.270	**501**	0	0	0	**0**
18	0.250	**710**	0	0	0	**0**
19	0.232	**693**	0	0	0	**0**
20	0.215	**676**	0	0	0	**0**
21	0.199	**658**	0	0	0	**0**
22	0.184	**874**	0	0	0	**0**
23	0.170	**852**	0	0	0	**0**
24	0.158	**829**	0	0	0	**0**
25	0.146	**806**	0	0	0	**0**
26	0.135	**1,030**	0	0	0	**0**
27	0.125	**1,001**	0	0	0	**0**
28	0.116	**973**	0	0	0	**0**
29	0.107	**944**	0	0	0	**0**
Total CPVRR =		**$3,582**	**$0**	**$0**	**$0**	**$0**

TABLE 5.13

(See color insert.) Variable Costs Calculation for Supply Only Resource Plan 1 (CC)

	(1)	(6)	(10)	(11)	(12)	(13)	(14)
		= Sum of Cols. (2) through (5)	= Col (7) + Col (8) − Col (9)				= Sum of Cols. (11) through (13)
Year	Annual Discount Factor 8.000%	Total Generation Fixed Costs (Nominal $, Millions)	DSM Net Costs (Nominal $, Millions)	Generation Variable O&M (Nominal $, Millions)	System Net Fuel (Nominal $, Millions)	System Environmental Compliance (Nominal $, Millions)	Total Variable Costs (Nominal $, Millions)
Current Year	1.000	**0**	**0**	0	1,660	275	**1,936**
1	0.926	**0**	**0**	0	1,726	390	**2,116**
2	0.857	**0**	**0**	0	1,794	507	**2,300**
3	0.794	**0**	**0**	0	1,863	625	**2,488**
4	0.735	**0**	**0**	0	1,935	746	**2,681**
5	0.681	**190**	**0**	7	1,928	845	**2,780**
6	0.630	**186**	**0**	7	2,002	968	**2,977**
7	0.583	**182**	**0**	7	2,079	1,092	**3,179**
8	0.540	**179**	**0**	7	2,158	1,219	**3,384**
9	0.500	**374**	**0**	15	2,151	1,315	**3,482**
10	0.463	**366**	**0**	15	2,233	1,444	**3,693**
11	0.429	**358**	**0**	16	2,317	1,574	**3,908**
12	0.397	**350**	**0**	16	2,404	1,707	**4,127**
13	0.368	**551**	**0**	25	2,399	1,800	**4,224**
14	0.340	**538**	**0**	25	2,489	1,935	**4,449**
15	0.315	**526**	**0**	26	2,581	2,072	**4,679**
16	0.292	**513**	**0**	26	2,677	2,211	**4,914**
17	0.270	**501**	**0**	27	2,775	2,352	**5,153**
18	0.250	**710**	**0**	36	2,771	2,441	**5,249**
19	0.232	**693**	**0**	37	2,873	2,585	**5,495**
20	0.215	**676**	**0**	38	2,978	2,730	**5,746**
21	0.199	**658**	**0**	38	3,085	2,878	**6,002**
22	0.184	**874**	**0**	49	3,083	2,964	**6,096**
23	0.170	**852**	**0**	50	3,195	3,114	**6,359**
24	0.158	**829**	**0**	51	3,310	3,266	**6,627**
25	0.146	**806**	**0**	52	3,428	3,421	**6,902**
26	0.135	**1,030**	**0**	64	3,428	3,503	**6,994**
27	0.125	**1,001**	**0**	65	3,551	3,660	**7,275**
28	0.116	**973**	**0**	66	3,677	3,819	**7,562**
29	0.107	**944**	**0**	68	3,807	3,981	**7,856**
Total CPVRR =		**$3,582**	**$0**	**$184**	**$27,271**	**$16,790**	**$44,246**

Furthermore, these annual Variable Costs escalate over time due to rising fuel and environmental compliance costs while the capital cost for an individual generating unit declines from year-to-year due to depreciation of the generating unit's capital cost.

Table 5.14 further condenses the tabular format to now summarize the three groups of cost components for Supply Only Resource Plan 1 (CC). This table carries over the total Fixed Costs (Column (6)), the total DSM Costs (Column (10)), and the total Variable Costs (Column (14)). The sum of these three cost components results in a total CPVRR cost for Supply Only Resource Plan 1 (CC) of $47,828 million CPVRR. This is indicated in each of the new Columns (15), (16), and (17) which, respectively, present annual nominal costs, annual NPV costs, and cumulative NPV (or CPVRR) costs.

The total costs for Supply Only Resource Plan 1 (CC) have now been presented in detail in Tables 5.11 through 5.13. In addition, a summary version of those total costs has been presented in Table 5.14. The resource plan-specific CPVRR total costs for the other two Supply Only resource plans are calculated in a similar way. We now provide the corresponding values for the other two Supply Only resource plans in condensed summary form in Table 5.15.

The values for the three types of cost components presented earlier for Supply Only Resource Plan 1 (CC) appear in the first row of the table in Columns (1) through (3). The total CPVRR cost for this resource plan is provided in Column (4). The corresponding values for the other two Supply Only resource plans, featuring the CT and PV options are then provided on the second and third rows, respectively.

A quick glance at the CPVRR values in Column (4) shows that Supply Only Resource Plan 1 (CC) is projected to have the lowest CPVRR costs of any of the three resource plans. In order to more clearly see the CPVRR cost difference between the resource plans, Column (5) has been added which presents the CPVRR cost difference between the resource plan with the lowest CPVRR cost, Supply Only Resource Plan 1 (CC), and the other two Supply Only resource plans.

The table shows us several things regarding these Supply Only resource plans. First, the lowest CPVRR cost plan is Supply Only Resource Plan 1 (CC). Its cost of $47,828 million CPVRR is $122 million CPVRR lower than Supply Only Resource Plan 2 (CT), and $397 million CPVRR lower than Supply Only Resource Plan 3 (PV). Therefore, from a CPVRR perspective, the CC unit option is the least expensive Supply option to add to our utility system. But why is this so? A further examination of the table provides the answer.

The second thing one can glean from the table is why one resource plan is better from a CPVRR perspective than another. We will start by comparing just Supply Only Resource Plan 1 (CC) and Supply Only Resource Plan 2 (CT). We see that in regard to the Fixed Costs category, the CT option (in Supply Only Resource Plan 2) has a cost advantage of $186 million CPVRR over the CC option. (In other words, $3,582 million CPVRR for Supply Only

TABLE 5.14

(See color insert.) Total Costs Calculation for Supply Only Resource Plan 1 (CC)

	(1)	(6)	(10)	(14)	(15)	(16)	(17)
		= Sum of Cols. (2) through (5)	= Col (7) + Col (8) − Col(9)	= Sum of Cols. (11) through (13)	= (6) + (10) + (14)	= (1) × (15)	
Year	Annual Discount Factor 8.000%	Total Generation Fixed Costs (Nominal $, Millions)	DSM Net Costs (Nominal $, Millions)	Total Variable Costs (Nominal $, Millions)	Total Annual Costs (Nominal $, Millions)	Total Annual NPV Costs (NPV $, Millions)	Cumulative Total NPV Costs(NPV $, Millions)
Current Year	1.000	0	0	1,936	1,936	1,936	1,936
1	0.926	0	0	2,116	2,116	1,959	3,895
2	0.857	0	0	2,300	2,300	1,972	5,867
3	0.794	0	0	2,488	2,488	1,975	7,842
4	0.735	0	0	2,681	2,681	1,970	9,813
5	0.681	190	0	2,780	2,970	2,022	11,834
6	0.630	186	0	2,977	3,164	1,994	13,828
7	0.583	182	0	3,179	3,361	1,961	15,789
8	0.540	179	0	3,384	3,563	1,925	17,714
9	0.500	374	0	3,482	3,856	1,929	19,643
10	0.463	366	0	3,693	4,058	1,880	21,522
11	0.429	358	0	3,908	4,266	1,829	23,352
12	0.397	350	0	4,127	4,477	1,778	25,130
13	0.368	551	0	4,224	4,774	1,755	26,885
14	0.340	538	0	4,449	4,987	1,698	28,583
15	0.315	526	0	4,679	5,204	1,641	30,224
16	0.292	513	0	4,914	5,427	1,584	31,808
17	0.270	501	0	5,153	5,654	1,528	33,336
18	0.250	710	0	5,249	5,959	1,491	34,827
19	0.232	693	0	5,495	6,187	1,434	36,261
20	0.215	676	0	5,746	6,421	1,378	37,639
21	0.199	658	0	6,002	6,660	1,323	38,962
22	0.184	874	0	6,096	6,970	1,282	40,244
23	0.170	852	0	6,359	7,210	1,228	41,472
24	0.158	829	0	6,627	7,456	1,176	42,648
25	0.146	806	0	6,902	7,708	1,125	43,773
26	0.135	1,030	0	6,994	8,023	1,085	44,858
27	0.125	1,001	0	7,275	8,276	1,036	45,894
28	0.116	973	0	7,562	8,535	989	46,883
29	0.107	944	0	7,856	8,800	945	47,828
Total CPVRR =		$3,582	$0	$44,246	$47,828	$47,828	

TABLE 5.15

(See color insert.) Economic Evaluation Results of Supply Only Resource Plans: CPVRR Costs

	(1)	(2)	(3)	(4)	(5)
				= Sum of Cols. (1) through (3)	
Resource Plan	**Fixed Costs (Millions, CPVRR)**	**DSM Costs (Millions, CPVRR)**	**Variable Costs (Millions, CPVRR)**	**Total Costs (Millions, CPVRR)**	**Difference from Lowest Cost Supply Only Plan (Millions, CPVRR)**
Supply Only Resource Plan 1 (CC)	3,582	0	44,246	**47,828**	0
Supply Only Resource Plan 2 (CT)	3,396	0	44,554	**47,950**	122
Supply Only Resource Plan 3 (PV)	3,894	0	44,331	**48,225**	397

Resource Plan 1 (CC) – $3,396 million CPVRR for Supply Only Resource Plan 2 (CT) = $186 million CPVRR.)

One might expect this by remembering several things including: (i) the CT option's cost per kW is $650/kW versus $1,000/kW for CC Unit A; (ii) the CT option's cost per kW is multiplied by 160 MW while the CC Unit A's cost per kW is multiplied by 500 MW; and (iii) the amount of firm gas required by the CT option (10,000 mmBTU/day) is much less than the amount of firm gas needed by the CC unit (100,000 mmBTU/day).

Each of these factors favor the CT option in Supply Only Resource Plan 2 in regard to Fixed Costs. However, we also know from our work in creating the resource plans that these advantages are offset to some degree by the fact that the 160 MW of the CT option, compared to the 500 MW of the CC option, results in our utility having to add one more filler unit, and to add filler units earlier, than if the larger CC unit had been chosen. However, after performing the economic analyses, we find that the Fixed Cost perspective still favors Supply Only Resource Plan 2 (CT).

However, the Variable Costs perspective provides a significantly different result. A comparison of the Variable Costs for these two resource plans shows that Supply Only Resource Plan 1 (CC) has an economic advantage of $308 million CPVRR over Supply Only Resource Plan 2 (CT) as shown by the calculation: $44,554 million CPVRR for Supply Only Resource Plan 2 (CT) – $44,246 million CPVRR for Supply Only Resource Plan 1 (CC) = $308 million CPVRR.

One might also expect a Variable Costs advantage for the CC unit featured in Supply Only Resource Plan 1 for two reasons. First, the CC unit is much more fuel efficient (6,600 BTU/kWh) than the CT option (10,400 BTU/kWh). Second, the more fuel efficient CC unit will be operated much more (a projected capacity factor of 80%) than the CT option (a projected capacity factor of 5%).

These two factors will result not only in reduced system fuel costs, but also in reduced system environmental compliance costs. This is because the operation of the CC unit will displace a much greater amount of energy that would otherwise have been produced by the utility's existing marginal generating units which we recall, from Chapter 2, are more expensive to operate (and which typically have higher emission rates) than a new CC unit.

As the values in Table 5.15 show, these advantages for the CC unit result in the significant Variable Cost advantage of $308 million CPVRR compared to the CT option. When the comparative results for Fixed Costs and Variable Costs are combined the net result is an economic advantage of $122 million CPVRR for Supply Only Resource Plan 1 (CC) as shown by the calculation: $308 million CPVRR advantage in Variable Costs – $186 million CPVRR disadvantage in Fixed Costs = $122 million CPVRR net cost advantage for the CC option.

Now, what about the PV option featured in Supply Only Resource Plan 3? A similar comparison to Supply Only Resource Plan 1 (CC) shows that the CC unit has an economic advantage compared to the PV option in both the Fixed Costs and Variable Costs categories.

In regard to Fixed Costs, the CC unit has an advantage of $312 million CPVRR as shown by the calculation: $3,894 million CPVRR for Supply Only Resource Plan 3 (PV) – $3,582 million CPVRR for Supply Only Resource Plan 1 (CC) = $312 million CPVRR. As you recall, the capital cost of the PV option was $5,000/kW compared to $1,000/kW for CC Unit A. This fact, combined with the fact that the PV option's 120 MW will result in the utility adding more filler units, and adding them earlier, than will be required with the 500 MW of the CC unit, results in a significant Fixed Costs disadvantage for the PV option.

However, the Variable Costs result is different with the CC unit having a smaller advantage of $85 million CPVRR over the PV option as shown by the calculation: $44,331 million CPVRR for Supply Only Resource Plan 3 (PV) – $44,246 million CPVRR for Supply Only Resource Plan 1 (CC) = $85 million CPVRR.

Upon seeing this result, one might react with a strong (but elegantly stated) response of: "Huh? How can this result be correct if the PV option has no fuel or emission costs?"

The answer to this question is driven by three factors. First, the CC unit is more fuel efficient than any of the existing oil- and gas-fired existing units on our utility system (i.e., the primary marginal generating units for our utility system). Second, this fuel efficiency results in the CC unit being operated a projected 80% of the hours in the year compared to the PV option maximum annual operation of 20% of the hours in the year. Third, the CC unit is a 500 MW resource compared to PV's 120 MW. These factors allow us to "reframe" the question of Variable Costs in the following way:

> "Which option will result in lower fuel and environmental compliance costs for the utility system as a whole: an option (PV) that has no fuel or environmental compliance costs itself and whose 120 MW operates 20% of the hours in the year, or an option (CC) that does have fuel and environmental compliance costs, but whose 500 MW operates 80% of the hours in the year?"*

In regard to our utility system and these specific Supply options, the answer is that the CC unit will lower fuel and environmental compliance costs for our utility system as a whole compared to the PV option. In addition, recall that the Variable Cost category also includes variable O&M costs for these Supply options. The CC unit's Variable Costs also include a variable O&M cost that the PV option doesn't have because its variable O&M cost has been assumed to be $0/MWh. This further points out that the fuel and emission cost savings from the CC unit is actually greater than $85 million CPVRR, but that the higher variable O&M costs of the CC unit reduces its net Variable Cost advantage to $85 million CPVRR.

Consequently, Supply Only Resource Plan 1 (CC) is projected to have a total net Variable Costs advantage of $85 million CPVRR compared to Supply Only Resource Plan 3 (PV).†

We have now established that Supply Only Resource Plan 1 (CC) is the best economic choice among the Supply Only resource plans from a total CPVRR cost perspective. We now turn our attention to examining an electric rate perspective of these three resource plans.

Electric Rate Perspective (Levelized System Average Electric Rate) for the Supply Only Resource Plans

In Chapter 3, we discussed that when conducting economic evaluations of Supply options only, once we knew what Supply option resulted in the lowest total cost (the CPVRR perspective), this Supply option would also be the Supply option that resulted in the lowest electric rates. We will now see whether this is true. (Of course it is or I wouldn't have told you earlier that it was.)

---

* This question sheds further light on the fundamental problems with screening curve analyses when the resource options being considered are not identical or very similar in regard to just two of the four key characteristics: capacity (MW) and capacity factor. Differences in these two characteristics have significant impacts on both system fuel and system environmental compliance costs that are not captured in a simple screening curve analysis approach.

† To further assist the reader, Appendix G provides a completely different look at how a resource option that burns fossil fuel can still result in lower fuel costs and environmental compliance costs for a system than does a resource option that burns no fossil fuel.

Table 5.16 presents a revised and expanded version of previously presented Table 5.14. That table provided the annual and total CPVRR revenue requirements for Supply Only Resource Plan 1 (CC). New Table 5.16 provides the electric rate impact perspective for this same resource plan.*

This table appears complicated at first glance, but we have seen some of it before and the remainder is straightforward. The table is constructed as follows:

- Previous Table 5.14 is the starting point for Table 5.16. However, in the interests of space, we have hidden all columns except the unnumbered column on the left-hand side that shows the year, Column (1) that shows the annual discount rate factors, and Column (15) that shows the annual costs and the total CPVRR cost.
- To this revised version of Table 5.14, we have added nine new columns and have labeled the new columns as Columns (18) through (26).
- Column (18) provides a projection of "Other System Costs" that are not affected by the selection of the resource plan. As previously discussed, these costs include costs of existing utility power plants, transmission and distribution lines, buildings, utility staff, etc.†
- Column (19) then adds the values in Columns (15) and (18) to derive a total utility cost per year.
- Columns (20) through (22) then present, respectively, the forecasted total annual sales (or Net Energy for Load values used by some utilities), the projected reduced energy usage from new DSM options (if any), and the resulting net energy usage by our utility's customers that the utility must serve. (Note that in this Supply Only resource plan there is no reduction in energy sales from additional DSM as shown by the zero values in Column (21)).
- Column (23) then calculates (for each year) an annual system average electric rate by dividing the total utility cost per year shown in Column (19) by the net energy usage for that year shown in Column (22).
- Column (24) calculates the present value of each annual system average electric rate, then sums these annual values at the bottom of the page.
- Column (25) presents the constant or levelized system average electric rate which, when present valued and summed in Column (26), results in an identical present valued sum as that shown at the bottom of Column (24).

---

* Note that the calculation method presented in Table 5.16 is one of several ways in which levelized system average rates for electric utilities may be calculated.

† The previously mentioned variable O&M for the utility's existing generating units are also accounted for in the cost values in Column (18).

This levelized system average electric rate value presented in Column (25) provides the electric rate perspective value for the resource plan in question.*

As we see from the table, the levelized electric rate for Supply Only Resource Plan 1 (CC) is 12.8116 cents/kWh. This value does not represent the annual electric rate for any one class of customer (i.e., for residential or business customers). As previously stated, this one value represents a system average electric rate (cents/kWh) value over the life of the time period covered in the analysis.†

Similar calculations have been performed for the remaining two Supply Only resource plans. We show these results in Table 5.17 which expands the previously presented Table 5.15 to include a new Column (6) that provides the corresponding levelized electric rate for each of the three Supply Only resource plans.

As shown in this table, a ranking of the three Supply Only resource plans based on the CPVRR values in Column (4), and a ranking based on the levelized electric rates in Column (6), would be identical. In either case, Supply Only Resource Plan 1 (CC) is the most economical Supply Only resource plan from either the CPVRR or electric rate perspective. This resource plan is followed, in economic ranking order, by Supply Only Resource Plan 2 (CT), then by Supply Only Resource Plan 3 (PV).

We walk away from the economic evaluation of the Supply Only resource plans considered by our utility system with (at least) four pieces of information. First, we now see why a utility may choose to stop its economic evaluation of Supply options once it has determined which Supply option results in the lowest CPVRR total cost. This is because the Supply option that has the lowest CPVRR total costs will also be the Supply option that results in the lowest levelized electric rate. Although taking the next step of calculating the levelized electric rate is absolutely necessary when Supply options are compared to DSM options, it is not necessary when comparing only Supply options.

Second, by examining the differences between the three Supply Only resource plans from both the CPVRR cost and levelized electric rate perspectives, we see that even large differences in resource plan costs (such as a $397 million CPVRR difference between Supply Only Resource Plan 1 (CC) and Supply Only Resource Plan 3 (PV)) will typically result in only seemingly small differences in levelized electric rate values (0.0583 cents/kWh)

---

* The levelized electric rate is also referred to by the term "levelized system average electric rate." Although this term is more descriptive of what the value actually represents, it is a mouthful to say/read. Therefore, we will often use the shorter terminology of "levelized electric rate" from this point forward.

† Electric utilities typically use multi-part electric rates for certain types of customers. An example is the use of separate demand ($/kW) and energy (cents/kWh) charges for larger commercial and industrial customers. The levelized electric rate is presented as a single "cents per kWh" rate value for ease in comparing the levelized electric rates for different resource plans.

TABLE 5.16

(See color insert.) Levelized Electric Rate Calculation for Supply Only Resource Plan 1 (CC)

	(1)	(15)	(18)	(19)	(20)	(21)	(22)	(23)	(24)	(25)	(26)
		= (6) + (10) + (14)		= (15) + (18)			= (20) − (21)	= ((19) × 100)/(22)	= (1) × (23)		= (1) × (25)
Year	Annual Discount Factor 8.000%	Total Annual Costs (Nominal $, Millions)	Other System Costs Not Affected by Plan (Nominal $, Millions)	Total Utility Costs (Nominal $, Millions)	Forecasted Utility Sales/NEL (GWh)	DSM Energy Reduction (GWh)	Net Utility Sales/ NEL (GWh)	System Average Electric Rate Nominal (Cents/ kWh)	System Average Electric Rate NPV (Cents/ kWh)	Levelized System Average Electric Rate (Cents/ kWh)	System Average Electric Rate NPV (Cents/ kWh)
Current Year	1.000	1,936	2,700	4,636	50,496	0	50,496	9.1800	9.1800	12.8116	12.8116
1	0.926	2,116	2,754	4,870	51,022	0	51,022	9.5447	8.8377	12.8116	11.8626
2	0.857	2,300	2,809	5,109	51,548	0	51,548	9.9116	8.4976	12.8116	10.9839
3	0.794	2,488	2,865	5,354	52,074	0	52,074	10.2809	8.1613	12.8116	10.1703
4	0.735	2,681	2,923	5,603	52,600	0	52,600	10.6528	7.8301	12.8116	9.4169
5	0.681	2,970	2,981	5,951	53,126	0	53,126	11.2024	7.6242	12.8116	8.7194
6	0.630	3,164	3,041	6,204	53,652	0	53,652	11.5638	7.2872	12.8116	8.0735
7	0.583	3,361	3,101	6,462	54,178	0	54,178	11.9282	6.9600	12.8116	7.4755
8	0.540	3,563	3,163	6,726	54,704	0	54,704	12.2958	6.6430	12.8116	6.9217
9	0.500	3,856	3,227	7,083	55,230	0	55,230	12.8237	6.4150	12.8116	6.4090
10	0.463	4,058	3,291	7,350	55,756	0	55,756	13.1819	6.1058	12.8116	5.9343
11	0.429	4,266	3,357	7,623	56,282	0	56,282	13.5437	5.8087	12.8116	5.4947
12	0.397	4,477	3,424	7,902	56,808	0	56,808	13.9093	5.5236	12.8116	5.0877
13	0.368	4,774	3,493	8,267	57,334	0	57,334	14.4189	5.3018	12.8116	4.7108
14	0.340	4,987	3,563	8,549	57,860	0	57,860	14.7761	5.0307	12.8116	4.3619

15	0.315	5,204	3,634	8,838	58,386	0	58,386	15.1375	4.7720	12.8116	4.0388
16	0.292	5,427	3,707	9,133	58,912	0	58,912	15.5033	4.5253	12.8116	3.7396
17	0.270	5,654	3,781	9,435	59,438	0	59,438	15.8735	4.2901	12.8116	3.4626
18	0.250	5,959	3,856	9,815	59,964	0	59,964	16.3686	4.0962	12.8116	3.2061
19	0.232	6,187	3,933	10,121	60,490	0	60,490	16.7314	3.8769	12.8116	2.9686
20	0.215	6,421	4,012	10,433	61,016	0	61,016	17.0992	3.6686	12.8116	2.7487
21	0.199	6,660	4,092	10,753	61,542	0	61,542	17.4720	3.4709	12.8116	2.5451
22	0.184	6,970	4,174	11,144	62,068	0	62,068	17.9551	3.3027	12.8116	2.3566
23	0.170	7,210	4,258	11,468	62,594	0	62,594	18.3212	3.1204	12.8116	2.1820
24	0.158	7,456	4,343	11,799	63,120	0	63,120	18.6930	2.9479	12.8116	2.0204
25	0.146	7,708	4,430	12,138	63,646	0	63,646	19.0704	2.7846	12.8116	1.8707
26	0.135	8,023	4,518	12,542	64,172	0	64,172	19.5439	2.6424	12.8116	1.7322
27	0.125	8,276	4,609	12,885	64,698	0	64,698	19.9153	2.4931	12.8116	1.6038
28	0.116	8,535	4,701	13,236	65,224	0	65,224	20.2930	2.3522	12.8116	1.4850
29	0.107	8,800	4,795	13,595	65,750	0	65,750	20.6770	2.2192	12.8116	1.3750
Total CPVRR =		**$47,828**							——		——
									155.7691		**155.7691**

TABLE 5.17

Economic Evaluation Results of Supply Only Resource Plans: CPVRR Costs and Levelized System Average Electric Rates

	(1)	(2)	(3)	(4)	(5)	(6)
				=Sum of Cols. (1) through (3)		
Resource Plan	Fixed Costs (Millions, CPVRR)	DSM Costs (Millions, CPVRR)	Variable Costs (Millions, CPVRR)	Total Costs (Millions, CPVRR)	Difference form Lowest Cost Supply Only Plan (Millions, CPVRR)	Levelized System Average Electric Rate (cents/kWh)
Supply Only Resource Plan 1 (CC)	3,582	0	44,246	47,828	0	**12.8116**
Supply Only Resource Plan 2 (CT)	3,396	0	44,554	47,950	122	**12.8283**
Supply Only Resource Plan 3 (PV)	3,894	0	44,331	48,225	397	**12.8699**

for these same two resource plans). The important point is that even seemingly small differences in levelized electric rate values are equivalent to very large CPVRR cost differences. This becomes especially important to remember when examining DSM options for which the electric rate perspective is a necessity.

Third, we have seen how even relatively small dissimilarities in just one of the four key characteristics of resource options, capacity (MW), can result in "downstream" changes in the number and timing of future resource additions (i.e., the filler units in our analyses). The capital and fixed costs associated with such changes are simply not captured in a screening curve analytical approach. In addition, we have seen that dissimilarities in both capacity (MW) and capacity factor will also result in changes in system fuel costs and system environmental costs. These cost impacts are also not captured if one attempts to use a screening curve analytical approach.

We can also now understand why the other two key characteristics (the percentage of the resource option's capacity (MW) that can be considered as firm capacity at the utility's system peak hours and the projected life of the resource option) are important. Differences in either of these two characteristics between two otherwise identical resource options would also result in differences in the multi-year resource plans for the two options in regard to the number and timing of filler unit additions. In turn, these resource plan differences would drive differences in both Fixed and Variable Costs between the resource plans over the analysis period.

Consequently, we now see why if even if one of the four key characteristics of two resource options is dissimilar, the use of a screening curve analytical approach is fundamentally flawed.

Fourth, our hypothetical utility system has now identified, from an economic perspective, its best Supply option: the CC unit. Our utility is now ready to turn its attention to examining the other type of resource option, DSM, which it is considering. We do so in the next chapter.

6

Resource Option Analyses for Our Utility System: DSM Options

In this chapter, we follow our utility through economic analyses of two DSM options it will consider to meet its projected resource need 5 years in the future. Just as we noted in our discussion of Supply options, in practice a utility would likely consider many more DSM options than the two DSM options that we will be examining here. However, in order to simplify the discussion, while still illustrating the important points regarding economic analyses of DSM options, we have condensed this list to two different DSM options that can reduce customers' demand for electricity.

Types of DSM Resource Options Under Consideration

All DSM options that would be considered for meeting a utility's resource needs by lowering electrical demand at the utility's peak hour will have three main characteristics. We will first briefly introduce these characteristics, then discuss them in more detail.

First, the DSM options will require certain types of expenditures. These expenditures typically include administrative costs and usually include incentive payment costs. The administrative costs include expenditures to market, advertise, operate, and monitor the DSM program on an on-going basis. Incentive costs are typically either a one-time payment to a participating customer to pay a portion of the cost that the participating customer will have to pay to install the DSM measure (such as higher levels of ceiling insulation or a higher efficiency air conditioner), or recurring (typically monthly) credits on the electric bill in exchange for the customer's continuing participation in a DSM program. In addition, for certain types of DSM options there will also be capital costs for utility-owned equipment, including equipment that is placed at the customer's premises.

Second, the DSM options will typically reduce demand at the utility's peak hour. This is commonly referred to as the kilowatt reduction at the peak hour (kW reduction) aspect of the DSM option or program.

The third characteristic that DSM options have in common is that the DSM options will typically reduce the annual electricity consumption of participating customers (and may also shift the timing during the day of when the electricity is used). This is referred to as the annual kilowatt-hour reduction (kWh reduction) aspect of the DSM program.

Now that these three characteristics of DSM options have been introduced, we will discuss them in more detail. This information should help in understanding the presentation of the economic and non-economic evaluations of DSM options that follow in the remainder of this chapter and in Chapter 7.

The first characteristic, DSM expenditures, is straightforward. The administrative costs and incentive payments directly associated with a specific DSM option are those expenditures required to implement and operate a DSM program. These costs are typically presented in terms of dollars per participating customer ($/participant) or dollars per peak hour kW reduction ($/kW). The administrative costs are typically a one-time cost that is incurred when a participating customer is signed up for the DSM program. As mentioned earlier, the incentive payment can either be a one-time payment when the customer signs up for the program (or when the customer pays for the DSM measure), or an on-going payment such as a monthly credit on the participating customer's monthly bill.

These $/participant, or $/kW, costs are typically small in comparison to most of the $/kW cost values we discussed in regard to the Supply options. However, a DSM option will usually only provide a relatively small amount of kW reduction for each participating customer compared to the MW output of Supply options. For example, the kW reduction may be 1 kW or less for any single residential or small commercial participating customer. Therefore, a very large number of participating customers must be signed up and retained for the DSM option to be able to compete with Supply options that will typically contribute capacity in terms of MW values (in which 1 MW = 1,000 kW).

For example, recall that our utility system has a resource need in 5 years of either 120 MW of new generating capacity or a peak load reduction of 100 MW. If a DSM option is projected to offer 1 kW of demand reduction per participating customer, the utility would need to sign up 100,000 participating customers in order to meet this 100 MW peak load reduction objective. (100,000 participating customers × 1 kW per customer = 100,000 kW or 100 MW.) One hundred thousand participating customers, each reducing load by 1 kW, equates to 100 MW of demand reduction on a utility system.

Therefore, although the DSM option's $/kW administrative costs and incentive payments are small relative to the $/kW values we have discussed for new Supply options, these "per kW" DSM costs will be multiplied by very large numbers of participating customers.

In regard to the second characteristic of DSM options, kW reduction at the utility's peak hour, this is the aspect of DSM options that drives most of the categories of benefits provided by DSM options. It is the kW reduction characteristic of DSM that results in avoiding the new generating unit (Supply option) that otherwise would have been built. In addition, the kW reduction characteristic will be the sole driver of a number of additional types of utility system cost savings, many of which would be the result of avoiding the construction and operation of a new Supply option if a DSM option is chosen instead.

The following list contains 10 DSM-related utility cost savings categories that are solely driven by the kW reduction characteristic of DSM options:

1. The avoided capital cost of the new generating unit;
2. The avoided transmission capital costs of both interconnecting the new generating unit to the existing transmission system (often referred to as transmission interconnection costs), and of any modifications to the existing transmission system to handle the new amount of power that will be supplied from the location of the new generating unit (often referred to as transmission integration costs);
3. The avoided fixed operating and maintenance (O&M) cost of the new generating unit;
4. The avoided capital replacement cost of the new generating unit;
5. The avoided firm gas transportation cost (if applicable) associated with the new generating unit;
6. The avoided variable O&M cost of the new generating unit;
7. The avoided capital costs of other transmission facilities throughout the utility system that otherwise would have been built if the peak load had not been lowered;
8. The avoided O&M costs for these other system transmission facilities that otherwise would have been built if the peak load had not been lowered;
9. The avoided capital costs of system distribution facilities that otherwise would have been built if the peak load had not been lowered; and,
10. The avoided O&M costs for system distribution facilities that otherwise would have been built if the peak load had not been lowered.

In addition, the kW reduction characteristic of DSM options will also drive two of three calculations that, in total, comprise the net changes in the utility system's fuel costs and environmental compliance costs from the addition of DSM options. In order to see this, we will examine the three specific types of changes in system fuel costs that will result from the introduction of a DSM option.

These three types of changes can be described as follows:

1. If the new generating unit is not built due to selection of the DSM option, the fuel that would have been burned in the new generating unit is not burned. This, by itself, will result in lower system fuel costs. (This calculation is driven by DSM's kW reduction characteristic because the kW reduction characteristic is what avoids the need for the new generating unit.)

2. However, if the new generating unit is not built and, therefore, does not operate, the amount of energy the new generating unit would have supplied to the utility system will now have to be provided by the existing generating units on the utility system. Because these existing generating units are typically not as fuel-efficient as most new generating units (recall that new generating units would typically not be operated unless they produced electricity at a lower cost than the existing generating units), the operation of these existing generating units to supply this same amount of energy will result in higher annual fuel costs for the utility system. We will refer to this as the "fuel penalty" from avoiding the new generating unit. (Because this fuel penalty is a result of avoiding the new generating unit, and the avoidance of the new generating unit is driven by DSM's kW reduction, the fuel penalty is also driven by DSM's kW reduction.)
3. The third calculation involves how much fuel is saved by the utility system not having to serve as much energy as it otherwise would if the DSM option is not selected. The reduction in the amount of system energy that the utility must serve is driven by the kWh reduction characteristic of the DSM option.

Therefore, the net impact of DSM on the utility system's fuel cost can be viewed as a three-part calculation involving both reduced and increased fuel costs*: ***Net system fuel cost impact from DSM*** = (i) fuel cost savings from fuel not being burned in the avoided new generating unit (driven by kW reduction), *minus* (ii) higher system fuel costs ("fuel penalty") from existing generating units now supplying energy that would have been supplied by the new unit (driven by kW reduction), *plus* (iii) system fuel cost savings from reduced energy that must be served by the utility (driven by kWh reduction).†

In similar fashion, the net environmental compliance (or emission) cost impact to the utility system from avoiding a new generating unit with DSM can be derived in three calculations:

1. By avoiding the new generating unit, the emissions that would have occurred from burning fuel in this new generating unit are also avoided. This results in emission cost savings for the utility system. This impact is solely driven by DSM's kW reduction because it is the kW reduction that results in avoiding the need for the new generating unit.

* In reality, these three impacts generally occur simultaneously. This is particularly true for the second and third calculations. We discuss these calculations separately to clarify how the overall net impact of DSM on system fuel is calculated. This information is also presented, in a slightly different format, in Appendix G.

† The result of this calculation can be either a net savings, or a net cost, in system fuel costs. This is dependent upon the individual utility system's existing generating units and the types of Supply and DSM options being considered.

2. Because the existing generating units on the utility system must now supply the same amount of energy the avoided new generating unit would have supplied, this increased output from the existing generating units typically results in increased emissions from the existing generating units. Because these existing generating units are typically less fuel-efficient than the avoided unit, the result is typically higher system emissions for this amount of energy than would have been the case if the avoided unit had been built and had supplied the energy. This "emission penalty" is also solely driven by DSM's kW reduction because it is the kW reduction characteristic that avoids the need for the new generating unit.
3. The lower amount of energy that the utility system must supply due to the selection of the DSM option will lower the utility system's emissions. This emission savings is solely driven by DSM's kWh reduction.

Therefore, the net impact of DSM on the utility system's environmental compliance costs can be viewed, similar to that for system fuel cost impacts discussed earlier, as a three-part calculation involving both reduced and increased environmental compliance costs:

Net system environmental compliance cost impact from DSM = (i) environmental compliance cost savings from fuel not being burned in the avoided new generating unit (driven by kW reduction), *minus* (ii) higher system environmental compliance costs ("environmental compliance cost penalty") from existing generating units now supplying energy that would have been supplied by the new unit (driven by kW reduction), *plus* (iii) system environmental compliance cost savings from reduced energy that must be served by the utility (driven by kWh reduction).*

In regard to the third characteristic of DSM, kWh reduction, we have seen that kWh reduction is responsible for only one of the three DSM-based impacts regarding either system fuel, or environmental compliance, cost savings. In other words, the kWh reduction characteristic of DSM options drives only two DSM-based cost savings categories. By comparison, the kW reduction characteristic of DSM options drives a total of 14 DSM cost savings categories: the original list of 10 cost savings categories we previously discussed, plus the two fuel cost impact categories, and the two environmental compliance cost impact categories, that were just discussed.

Therefore, we see that the second characteristic of DSM options, kW reduction, is responsible for driving many more categories (14) of DSM-related utility benefits than does the third characteristic, kWh reduction (2 categories). However, the two cost savings categories that are driven by the kWh reduction characteristic usually involve relatively large cost savings values. The key point

* The result of this calculation can also be either a net savings, or a net cost, in system environmental compliance costs. This is again dependent upon the individual utility system's existing generating units and the types of Supply and DSM options being considered.

is that there are 16 utility cost impact, or net benefit, categories and that neither the kW reduction characteristic, nor the kWh reduction characteristic, of DSM options should be overlooked in determining the true impact of DSM options.*

However, we are not quite through with the utility impacts resulting from the DSM kW reduction and kWh reduction characteristics. Although we have examined 16 categories of utility net benefits that result from selecting a DSM option, there is one more impact of DSM that must be considered.

That impact is the reduction in monies the utility receives from its customers due to the introduction of a DSM option. This is due to the DSM option's kW and kWh reduction characteristics. This impact is often referred to as "lost revenues," but is more correctly referred to as "revenue requirements not received" or as "unrecovered revenue requirements."

Without going into too much detail into how electric rates are set for a utility by its regulatory authority, here is how electric rates are basically set. A projection of the total demand (kW) and total energy (kWh) the utility will be expected to serve over a given time period is developed. The utility's projected operating costs over this same time period are also developed. The regulatory authority then develops an allowed rate of return (simplistically, think "return on the investments made in the utility by investors.") These factors lead to a determination of the total amount of money per year the utility must take in to meet both its operating costs and its allowed rate of return. This amount of money per year that the utility must take in is referred to as the utility's annual "revenue requirements." Electric rates are then set on both a cents/kWh basis and a \$/kW basis† that will allow the recovery of this amount of revenue requirements assuming the projected amounts of kW and kWh are served.

For example, using a very simplistic example, let's assume that some (again, very small) utility's projected total revenue requirements are \$1 million per year and that its projected amount of energy it must serve (its sales) are 10 million kWh. Therefore, if this utility's electric rates are set solely on a cents/kWh basis, its electric rates will be set as follows:

$$\text{Electric rate (cents/kWh)} = (\$1{,}000{,}000 \times 100 \text{ cents/\$1})/10{,}000{,}000 \text{ kWh}$$

$$= 10.0 \text{ cents/kWh}.$$

* In the years leading up to the writing of this book, there has been a trend by certain parties to focus solely on the kWh reduction characteristic of DSM options. Such a focus is ill advised because, as discussed above, there are many more categories of cost savings that are driven by kW reduction than are driven by kWh reduction. Therefore, a focus solely on kWh reduction does not provide a complete picture of the impacts, both positive and negative, of DSM options.

† All customers served by a traditional regulated utility typically pay a cents/kWh charge for each kWh they use. In addition, non-residential customers (i.e., commercial and industrial customers) whose highest monthly demand is above a given threshold (for example, 20 kW), also typically pay a \$/kW charge based on their highest monthly demand.

This electric rate of 10.0 cents/kWh will allow the utility to recover the desired $1 million of revenue requirements if its sales are, as projected, 10,000,000 kWh. But what happens if the utility's projected sales drop by 5% (or 500,000 kWh) to 9,500,000 kWh? The utility will recover only $950,000 in revenue requirements instead of $1,000,000. In order to recover the desired $1,000,000 in revenue requirements, the utility's electric rates would have to be raised to approximately 10.53 cents/kWh. ($1,000,000 in revenue requirements divided by 9,500,000 kWh = 10.53 cents/kWh.)

Now, switching our perspective from that of the utility system as a whole to the perspective of an individual customer, let's see what happens. Let's assume you are that customer. If your electric usage was 1,000 kWh per month, you originally would be paying $100 per month with the electric rate at 10.0 cents/kWh. But if the electric rate increased to 10.53 cents/kWh, and your usage had not changed, you would be paying $105.30 per month. (You might not be pleased with this outcome.)

The purpose of this simple example is to show that DSM options which result in kW and kWh reductions will have impacts on both the numerator (the projected revenue requirements or costs of the utility due to the costs and benefits of the DSM option) and the denominator (the projected number of sales in kWh). Both of these impacts affect the utility's electric rates and affect customers' bills.*

We have discussed which categories of utility costs (revenue requirements) are driven by either the kW reduction and the kWh reduction characteristics of the DSM option. Thus it is clear that DSM impacts the numerator value in an electric rate calculation. However, because DSM impacts the number of kWh that the utility system will serve, the denominator in an electric rate calculation is also impacted by DSM. We will return later in this chapter to examine the electric rate impact on our hypothetical utility system of the two DSM options it will be evaluating.

With this discussion of the three basic characteristics of DSM options, we will now introduce the two DSM options our utility is considering. Our assumptions for the key inputs for these two DSM options are presented in Table 6.1. The values shown for these inputs are assumed to be average values for each participating customer.

As shown from Row (1) of this table, the two DSM options differ in regard to the projected demand (kW) reduction per participant value. DSM Option 1 reduces the peak hour demand of each participating customer by 1.00 kW while DSM Option 2 reduces the peak hour demand of each participating customer by 0.50 kW.

---

* Obviously, the amount of electricity that a customer uses also affects the customer's bill. However, some customers (such as those with special medical conditions, limited income, etc.) may not be able to easily reduce their usage enough to offset any increase in electric rates resulting from the selection of certain DSM options (or of too many DSM options). Therefore, a utility should be very careful in its selection of DSM options.

TABLE 6.1

Key Inputs and Descriptors for Economic Evaluation: DSM Options

Input	Units of Measurement	DSM Option 1	DSM Option 2
1. Demand reduction (peak hour)	kW reduction/ participant	1.00	0.50
2. Energy reduction (annual)	kWh reduction/ participant	2,000	3,000
3. Administrative cost	$/participant	$100	$100
4. Incentive payment	$/participant	$200	$200
5. Participant equipment cost	$/participant	$800	$800
6. Life of DSM measure	Years	15	5
7. Energy-to-demand reduction ratio	Equivalent hours	2,000	6,000
8. Equivalent capacity factor	% of annual hours	23%	68%

The alert reader will quickly realize that, because DSM Option 2 will provide only 0.50 kW per participating customer compared to 1.00 kW per participating customer, twice as many participating customers (200,000), will need to be signed up for DSM Option 2 compared to DSM Option 1 (100,000) in order to achieve the same total desired demand reduction level (100 MW) to meet the projected resource needs of our utility system 5 years in the future.

There are other differences in the two DSM options as can be seen from Rows (2) through (6) in the table which present other key characteristics for the DSM options. In Row (2), the annual energy (kWh) reduction per participant is presented. DSM Option 1 will result in 2,000 kWh of energy use being reduced by each participant and DSM Option 2 will result in a 3,000 kWh reduction per participant.

Rows (3) and (4) present, respectively, the "per participant" administrative cost and incentive payment. These values for the two DSM options are identical: administrative costs of $100 and incentive payments of $200.

However, a note regarding the *total* administrative and incentive payment costs for the two DSM options is in order. Despite the fact that the per participant administrative and incentive costs are identical for the two DSM options, the *total* administrative and incentive costs that will be incurred for DSM Option 2 will be at least twice as high as these costs for DSM Option 1 in meeting the 100 MW objective for DSM. This is because the kW reduction per participant for DSM Option 2 (0.50 kW) is only half that of DSM Option 1 (1.00 kW). Therefore, twice as many total participants will need to be signed up for DSM Option 2 than for DSM Option 1 in order to achieve the same level of demand reduction (100 MW). As a consequence, the total administrative and incentive payment costs for DSM Option 2 will be at least twice as high as for DSM Option 1. (You may ask why these costs for DSM Option 2 are not exactly twice as high as for DSM Option 1. The answer is the "life expectancy" of the equipment installed as part of either DSM option. We will get to that factor momentarily as we discuss Row (6).)

Row (5) presents the incremental equipment cost that will be paid by the DSM participant for each option: $800. For example, if a DSM option were a much more efficient air conditioning unit, a customer might have to pay $800 more to install this higher efficiency air conditioner instead of a more standard efficiency air conditioner. Row (6) then presents the life expectancy of the DSM equipment before it needs to be replaced: 15 years for DSM Option 1 and 5 years for DSM Option 2. The life expectancy of the DSM equipment installed for DSM Option 2 is only 1/3 as long as the life expectancy is for the different equipment installed for DSM Option 1. Therefore, it will be necessary to replace that DSM equipment much earlier for DSM Option 2 than for DSM Option 1.

For purposes of this discussion, we will assume that we perform this replacement for either DSM option by signing up new customers and paying an incentive to the new customers. Because we are doing this every 5 years for DSM Option 2, or three times in the same 15 year period that constitutes the life expectancy of DSM Option 1, the *total* administrative and incentive costs over the life of the analyses are further increased for DSM Option 2 in comparison to those costs for DSM Option 1. This explains why the total costs over the analysis period will actually be more than twice as high for DSM Option 2 as for DSM Option 1.

Returning to Table 6.1, Rows (1) through (6) present all of the key characteristics of the two DSM options in the manner they are normally discussed. However, I have found that an examination of the information presented on these 6 rows often fails to provide a clear picture of how DSM options really compare with each other, especially in regard to how they will "operate" once they are implemented on a utility system. Consequently, Rows (7) and (8) have been added.

Rows (7) and (8) are not actually inputs for the DSM options themselves, but are "descriptors" of DSM options that I have found useful in helping to explain differences between DSM options and the impact that these different DSM options have on utility systems. Row (7) calculates an energy-to-demand reduction ratio by dividing the annual kWh reduction value in Row (2) by the kW reduction value in Row (1). The resulting value represents the equivalent number of hours the DSM option will impact the utility system assuming its kW reduction value remains constant during all of these hours (which is a simplistic assumption). When two DSM options have different kW reduction values and/or different kWh reduction values, this descriptor allows one to better understand the differences between the DSM options and how the utility system will be impacted by the different DSM options.

For example, Row (7) helps shed more light on how much energy reduction will actually result from each of the two DSM options. A glance at the information presented in Row (2) shows that, on a per participant basis, DSM

Option 1 will reduce 2,000 kWh and DSM Option 2 will reduce 3,000 kWh. However, recall that DSM options are typically utilized as a resource option to meet a pre-determined resource need (100 MW for our utility if DSM is selected to meet its resource needs 5 years in the future).

Also recall that the per participant kW reduction for DSM Option 2 is 0.50 kW compared to 1.00 kW for DSM Option 1, which leads to the fact that twice as many participants will need to be signed up for DSM Option 2 than for DSM Option 1 to meet the pre-determined resource need of 100 MW.

Therefore, Row (7) shows that DSM Option 2 will actually reduce 6,000 kWh for every 1 kW reduction (3,000 kWh/0.5 kW = 6,000 kWh/kW) versus a 2,000 kWh reduction per kW for DSM Option 1. Thus, implementing the desired 100 MW of either DSM option will result in DSM Option 2 reducing three times more energy than DSM Option 1.

Row (8) then calculates an "equivalent capacity factor" by dividing the equivalent hour (kWh/kW) value in Row (7) by 8,760 hours in a year. The resulting ratio can be thought of as the DSM equivalent of the capacity factor values for generating units such as the four Supply options previously evaluated in Chapter 5. This "equivalent capacity factor" information is also useful in understanding and discussing how the DSM options will impact the utility system in comparison to the Supply options that DSM will be competing with.

Recall that our four Supply options had capacity factors, in ascending order, of 5% for the CT unit, 20% for the PV option, and 80% for the two CC units. By comparing these capacity factors with the equivalent capacity factors for the two DSM options of 23% for DSM Option 1 and 68% for DSM Option 2, we see that both of the DSM options' equivalent capacity factors are higher than the capacity factor for the CT and PV options, but lower than the capacity factor for the CC options. Thus, these DSM options can be seen as impacting the utility system—or "operating"—in a range that would be similar to how a low-level intermediate generating unit (DSM Option 1) to a high-level intermediate generating unit (DSM Option 2) would operate on our utility system.* This information is important not only in regard to the economic impacts the resource options will have, but also how the resource options will affect system fuel use and system emissions (i.e., environmental compliance costs).

This completes the introduction of the DSM option-specific inputs that will be used in the economic evaluation of these options. Before we leave this section, let's look back (just as we did for the Supply options) and see what appears to be obvious, and perhaps more importantly, what is not obvious, from looking at these inputs.

* The vast majority of DSM options will "operate" on a utility system in a range similar to a peaking-to-intermediate level generating unit. DSM options that operate similar to baseload generating units are less common.

First, from looking at the inputs, the following statements about these two DSM options appear to be obvious:

- Because twice as many participating customers will have to be signed up for DSM Option 2 than for DSM Option 1 in order to reach the same 100 MW objective, and because the life expectancy of the DSM equipment for DSM Option 2 is only 1/3 as long as for DSM Option 1, DSM Option 2 will have significantly higher total administrative and incentive costs over the period of the analysis than will DSM Option 1.
- Assuming that the same 100 MW level is implemented with either DSM option, all of the kW reduction-driven impacts to the utility system will be identical for DSM Option 1 and DSM Option 2.
- Conversely, the kWh reduction-driven impacts to the utility system will be significantly greater for DSM Option 2 than for DSM Option 1.

However, two outcomes are not clear. First, it is unclear which of the two combinations: DSM Option 1's lower total administrative and incentive costs, plus lower total kWh reduction impacts, or DSM Option 2's higher total administrative and incentive costs, plus higher total kWh reduction impacts, will result in a lower total cost (CPVRR) for the utility.

Second, and much more importantly, in regard to the electric rate perspective, this CPVRR value will represent only one half of the picture; i.e., the numerator in the electric rate calculation. While the aforementioned determination of CPVRR will identify the numerator of an electric rate calculation, the denominator (the amount of total energy to be served by the utility if a particular DSM option is selected) must also be accounted for. It is only after both the numerator and denominator are determined that we will know which DSM option will result in the lower electric rate. And it is only by calculating the electric rate impact that we will have a full picture of the economic impacts of both of the DSM options.

We will see how all of this actually plays out in the next sections of this chapter. We will start by performing preliminary economic screening evaluations of the two DSM options.

Preliminary Economic Screening Evaluation of DSM Options: Understanding the Cost-Effectiveness Screening Tests

As mentioned in Chapter 3, preliminary economic screening evaluations of DSM options typically utilize specific "cost-effectiveness screening tests" that compare the DSM option with a comparably sized Supply option.* Therefore, we will next perform a preliminary economic screening evaluation of these two DSM options and we will use the Participant test, RIM test, and TRC test to perform these preliminary economic screening evaluations of the two DSM options. The Supply option to which the two DSM options will be compared will be the most economic Supply option for our hypothetical utility system: CC Unit A.

We will begin by providing more detail regarding these tests. These three tests are comparisons of DSM-driven benefits and costs from the perspective of either a potential participating customer (as the Participant test is designed to do) or all of the utility's customers (as the RIM and TRC tests are purported to do). The test results are typically presented in a benefit-to-cost ratio format in which the present value of the benefits derived from DSM is divided by the present value of the DSM costs. A benefits-divided-by-costs resulting value of 1.00 means that the present value of the benefits is exactly equal to the present value of the costs. A ratio greater than 1.00 indicates that the benefits exceed the costs. In such a case, the DSM option is said to have passed that particular cost-effectiveness screening test. Conversely, a ratio of less than 1.00 indicates that benefits are less than the costs. In this case, the DSM option is said to have failed that particular cost-effectiveness test.

In order to better understand the results of these cost-effectiveness screening tests when our two DSM options are evaluated, we need to better understand the specific cost-effectiveness tests. We will start with the Participant test.

Table 6.2 first indicates, in the shaded column, the types of economic impacts, or benefits, that a participating customer would actually receive from a DSM option or program. These include: the bill savings from the DSM-induced reduction in energy usage by the participant, the incentive payments received from the utility, and tax credits that may be available for

TABLE 6.2

Economic Elements Included in the Participant Test: Benefits Only

Economic Elements	Participant Incurred Economic Impacts	Included in the Participant Test?
Benefits of DSM		
(=avoided costs and/or direct benefits)		
Bill savings by participants	Yes	Yes
Incentives received by participants	Yes	Yes
Tax credits received by participants	Yes	Yes

* In Chapter 3, it was mentioned that this approach to preliminary economic screening of DSM options was a meaningful approach. However, following up on previous discussions of the inherent problems with using a screening curve analysis approach in which even one of four key characteristics are dissimilar for Supply options, we see that using such an approach for DSM options (such as the two discussed here) is again flawed due to significant differences in DSM and Supply options in regard to at least two key characteristics: capacity factor and life of the measure.

some DSM measures. The unshaded column on the right-hand side of the table then indicates whether the Participant test actually includes these categories of benefits in its calculations.

As shown in this table, the Participant test does include all of these categories of benefits that a participating customer would actually receive from a DSM program.

Now we turn our attention to the costs that a participating customer would incur from his/her participation in the DSM program. This information is presented in Table 6.3 which is an expansion of the previous table.

As shown in the shaded column of Table 6.3, a participating customer would expect to incur initial purchase (i.e., capital) costs as well as on-going operation and/or maintenance costs. As the unshaded column on the right-hand side of the table also shows, the Participant test does include these DSM-related costs in its calculation.

Therefore, the Participant test correctly includes all of the DSM-related benefits and costs that a potential participating customer should consider when deciding whether to participate in a DSM program offered by a utility. It is a meaningful test.

We now turn our attention to the RIM and TRC tests. We again begin by looking at the benefits side of the calculation for these two tests in Table 6.4.

This table shows that the DSM-related benefits of DSM include the following costs that will be avoided by DSM's kW and kWh reduction characteristics: (i) generation capital and O&M, (ii) transmission capital and O&M, and (iii) distribution capital and O&M. In addition, these two tests include: (iv) the net fuel impacts to the utility system, and (v) the net environmental compliance impacts to the utility system.*

TABLE 6.3

Economic Elements Included in the Participant Test: Benefits and Costs

Economic Elements	Participant Incurred Economic Impacts	Included in the Participant Test?
Benefits of DSM		
(= avoided costs and/or direct benefits)		
Bill savings by participants	Yes	Yes
Incentives received by participants	Yes	Yes
Tax credits received by participants	Yes	Yes
Costs of DSM (= incurred costs)		
Participants' capital and O&M costs	Yes	Yes

* These five categories of DSM benefits is a "collapsed" version of the 16 types of DSM impacts that were discussed earlier. This collapsed set of types of impacts is used here solely to make the RIM and TRC tests easier to understand. The use of the word "net" in regard to system fuel and environmental compliance impacts denotes that DSM options can have both positive and negative impacts in these areas as previously discussed.

TABLE 6.4

Economic Elements Included in the RIM and TRC Tests: Benefits Only

Economic Elements	Utility-Incurred Economic Impacts	Included in the RIM Test?	Included in the TRC Test?
Benefits of DSM			
(= avoided costs and/or cost impacts)			
Generation capital and O&M	Yes	Yes	Yes
Transmission capital and O&M	Yes	Yes	Yes
Distribution capital and O&M	Yes	Yes	Yes
Net system fuel impacts	Yes	Yes	Yes
Net system environmental compliance impacts	Yes	Yes	Yes

And, similar to what we observed with the Participant test, all of the benefits that will result from the implementation of DSM options are included in the benefits calculation of both the RIM and TRC tests. In fact, assuming all else equal, the RIM and TRC tests will calculate identical benefit values for a given DSM program.

Two logical questions remain. First, as was the case with the Participant test, are all of the relevant costs of DSM included in the RIM and TRC tests? Second, in regard to the RIM and TRC tests calculating identical benefits from DSM, do the RIM and TRC tests also calculate identical DSM costs?

The answer is "no" to both of these questions. This is seen from Table 6.5. This table shows that we have listed four types of DSM-related costs: (i) utility equipment and administration costs, (ii) incentives paid to participating customers, (iii) unrecovered revenue requirements, and (iv) capital and O&M costs that participating customers pay for themselves.

As shown by the bottom portion of the shaded column in Table 6.5, the rows marked with a "Yes" indicate that there are three types of DSM-related costs that are incurred by the utility and which are passed on to, or otherwise impact, all of a utility's customers. These three types of DSM costs are: (i) utility equipment and administration, (ii) incentives paid to participants, and (iii) unrecovered revenue requirements.

As Table 6.5 shows, both the RIM and TRC tests include the DSM-related costs of utility equipment and administration. However, at this point the RIM and TRC tests significantly diverge in regard to the accounting of DSM costs.

The RIM test includes the remaining two DSM costs that will be incurred by the utility and which will be passed on to, or otherwise impact, all of the utility's customers: incentives paid to participants and unrecovered revenue requirements. Conversely, the TRC test does not include either of these two utility-incurred DSM costs. Furthermore, the TRC test includes the participants' capital and O&M costs even though these costs are not incurred by the

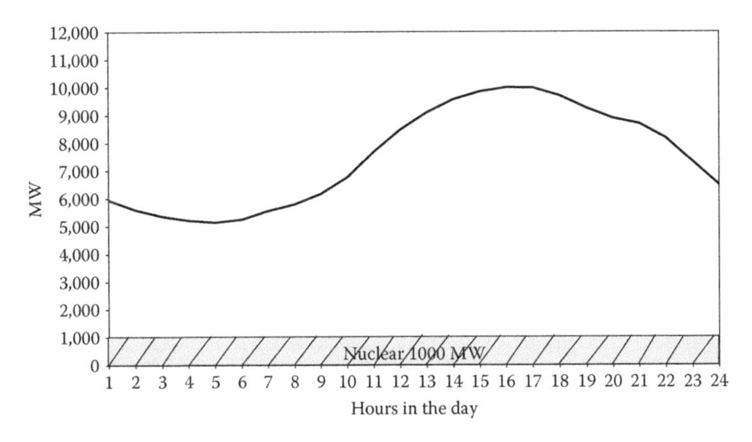

FIGURE 2.3
The potential contribution from nuclear generation during the summer peak day.

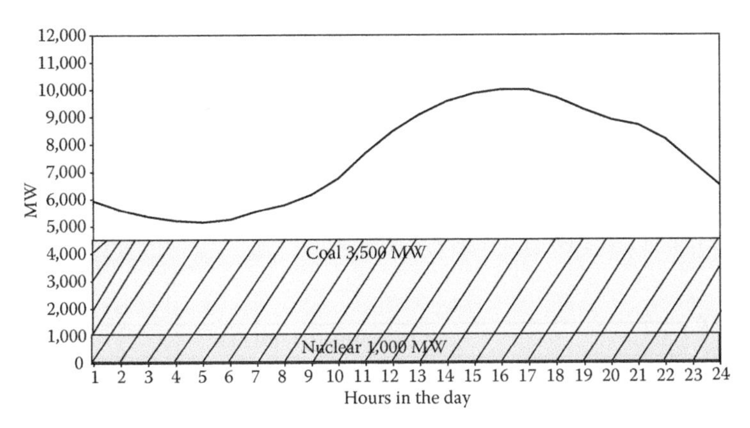

FIGURE 2.4
The potential contribution from nuclear and coal generation during the summer peak day.

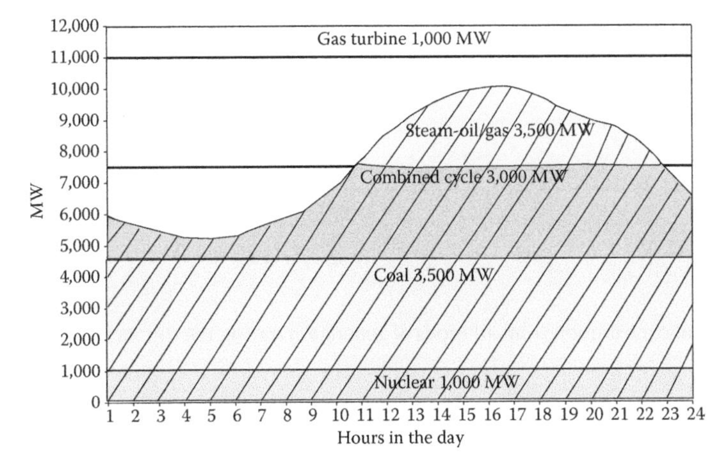

FIGURE 2.5
The potential contribution from all types of generation during the summer peak day.

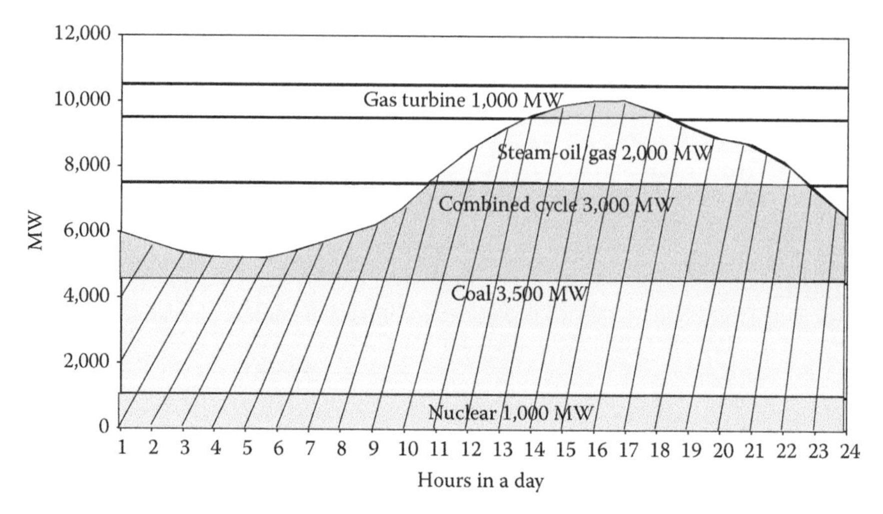

FIGURE 2.6
The potential contribution from all types of generation during the summer peak day (assuming a reduction of 1,500 MW of steam-oil/gas capacity).

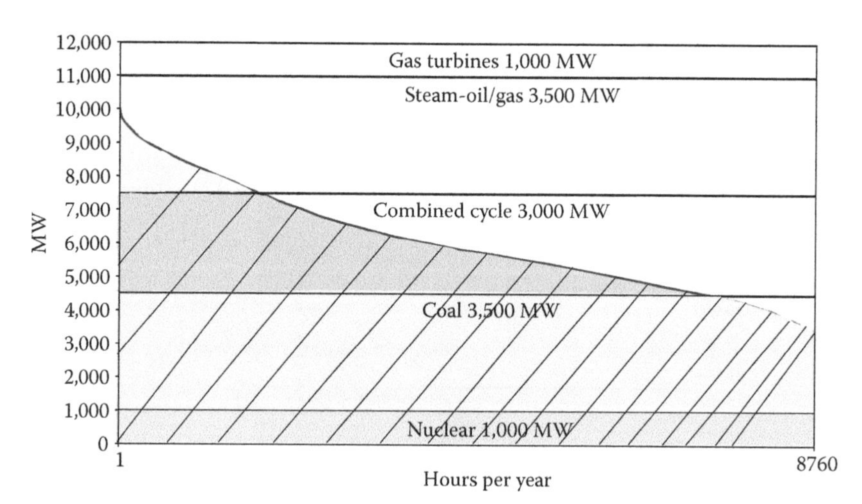

FIGURE 2.7
The potential contribution from all types of generation during the course of a year.

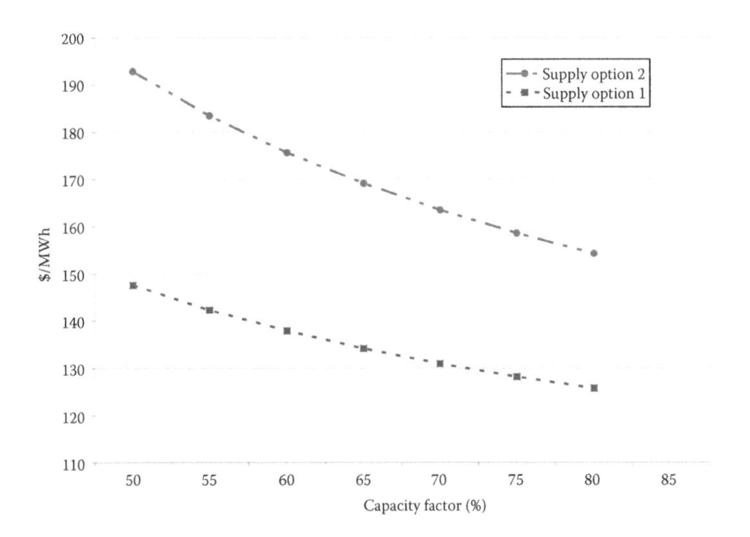

FIGURE 3.1
Preliminary economic analysis: screening curve approach levelized $/MWh costs for two supply options.

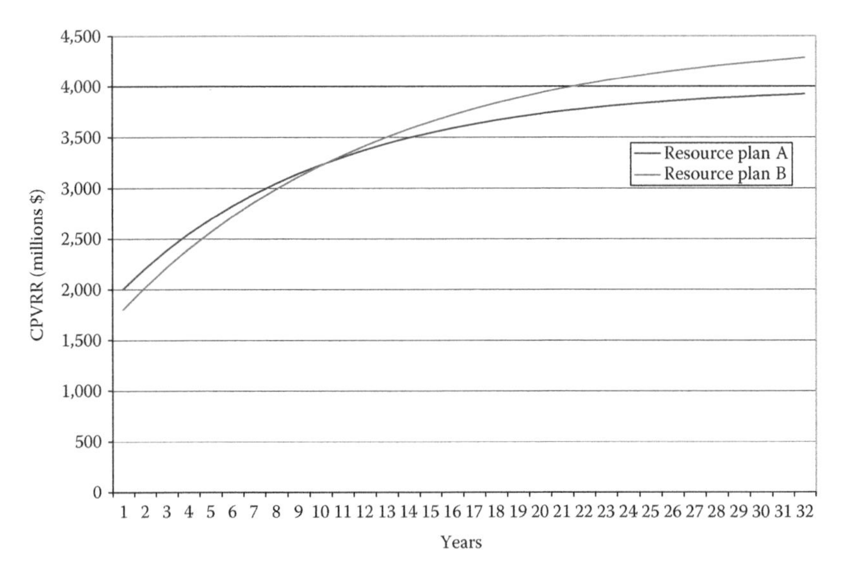

FIGURE 3.2
"Cross over" graph of two hypothetical resource plans: cross over in 10 years.

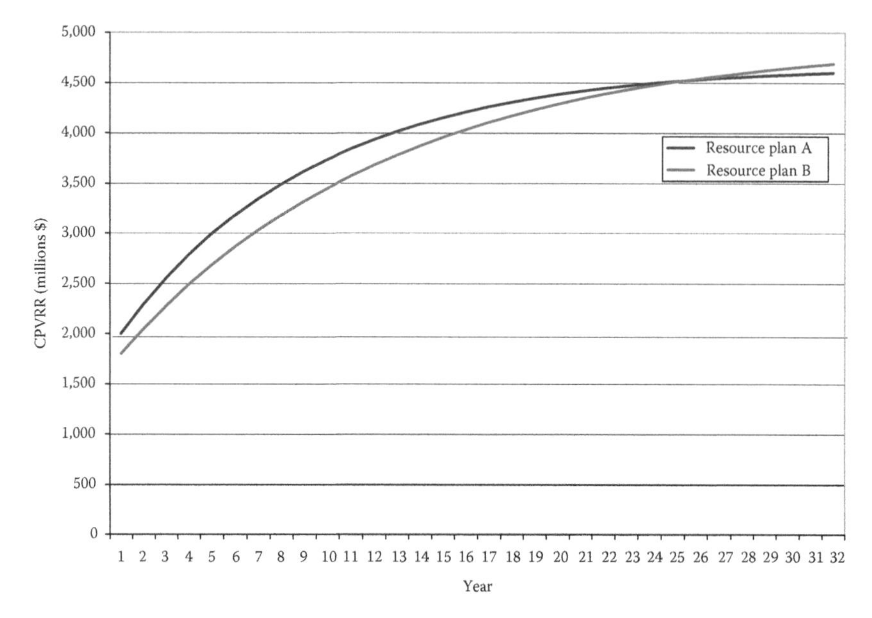

FIGURE 3.3
"Cross over" graph of two hypothetical resource plans: cross over in 25 years.

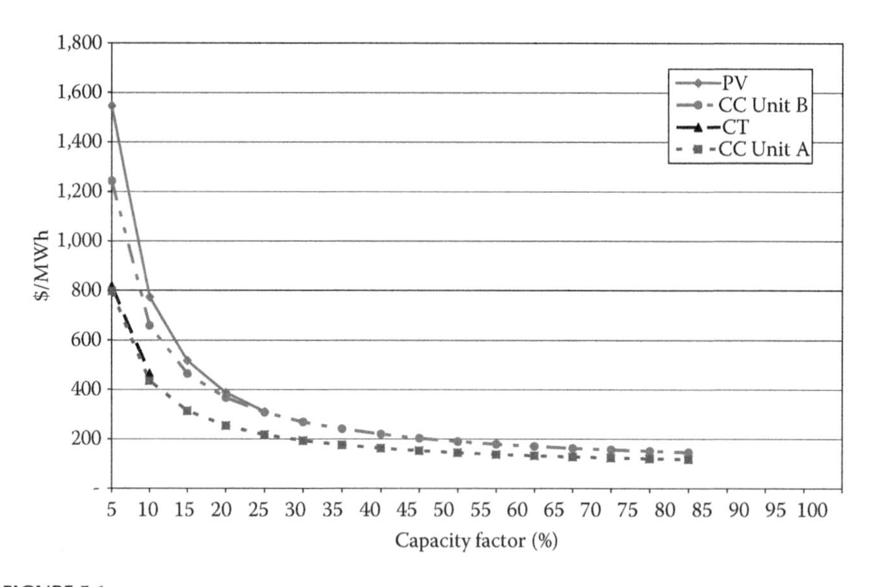

FIGURE 5.1
Preliminary economic screening analysis: Screening approach levelized $/MWH costs for all four supply options.

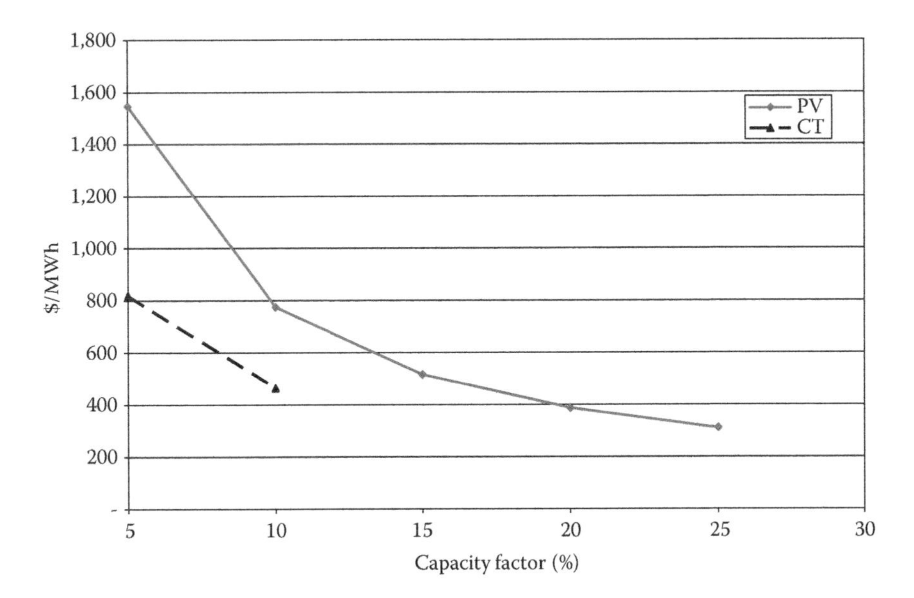

FIGURE 5.2
Preliminary economic screening analysis: Screening approach levelized $/MWH costs for CT and PV options.

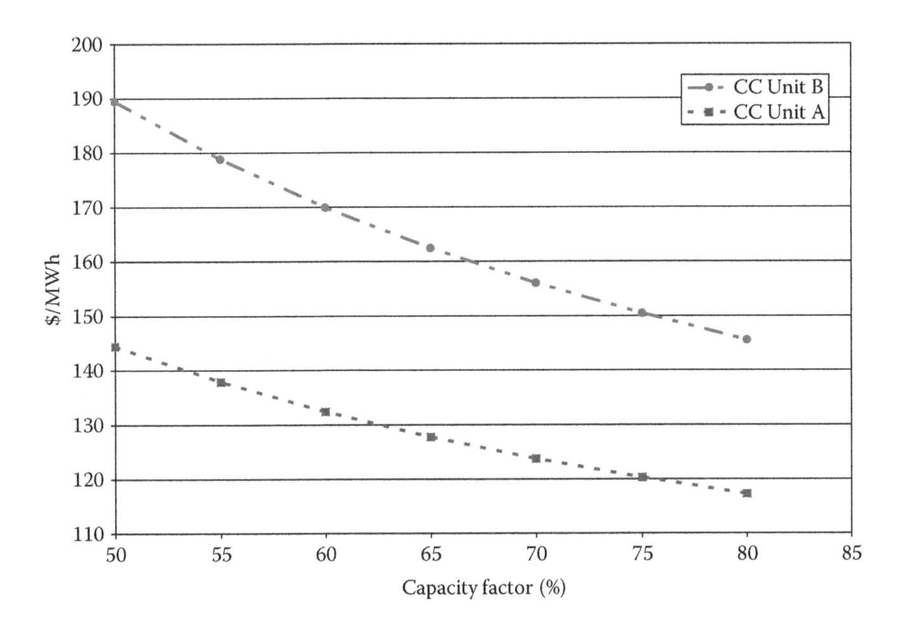

FIGURE 5.3
Preliminary economic screening analysis: Screening approach levelized $/MWH costs for CC Unit A and CC Unit B.

TABLE 5.11

Fixed Costs Calculation for Supply Only Resource Plan 1 (CC)

	(1)	(2)	(3)	(4)	(5)	(6) = Sum of Cols. (2) through (5)
Year	Annual Discount Factor 8.000%	Generation Capital (Nominal $, Millions)	Generation Fixed O & M (Nominal $, Millions)	Generation Capital Replacement (Nominal $, Millions)	Firm Gas Transportation (Nominal $, Millions)	**Total Generation Fixed Costs** (Nominal $, Millions)
Current Year	1.000	0	0	0	0	**0**
1	0.926	0	0	0	0	**0**
2	0.857	0	0	0	0	**0**
3	0.794	0	0	0	0	**0**
4	0.735	0	0	0	0	**0**
5	0.681	100	3	5	82	**190**
6	0.630	96	3	5	82	**186**
7	0.583	92	3	5	82	**182**
8	0.540	88	3	5	82	**179**
9	0.500	192	6	11	164	**374**
10	0.463	184	7	11	164	**366**
11	0.429	176	7	11	164	**358**
12	0.397	167	7	11	164	**350**
13	0.368	276	11	18	246	**551**
14	0.340	263	11	18	246	**538**
15	0.315	250	11	18	246	**526**
16	0.292	237	11	19	246	**513**
17	0.270	224	11	19	246	**501**
18	0.250	340	16	26	329	**710**
19	0.232	322	16	26	329	**693**
20	0.215	304	16	27	329	**676**
21	0.199	286	16	27	329	**658**
22	0.184	408	21	35	411	**874**
23	0.170	384	21	36	411	**852**
24	0.158	360	22	36	411	**829**
25	0.146	336	22	37	411	**806**
26	0.135	464	27	45	493	**1,030**
27	0.125	434	28	46	493	**1,001**
28	0.116	404	28	47	493	**973**
29	0.107	374	29	48	493	**944**
Total CPVRR =		**$1,685**	**$79**	**$131**	**$1,687**	**$3,582**

TABLE 5.12

DSM Costs Calculation for Supply Only Resource Plan 1 (CC)

	(1)	(6) = Sum of Cols. (2) through (5)	(7)	(8)	(9)	(10) = Col (7) + Col (8) – Col (9)
Year	Annual Discount Factor 8.000%	**Total Generation Fixed Costs** (Nominal $, Millions)	DSM Administrative Costs (Nominal $, Millions)	DSM Incentive Payments (Nominal $, Millions)	T & D Costs Avoided by DSM (Nominal $, Millions)	**DSM Net Costs** (Nominal $, Millions)
Current Year	1.000	**0**	0	0	0	**0**
1	0.926	**0**	0	0	0	**0**
2	0.857	**0**	0	0	0	**0**
3	0.794	**0**	0	0	0	**0**
4	0.735	**0**	0	0	0	**0**
5	0.681	**190**	0	0	0	**0**
6	0.630	**186**	0	0	0	**0**
7	0.583	**182**	0	0	0	**0**
8	0.540	**179**	0	0	0	**0**
9	0.500	**374**	0	0	0	**0**
10	0.463	**366**	0	0	0	**0**
11	0.429	**358**	0	0	0	**0**
12	0.397	**350**	0	0	0	**0**
13	0.368	**551**	0	0	0	**0**
14	0.340	**538**	0	0	0	**0**
15	0.315	**526**	0	0	0	**0**
16	0.292	**513**	0	0	0	**0**
17	0.270	**501**	0	0	0	**0**
18	0.250	**710**	0	0	0	**0**
19	0.232	**693**	0	0	0	**0**
20	0.215	**676**	0	0	0	**0**
21	0.199	**658**	0	0	0	**0**
22	0.184	**874**	0	0	0	**0**
23	0.170	**852**	0	0	0	**0**
24	0.158	**829**	0	0	0	**0**
25	0.146	**806**	0	0	0	**0**
26	0.135	**1,030**	0	0	0	**0**
27	0.125	**1,001**	0	0	0	**0**
28	0.116	**973**	0	0	0	**0**
29	0.107	**944**	0	0	0	**0**
Total CPVRR =		**$3,582**	**$0**	**$0**	**$0**	**$0**

TABLE 5.13

Variable Costs Calculation for Supply Only Resource Plan 1 (CC)

	(1)	(6) = Sum of Cols. (2) through (5)	(10) = Col (7)+ Col (8) – Col (9)	(11)	(12)	(13)	(14) = Sum of Cols. (11) through (13)
Year	Annual Discount Factor 8.000%	**Total Generation Fixed Costs** (Nominal $, Millions)	**DSM Net Costs** (Nominal $, Millions)	Generation Variable O & M (Nominal $, Millions)	System Net Fuel (Nominal $, Millions)	System Environmental Compliance (Nominal $, Millions)	**Total Variable Costs** (Nominal $, Millions)
Current Year	1.000	**0**	**0**	0	1,660	275	**1,936**
1	0.926	**0**	**0**	0	1,726	390	**2,116**
2	0.857	**0**	**0**	0	1,794	507	**2,300**
3	0.794	**0**	**0**	0	1,863	625	**2,488**
4	0.735	**0**	**0**	0	1,935	746	**2,681**
5	0.681	**190**	**0**	7	1,928	845	**2,780**
6	0.630	**186**	**0**	7	2,002	968	**2,977**
7	0.583	**182**	**0**	7	2,079	1,092	**3,179**
8	0.540	**179**	**0**	7	2,158	1,219	**3,384**
9	0.500	**374**	**0**	15	2,151	1,315	**3,482**
10	0.463	**366**	**0**	15	2,233	1,444	**3,693**
11	0.429	**358**	**0**	16	2,317	1,574	**3,908**
12	0.397	**350**	**0**	16	2,404	1,707	**4,127**
13	0.368	**551**	**0**	25	2,399	1,800	**4,224**
14	0.340	**538**	**0**	25	2,489	1,935	**4,449**
15	0.315	**526**	**0**	26	2,581	2,072	**4,679**
16	0.292	**513**	**0**	26	2,677	2,211	**4,914**
17	0.270	**501**	**0**	27	2,775	2,352	**5,153**
18	0.250	**710**	**0**	36	2,771	2,441	**5,249**
19	0.232	**693**	**0**	37	2,873	2,585	**5,495**
20	0.215	**676**	**0**	38	2,978	2,730	**5,746**
21	0.199	**658**	**0**	38	3,085	2,878	**6,002**
22	0.184	**874**	**0**	49	3,083	2,964	**6,096**
23	0.170	**852**	**0**	50	3,195	3,114	**6,359**
24	0.158	**829**	**0**	51	3,310	3,266	**6,627**
25	0.146	**806**	**0**	52	3,428	3,421	**6,902**
26	0.135	**1,030**	**0**	64	3,428	3,503	**6,994**
27	0.125	**1,001**	**0**	65	3,551	3,660	**7,275**
28	0.116	**973**	**0**	66	3,677	3,819	**7,562**
29	0.107	**944**	**0**	68	3,807	3,981	**7,856**
Total CPVRR =		**$3,582**	**$0**	**$184**	**$27,271**	**$16,790**	**$44,246**

TABLE 5.14

Total Costs Calculation for Supply Only Resource Plan 1 (CC)

	(1)	(6) = Sum of Cols. (2) through (5)	(10) = Col (7)+ Col (8) – Col (9)	(14) = Sum of Cols. (11) through (13)	(15) – (6) + (10) + (14)	(16) = (1) × (15)	(17)
Year	Annual Discount Factor 8.000%	**Total Generation Fixed Costs** (Nominal $, Millions)	**DSM Net Costs** (Nominal $, Millions)	**Total Variable Costs** (Nominal $, Millions)	Total Annual Costs (Nominal $, Millions)	Total Annual NPV Costs (NPV $, Millions)	Cumulative Total NPV Costs (NPV $, Millions)
Current Year	1.000	**0**	**0**	**1,936**	1,936	1,936	1,936
1	0.926	**0**	**0**	**2,116**	2,116	1,959	3,895
2	0.857	**0**	**0**	**2,300**	2,300	1,972	5,867
3	0.794	**0**	**0**	**2,488**	2,488	1,975	7,842
4	0.735	**0**	**0**	**2,681**	2,681	1,970	9,813
5	0.681	**190**	**0**	**2,780**	2,970	2,022	11,834
6	0.630	**186**	**0**	**2,977**	3,164	1,994	13,828
7	0.583	**182**	**0**	**3,179**	3,361	1,961	15,789
8	0.540	**179**	**0**	**3,384**	3,563	1,925	17,714
9	0.500	**374**	**0**	**3,482**	3,856	1,929	19,643
10	0.463	**366**	**0**	**3,693**	4,058	1,880	21,522
11	0.429	**358**	**0**	**3,908**	4,266	1,829	23,352
12	0.397	**350**	**0**	**4,127**	4,477	1,778	25,130
13	0.368	**551**	**0**	**4,224**	4,774	1,755	26,885
14	0.340	**538**	**0**	**4,449**	4,987	1,698	28,583
15	0.315	**526**	**0**	**4,679**	5,204	1,641	30,224
16	0.292	**513**	**0**	**4,914**	5,427	1,584	31,808
17	0.270	**501**	**0**	**5,153**	5,654	1,528	33,336
18	0.250	**710**	**0**	**5,249**	5,959	1,491	34,827
19	0.232	**693**	**0**	**5,495**	6,187	1,434	36,261
20	0.215	**676**	**0**	**5,746**	6,421	1,378	37,639
21	0.199	**658**	**0**	**6,002**	6,660	1,323	38,962
22	0.184	**874**	**0**	**6,096**	6,970	1,282	40,244
23	0.170	**852**	**0**	**6,359**	7,210	1,228	41,472
24	0.158	**829**	**0**	**6,627**	7,456	1,176	42,648
25	0.146	**806**	**0**	**6,902**	7,708	1,125	43,773
26	0.135	**1,030**	**0**	**6,994**	8,023	1,085	44,858
27	0.125	**1,001**	**0**	**7,275**	8,276	1,036	45,894
28	0.116	**973**	**0**	**7,562**	8,535	989	46,883
29	0.107	**944**	**0**	**7,856**	8,800	945	47,828
Total CPVRR =		**$3,582**	**$0**	**$44,246**	**$47,828**	**$47,828**	

TABLE 5.15

Economic Evaluation Results of Supply Only Resource Plans: CPVRR Costs

	(1)	(2)	(3)	(4) = Sum of Cols (1) through (3)	(5)
Resource Plan	Fixed Costs (Millions, CPVRR)	DSM Costs (Millions, CPVRR)	Variable Costs (Millions, CPVRR)	Total Costs (Millions, CPVRR)	Difference from Lowest Cost Supply Only Plan (Millions, CPVRR)
-----	-----	-----	-----	-----	-----
Supply Only Resource Plan 1 (CC)	3,582	0	44,246	**47,828**	0
Supply Only Resource Plan 2 (CT)	3,396	0	44,554	**47,950**	122
Supply Only Resource Plan 3 (PV)	3,894	0	44,331	**48,225**	397

TABLE 5.16

Levelized Electric Rate Calculation for Supply Only Resource Plan 1 (CC)

	(1)	(15) = (6)+(10) + (14)	(18)	(19) = (15) + (18)	(20)	(21)	(22) =(20) – (21)	(23) =((19) × 100) / (22)	(24) =(1) × (23)	(25)	(26) = (1) × (25)
Year	Annual Discount Factor 8.000%	Total Annual Costs (Nominal $, Millions)	Other System Costs Not Affected by Plan (Nominal $, Millions)	Total Utility Costs (Nominal $, Millions)	Forecasted Utility Sales/NEL (GWh)	DSM Energy Reduction (GWh)	Net Utility Sales/NEL (GWh)	System Average Electric Rate Nominal (Cents/kWh)	System Average Electric Rate NPV (Cents/kWh)	Levelized System Average Electric Rate (Cents/kWh)	System Average Electric Rate Nominal (Cents/kWh)
Current Year	1.000	1,936	2,700	4,636	50,496	0	50,496	9.1800	9.1800	12.8116	12.8116
1	0.926	2,116	2,754	4,870	51,022	0	51,022	9.5447	8.8377	12.8116	11.8626
2	0.857	2,300	2,809	5,109	51,548	0	51,548	9.9116	8.4976	12.8116	10.9839
3	0.794	2,488	2,865	5,354	52,074	0	52,074	10.2809	8.1613	12.8116	10.1703
4	0.735	2,681	2,923	5,603	52,600	0	52,600	10.6528	7.8301	12.8116	9.4169
5	0.681	2,970	2,981	5,951	53,126	0	53,126	11.2024	7.6242	12.8116	8.7194
6	0.630	3,164	3,041	6,204	53,652	0	53,652	11.5638	7.2872	12.8116	8.0735
7	0.583	3,361	3,101	6,462	54,178	0	54,178	11.9282	6.9600	12.8116	7.4755
8	0.540	3,563	3,163	6,726	54,704	0	54,704	12.2958	6.6430	12.8116	6.9217
9	0.500	3,856	3,227	7,083	55,230	0	55,230	12.8237	6.4150	12.8116	6.4090
10	0.463	4,058	3,291	7,350	55,756	0	55,756	13.1819	6.1058	12.8116	5.9343
11	0.429	4,266	3,357	7,623	56,282	0	56,282	13.5437	5.8087	12.8116	5.4947
12	0.397	4,477	3,424	7,902	56,808	0	56,808	13.9093	5.5236	12.8116	5.0877
13	0.368	4,774	3,493	8,267	57,334	0	57,334	14.4189	5.3018	12.8116	4.7108
14	0.340	4,987	3,563	8,549	57,860	0	57,860	14.7761	5.0307	12.8116	4.3619
15	0.315	5,204	3,634	8,838	58,386	0	58,386	15.1375	4.7720	12.8116	4.0388
16	0.292	5,427	3,707	9,133	58,912	0	58,912	15.5033	4.5253	12.8116	3.7396
17	0.270	5,654	3,781	9,435	59,438	0	59,438	15.8735	4.2901	12.8116	3.4626
18	0.250	5,959	3,856	9,815	59,964	0	59,964	16.3686	4.0962	12.8116	3.2061
19	0.232	6,187	3,933	10,121	60,490	0	60,490	16.7314	3.8769	12.8116	2.9686
20	0.215	6,421	4,012	10,433	61,016	0	61,016	17.0992	3.6686	12.8116	2.7487
21	0.199	6,660	4,092	10,753	61,542	0	61,542	17.4720	3.4709	12.8116	2.5451
22	0.184	6,970	4,174	11,144	62,068	0	62,068	17.9551	3.3027	12.8116	2.3566
23	0.170	7,210	4,258	11,468	62,594	0	62,594	18.3212	3.1204	12.8116	2.1820
24	0.158	7,456	4,343	11,799	63,120	0	63,120	18.6930	2.9479	12.8116	2.0204
25	0.146	7,708	4,430	12,138	63,646	0	63,646	19.0704	2.7846	12.8116	1.8707
26	0.135	8,023	4,518	12,542	64,172	0	64,172	19.5439	2.6424	12.8116	1.7322
27	0.125	8,276	4,609	12,885	64,698	0	64,698	19.9153	2.4931	12.8116	1.6038
28	0.116	8,535	4,701	13,236	65,224	0	65,224	20.2930	2.3522	12.8116	1.4850
29	0.107	8,800	4,795	13,595	65,750	0	65,750	20.6770	2.2192	12.8116	1.3750
	Total CPVRR =	**$47,828**							------		------
									155.7691		**155.7691**

TABLE 6.9

Fixed Costs Calculation for With DSM Resource Plan 1

	(1)	(2)	(3)	(4)	(5)	(6) = Sum of Cols. (2) through (5)
Year	Annual Discount Factor 8.000%	Generation Capital (Nominal $, Millions)	Generation Fixed O & M (Nominal $, Millions)	Generation Capital Replacement (Nominal $, Millions)	Firm Gas Transportation (Nominal $, Millions)	**Total Generation Fixed Costs** (Nominal $, Millions)
Current Year	1.000	0	0	0	0	**0**
1	0.926	0	0	0	0	**0**
2	0.857	0	0	0	0	**0**
3	0.794	0	0	0	0	**0**
4	0.735	0	0	0	0	**0**
5	0.681	0	0	0	0	**0**
6	0.630	102	3	5	82	**192**
7	0.583	98	3	5	82	**188**
8	0.540	94	3	5	82	**184**
9	0.500	90	3	5	82	**181**
10	0.463	196	7	11	164	**378**
11	0.429	188	7	11	164	**370**
12	0.397	179	7	11	164	**362**
13	0.368	171	7	12	164	**354**
14	0.340	282	11	18	246	**557**
15	0.315	268	11	18	246	**544**
16	0.292	255	11	19	246	**531**
17	0.270	242	11	19	246	**519**
18	0.250	229	12	19	246	**506**
19	0.232	347	16	26	329	**718**
20	0.215	329	16	27	329	**700**
21	0.199	310	16	27	329	**682**
22	0.184	292	17	28	329	**665**
23	0.170	416	21	36	411	**884**
24	0.158	392	22	36	411	**860**
25	0.146	367	22	37	411	**837**
26	0.135	343	23	38	411	**814**
27	0.125	473	28	46	493	**1,040**
28	0.116	443	28	47	493	**1,011**
29	0.107	412	29	48	493	**982**
Total CPVRR =		**$1553**	**$72**	**$119**	**$1513**	**$3257**

TABLE 6.10

DSM Costs Calculation for With DSM Resource Plan 1

	(1)	(6) = Sum of Cols. (2) through (5)	(7)	(8)	(9)	(10) = Col (7) + Col (8) – Col (9)
Year	Annual Discount Factor 8.000%	**Total Generation Fixed Costs** (Nominal $, Millions)	DSM Administrative Costs (Nominal $, Millions)	DSM Incentive Payments (Nominal $, Millions)	T & D Costs Avoided by DSM (Nominal $, Millions)	**DSM Net Costs** (Nominal $, Millions)
Current Year	1.000	**0**	0	0	0	**0**
1	0.926	**0**	2	4	0	**6**
2	0.857	**0**	2	4	1	**5**
3	0.794	**0**	2	4	3	**3**
4	0.735	**0**	2	4	4	**2**
5	0.681	**0**	2	4	5	**1**
6	0.630	**192**	0	0	7	**(7)**
7	0.583	**188**	0	0	6	**(6)**
8	0.540	**184**	0	0	6	**(6)**
9	0.500	**181**	0	0	6	**(6)**
10	0.463	**378**	0	0	6	**(6)**
11	0.429	**370**	0	0	5	**(5)**
12	0.397	**362**	0	0	5	**(5)**
13	0.368	**354**	0	0	5	**(5)**
14	0.340	**557**	0	0	5	**(5)**
15	0.315	**544**	0	0	4	**(4)**
16	0.292	**531**	3	4	4	**3**
17	0.270	**519**	3	4	4	**3**
18	0.250	**506**	3	4	4	**3**
19	0.232	**718**	3	4	3	**4**
20	0.215	**700**	3	4	3	**4**
21	0.199	**682**	0	0	3	**(3)**
22	0.184	**665**	0	0	3	**(3)**
23	0.170	**884**	0	0	2	**(2)**
24	0.158	**860**	0	0	2	**(2)**
25	0.146	**837**	0	0	2	**(2)**
26	0.135	**814**	0	0	2	**(2)**
27	0.125	**1,040**	0	0	1	**(1)**
28	0.116	**1,011**	0	0	1	**(1)**
29	0.107	**982**	0	0	1	**(1)**
Total CPVRR =		**$3,257**	**$12**	**$21**	**$43**	**($10)**

TABLE 6.11

Variable Costs Calculation for With DSM Resource Plan 1

	(1)	(6) = Sum of Cols. (2) through (5)	(10) = Col (7) + Clo (8) - Col (9)	(11)	(12)	(13)	(14) = Sum of Cols. (11) through (13)
Year	Annual Discount Factor 8.000%	**Total Generation Fixed Costs** (Nominal $, Millions)	**DSM Net Costs** (Nominal $, Millions)	Generation Variable O & M (Nominal $, Millions)	System Net Fuel (Nominal $, Millions)	System Environmental Compliance (Nominal $, Millions)	**Total Variable Costs** (Nominal $, Millions)
Current Year	1.000	**0**	**0**	0	1,660	275	**1,936**
1	0.926	**0**	**6**	0	1,725	390	**2,115**
2	0.857	**0**	**5**	0	1,790	506	**2,296**
3	0.794	**0**	**3**	0	1,857	624	**2,481**
4	0.735	**0**	**2**	0	1,926	744	**2,670**
5	0.681	**0**	**1**	0	1,997	866	**2,863**
6	0.630	**192**	**(7)**	7	1,989	965	**2,961**
7	0.583	**188**	**(6)**	7	2,065	1,089	**3,161**
8	0.540	**184**	**(6)**	7	2,144	1,215	**3,366**
9	0.500	**181**	**(6)**	8	2,224	1,343	**3,576**
10	0.463	**378**	**(6)**	15	2,219	1,439	**3,673**
11	0.429	**370**	**(5)**	16	2,302	1,570	**3,888**
12	0.397	**362**	**(5)**	16	2,389	1,702	**4,107**
13	0.368	**354**	**(5)**	16	2,478	1,836	**4,331**
14	0.340	**557**	**(5)**	25	2,473	1,929	**4,427**
15	0.315	**544**	**(4)**	26	2,565	2,065	**4,656**
16	0.292	**531**	**3**	26	2,660	2,204	**4,890**
17	0.270	**519**	**3**	27	2,758	2,345	**5,129**
18	0.250	**506**	**3**	27	2,858	2,488	**5,374**
19	0.232	**718**	**4**	37	2,855	2,577	**5,469**
20	0.215	**700**	**4**	38	2,960	2,722	**5,720**
21	0.199	**682**	**(3)**	38	3,067	2,870	**5,975**
22	0.184	**665**	**(3)**	39	3,178	3,020	**6,237**
23	0.170	**884**	**(2)**	50	3,176	3,105	**6,330**
24	0.158	**860**	**(2)**	51	3,290	3,257	**6,598**
25	0.146	**837**	**(2)**	52	3,409	3,411	**6,872**
26	0.135	**814**	**(2)**	53	3,530	3,568	**7,151**
27	0.125	**1,040**	**(1)**	65	3,530	3,649	**7,244**
28	0.116	**1,011**	**(1)**	66	3,656	3,808	**7,530**
29	0.107	**982**	**(1)**	68	3,785	3,970	**7,823**
Total CPVRR =		**$3,257**	**($10)**	**$167**	**$27,328**	**$16,826**	**$44,322**

TABLE 6.12

Total Costs Calculation for With DSM Resource Plan 1

	(1)	(6) = Sum of Cols. (2) through (5)	(10) = Col (7) + Col (8) − Col (9)	(14) = Sum of Cols. (11) through (13)	(15) = (6) + (10) + (14)	(16) = (1) × (15)	(17)
Year	Annual Discount Factor 8.000%	**Total Generation Fixed Costs** (Nominal $, Millions)	**DSM Net Costs** (Nominal $, Millions)	**Total Variable Costs** (Nominal $, Millions)	Total Annual Costs (Nominal $, Millions)	Total Annual NPV Costs (NPV $, Millions)	Cumulative Total NPV Costs (NPV $, Millions)
Current Year	1.000	**0**	**0**	**1,936**	1,936	1,936	1,936
1	0.926	**0**	**6**	**2,115**	2,121	1,963	3,899
2	0.857	**0**	**5**	**2,296**	2,301	1,972	5,871
3	0.794	**0**	**3**	**2,481**	2,484	1,972	7,844
4	0.735	**0**	**2**	**2,670**	2,672	1,964	9,808
5	0.681	**0**	**1**	**2,863**	2,864	1,949	11,757
6	0.630	**192**	**(7)**	**2,961**	3,146	1,983	13,739
7	0.583	**188**	**(6)**	**3,161**	3,343	1,951	15,690
8	0.540	**184**	**(6)**	**3,366**	3,545	1,915	17,605
9	0.500	**181**	**(6)**	**3,576**	3,750	1,876	19,481
10	0.463	**378**	**(6)**	**3,673**	4,046	1,874	21,355
11	0.429	**370**	**(5)**	**3,888**	4,252	1,824	23,179
12	0.397	**362**	**(5)**	**4,107**	4,464	1,773	24,951
13	0.368	**354**	**(5)**	**4,331**	4,679	1,721	26,672
14	0.340	**557**	**(5)**	**4,427**	4,979	1,695	28,367
15	0.315	**544**	**(4)**	**4,656**	5,196	1,638	30,005
16	0.292	**531**	**3**	**4,890**	5,424	1,583	31,588
17	0.270	**519**	**3**	**5,129**	5,651	1,527	33,116
18	0.250	**506**	**3**	**5,374**	5,883	1,472	34,588
19	0.232	**718**	**4**	**5,469**	6,191	1,434	36,022
20	0.215	**700**	**4**	**5,720**	6,424	1,378	37,400
21	0.199	**682**	**(3)**	**5,975**	6,655	1,322	38,722
22	0.184	**665**	**(3)**	**6,237**	6,899	1,269	39,991
23	0.170	**884**	**(2)**	**6,330**	7,212	1,228	41,220
24	0.158	**860**	**(2)**	**6,598**	7,456	1,176	42,396
25	0.146	**837**	**(2)**	**6,872**	7,707	1,125	43,521
26	0.135	**814**	**(2)**	**7,151**	7,964	1,077	44,598
27	0.125	**1,040**	**(1)**	**7,244**	8,283	1,037	45,634
28	0.116	**1,011**	**(1)**	**7,530**	8,540	990	46,624
29	0.107	**982**	**(1)**	**7,823**	8,804	945	47,569
Total CPVRR =		**$3,257**	**($10)**	**$44,322**	**$47,569**	**$47,569**	

TABLE 6.13

Levelized Electric Rate Calculation for With DSM Resource Plan 1

	(1)	(15) = (6) + (10) + (14)	(18)	(19) = (15) + (18)	(20)	(21)	(22) = (20) − (21)	(23) = ((19)× 100)/(22)	(24) = (1) × (23)	(25)	(26) = (1) × (25)
Year	Annual Discount Factor 8.000%	Total Annual Costs (Nominal $, Millions)	Other System Costs Not Affected by Plan (Nominal $, Millions)	Total Utility Costs (Nominal $, Millions)	Forecasted Utility Sales/NEL (GWh)	DSM Energy Reduction (GWh)	Net Utility Sales/NEL (GWh)	System Avg. Electric Rate Nominal (¢/kWh)	System Avg. Electric Rate NPV (¢/kWh)	Levelized System Avg Electric Rate (¢/kWh)	System Avg. Electric Rate Nominal (¢/kWh)
Current Year	1.000	1,936	2,700	4,636	50,496	0	50,496	9.1800	9.1800	12.8095	12.8095
1	0.926	2,121	2,754	4,875	51,022	20	51,002	9.5576	8.8496	12.8095	11.8607
2	0.857	2,301	2,809	5,110	51,548	60	51,488	9.9240	8.5082	12.8095	10.9821
3	0.794	2,484	2,865	5,350	52,074	100	51,974	10.2930	8.1709	12.8095	10.1686
4	0.735	2,672	2,923	5,595	52,600	140	52,460	10.6646	7.8388	12.8095	9.4154
5	0.681	2,864	2,981	5,845	53,126	180	52,946	11.0390	7.5130	12.8095	8.7179
6	0.630	3,146	3,041	6,187	53,652	200	53,452	11.5747	7.2940	12.8095	8.0722
7	0.583	3,343	3,101	6,445	54,178	200	53,978	11.9395	6.9666	12.8095	7.4742
8	0.540	3,545	3,163	6,708	54,704	200	54,504	12.3074	6.6493	12.8095	6.9206
9	0.500	3,750	3,227	6,977	55,230	200	55,030	12.6786	6.3425	12.8095	6.4079
10	0.463	4,046	3,291	7,337	55,756	200	55,556	13.2065	6.1172	12.8095	5.9333
11	0.429	4,252	3,357	7,609	56,282	200	56,082	13.5684	5.8193	12.8095	5.4938
12	0.397	4,464	3,424	7,888	56,808	200	56,608	13.9340	5.5334	12.8095	5.0868
13	0.368	4,679	3,493	8,172	57,334	200	57,134	14.3035	5.2594	12.8095	4.7100
14	0.340	4,979	3,563	8,541	57,860	200	57,660	14.8135	5.0434	12.8095	4.3611
15	0.315	5,196	3,634	8,830	58,386	200	58,186	15.1747	4.7837	12.8095	4.0381
16	0.292	5,424	3,707	9,131	58,912	200	58,712	15.5517	4.5394	12.8095	3.7390
17	0.270	5,651	3,781	9,432	59,438	200	59,238	15.9216	4.3031	12.8095	3.4620
18	0.250	5,883	3,856	9,739	59,964	200	59,764	16.2962	4.0781	12.8095	3.2056
19	0.232	6,191	3,933	10,124	60,490	200	60,290	16.7922	3.8910	12.8095	2.9681
20	0.215	6,424	4,012	10,436	61,016	200	60,816	17.1593	3.6815	12.8095	2.7483
21	0.199	6,655	4,092	10,747	61,542	200	61,342	17.5201	3.4805	12.8095	2.5447
22	0.184	6,899	4,174	11,073	62,068	200	61,868	17.8976	3.2921	12.8095	2.3562
23	0.170	7,212	4,258	11,469	62,594	200	62,394	18.3819	3.1307	12.8095	2.1817
24	0.158	7,456	4,343	11,799	63,120	200	62,920	18.7528	2.9573	12.8095	2.0201
25	0.146	7,707	4,430	12,137	63,646	200	63,446	19.1293	2.7932	12.8095	1.8704
26	0.135	7,964	4,518	12,482	64,172	200	63,972	19.5117	2.6380	12.8095	1.7319
27	0.125	8,283	4,609	12,891	64,698	200	64,498	19.9869	2.5021	12.8095	1.6036
28	0.116	8,540	4,701	13,241	65,224	200	65,024	20.3633	2.3604	12.8095	1.4848
29	0.107	8,804	4,795	13,599	65,750	200	65,550	20.7459	2.2266	12.8095	1.3748
	Total CPVRR =	**$47,569**							-----		------
									155.7432		**155.7432**

TABLE D.1

Typical Screening Curve Results for a CC Unit: With No System Impacts

(1)	(2)	(3)	(4)	(5)	(6)	(7)	(8)	(9)	(10)	(11)	(12)	(13)	(14)
Capacity Factor (%)	Levelized Cost ($/kW)	Levelized Cost ($/MWh)	In-Service Year	Capital $000	Fixed O&M $000	Capital Repl $000	Firm Gas Transportation $000	NO_x Emission $000	SO_2 Emission $000	CO_2 Emission $000	Fuel Costs $000	Variable O&M $000	Total $000
				0	0	0	0	0	0	0	0	0	0
				0	0	0	0	0	0	0	0	0	0
				0	0	0	0	0	0	0	0	0	0
0	172	0		0	0	0	0	0	0	0	0	0	0
5	203	464	1	258,093	8,106	13,319	161,056	694	610	103,768	865,447	14,556	1,425,649
10	234	267	2	248,821	8,309	13,652	161,056	760	667	112,272	912,227	14,920	1,472,683
15	265	202	3	238,528	8,516	13,994	161,056	832	731	121,135	930,501	15,293	1,490,586
20	296	169	4	228,618	8,729	14,343	161,056	911	800	136,580	949,141	15,675	1,515,855
25	327	149	5	219,061	8,947	14,702	161,056	998	877	146,358	968,154	16,067	1,536,221
30	357	136	6	209,831	9,171	15,070	161,056	1,093	960	163,062	987,547	16,469	1,564,259
35	388	127	7	200,889	9,400	15,446	161,056	1,198	1,052	180,509	1,007,328	16,881	1,593,760
40	419	120	8	192,194	9,635	15,832	161,056	1,022	1,028	191,875	1,027,505	17,303	1,617,450
45	450	114	9	183,630	9,876	16,228	161,056	872	1,005	210,720	1,048,085	17,735	1,649,207
50	481	110	10	175,085	10,123	16,634	161,056	745	982	230,387	1,069,076	18,179	1,682,266
55	512	106	11	166,541	10,376	17,050	161,056	636	960	258,285	1,090,487	18,633	1,724,024
60	543	103	12	157,997	10,636	17,476	161,056	543	939	279,871	1,112,327	19,099	1,759,943
65	574	101	13	149,455	10,902	17,913	161,056	407	911	304,610	1,134,603	19,576	1,799,433
70	604	99	14	140,914	11,174	18,361	161,056	264	881	331,183	1,157,325	20,066	1,841,223
75	635	97	15	132,374	11,454	18,820	161,056	113	849	359,684	1,180,501	20,567	1,885,418
80	666	95	16	123,939	11,740	19,290	161,056	0	815	390,214	1,204,140	21,082	1,932,276
85	697	94	17	115,716	12,033	19,773	161,056	0	778	422,875	1,228,252	21,609	1,982,091
90	728	92	18	107,598	12,334	20,267	161,056	0	740	457,776	1,252,847	22,149	2,034,765
95	759	91	19	99,481	12,643	20,774	161,056	0	698	495,030	1,277,933	22,703	2,090,316
100	790	90	20	91,365	12,959	21,293	161,056	0	655	534,754	1,303,521	23,270	2,148,872
			21	83,933	13,283	21,825	161,056	0	608	577,072	1,329,621	23,852	2,211,249
			22	77,866	13,615	22,371	161,056	0	559	622,110	1,356,242	24,448	2,278,268
			23	72,484	13,955	22,930	161,056	0	507	670,002	1,383,396	25,059	2,349,390
			24	67,102	14,304	23,503	161,056	0	453	720,887	1,411,093	25,686	2,424,085
			25	61,722	14,661	24,091	161,056	0	395	774,909	1,439,344	26,328	2,502,507
			NPV total costs =	982,398	53,769	88,350	860,046	3,396	4,350	1,357,990	5,658,207	96,554	9,105,059
			NPV $/kW at 100% Capacity Factor =	85	5	8	75	0	0	118	491	8	790

TABLE D.2

Modified Screening Curve Results for a CC Unit: With Two System Impacts

(1)	(2)	(3)	(4)	(5)	(6)	(7)	(8)	(9)	(10)	(11)	(12)	(13)	(14)
Capacity Factor (%)	Levelized Cost ($/kW)	Levelized Cost ($/MWh)	In-Service Year	Capital $000	Fixed O&M $000	Capital Repl $000	Firm Gas Transportation $000	NO_x Emission $000	SO_2 Emission $000	CO_2 Emission $000	Fuel Costs $000	Variable O&M $000	Total $000
				0	0	0	0	0	0	0	0	0	0
				0	0	0	0	0	0	0	0	0	0
				0	0	0	0	0	0	0	0	0	0
0	172	0		0	0	0	0	0	0	0	0	0	0
5	170	387	1	258,093	8,106	13,319	161,056	(69)	(61)	(10,377)	(86,545)	14,556	358,079
10	167	191	2	248,821	8,309	13,652	161,056	(76)	(67)	(11,227)	(91,223)	14,920	344,165
15	164	125	3	238,528	8,516	13,994	161,056	(83)	(73)	(12,114)	(93,050)	15,293	332,067
20	162	92	4	228,618	8,729	14,343	161,056	(91)	(80)	(13,658)	(94,914)	15,675	319,679
25	159	73	5	219,061	8,947	14,702	161,056	(100)	(88)	(14,636)	(96,815)	16,067	308,195
30	156	60	6	209,831	9,171	15,070	161,056	(109)	(96)	(16,306)	(98,755)	16,469	296,330
35	154	50	7	200,889	9,400	15,446	161,056	(120)	(105)	(18,051)	(100,733)	16,881	284,664
40	151	43	8	192,194	9,635	15,832	161,056	(102)	(103)	(19,187)	(102,750)	17,303	273,878
45	148	38	9	183,630	9,876	16,228	161,056	(87)	(101)	(21,072)	(104,808)	17,735	262,457
50	146	33	10	175,085	10,123	16,634	161,056	(74)	(98)	(23,039)	(106,908)	18,179	250,957
55	143	30	11	166,541	10,376	17,050	161,056	(64)	(96)	(25,829)	(109,049)	18,633	238,619
60	141	27	12	157,997	10,636	17,476	161,056	(54)	(94)	(27,987)	(111,233)	19,099	226,896
65	138	24	13	149,455	10,902	17,913	161,056	(41)	(91)	(30,461)	(113,460)	19,576	214,849
70	135	22	14	140,914	11,174	18,361	161,056	(26)	(88)	(33,118)	(115,732)	20,066	202,605
75	133	20	15	132,374	11,454	18,820	161,056	(11)	(85)	(35,968)	(118,050)	20,567	190,156
80	130	19	16	123,939	11,740	19,290	161,056	0	(81)	(39,021)	(120,414)	21,082	177,590
85	127	17	17	115,716	12,033	19,773	161,056	0	(78)	(42,287)	(122,825)	21,609	164,996
90	125	16	18	107,598	12,334	20,267	161,056	0	(74)	(45,778)	(125,285)	22,149	152,267
95	122	15	19	99,481	12,643	20,774	161,056	0	(70)	(49,503)	(127,793)	22,703	139,289
100	120	14	20	91,365	12,959	21,293	161,056	0	(65)	(53,475)	(130,352)	23,270	126,050
			21	83,933	13,283	21,825	161,056	0	(61)	(57,707)	(132,962)	23,852	113,218
			22	77,866	13,615	22,371	161,056	0	(56)	(62,211)	(135,624)	24,448	101,465
			23	72,484	13,955	22,930	161,056	0	(51)	(67,000)	(138,340)	25,059	90,094
			24	67,102	14,304	23,503	161,056	0	(45)	(72,089)	(141,109)	25,686	78,408
			25	61,722	14,661	24,091	161,056	0	(39)	(77,491)	(143,934)	26,328	66,394
NPV total costs =				982,398	53,769	88,350	860,046	(340)	(435)	(135,799)	(565,821)	96,554	1,378,722
NPV $/kW at 100% Capacity Factor =				85	5	8	75	(0)	(0)	(12)	(49)	8	120

TABLE 6.5

Economic Elements Included in the RIM and TRC Tests: Benefits and Costs

Economic Elements	Utility-Incurred Economic Impacts	Included in the RIM Test?	Included in the TRC Test?
Benefits of DSM (=avoided costs and/or cost impacts)			
Generation capital and O&M	Yes	Yes	Yes
Transmission capital and O&M	Yes	Yes	Yes
Distribution capital and O&M	Yes	Yes	Yes
Net system fuel impacts	Yes	Yes	Yes
Net system environmental compliance impacts	Yes	Yes	Yes
Costs of DSM (=incurred costs)			
Utility equipment and administration	Yes	Yes	Yes
Incentives paid to participants	Yes	Yes	No
Unrecovered revenue requirements	Yes	Yes	No
Participants' capital and O&M	No	No	Yes

utility and are not passed on to all of its customers. (Recall that these costs, which are incurred only by the participant are already properly accounted for in the Participant test.)

Therefore, the RIM test correctly accounts for all of the DSM-related benefits and costs that will be incurred by the utility and passed on to, or otherwise impact, all of the utility's customers. However, although the TRC test does account for all of the DSM-related benefits, it falls far short of accounting for all of the DSM-related costs that will impact all of the utility's customers. If one's objective is to perform preliminary economic screening analyses of DSM options which account for all relevant cost impacts for all of a utility's customers, then the TRC test falls far short; i.e., it is a fundamentally flawed tool.

Because of the fundamental flaw in the TRC test, the TRC test will provide—in almost every case—different benefit-to-cost ratios from those provided by the RIM test when evaluating the same DSM option.* This shall be evident in the next section as we present the actual results of the RIM and TRC tests for DSM Options 1 and 2.†

* As one might expect from the fact that the TRC test does not account for all relevant cost impacts, the resulting benefit-to-cost values of the TRC test are generally higher than (and often considerably higher than) the benefit-to-cost values resulting from the RIM test that does account for all relevant cost impacts.

† A further discussion of the RIM and TRC tests is presented in Appendix E.

Preliminary Economic Screening Analyses of DSM Options: Results

The benefit-to-cost ratio results of the three DSM cost-effectiveness screening tests for our hypothetical utility, when applied to DSM Options 1 and 2 versus the most economic Supply Option (CC Unit A), are presented in Table 6.6.

We will start with the Participant test results shown in Row (1). Both of the DSM options are projected to have benefit-to-cost ratios significantly higher than 1.00; i.e., the CPVRR value of the benefits exceeds the CPVRR value of the costs. Therefore, the two DSM options are projected to be cost-effective from the perspective of a potential participant and our utility's customers should find either of these two DSM options attractive. The question now becomes: should our utility attempt to sign up its customers for either of these two DSM options? In other words, should our utility offer either of these DSM options?

In an attempt to answer that question, we turn to the RIM and TRC test results that are shown, respectively, on Rows (2) and (3). The RIM test results show that DSM Option 1 has a benefit-to-cost ratio of 1.07. Thus DSM Option 1 is projected to be cost-effective from the perspective of the RIM test. However, DSM Option 2 has a RIM test benefit-to-cost ratio of 0.78; i.e., less than 1.00. Therefore, DSM Option 2 is *not* projected to be cost-effective from the perspective of the RIM test.

However, the TRC test results show that DSM Option 1 has a TRC benefit-to-cost ratio of 2.69 and DSM Option 2 has a TRC benefit-to-cost ratio of 2.40. Thus both DSM options are projected to be cost-effective from a TRC test perspective (and, as previously mentioned, the benefits-to-costs ratio for the TRC test are significantly higher than the resulting RIM test values).

Our utility now has a dilemma based on the results of its preliminary economic screening analyses. Both DSM options are projected to be cost-effective from a participant's perspective (the Participant test) and both DSM options are projected to be cost-effective under at least one of the two cost-effectiveness tests that are intended to provide a perspective of the utility system as a whole.

Unless a utility has previously determined which of the so-called utility-perspective cost-effectiveness screening tests (RIM or TRC) it believes provides the more meaningful perspective, it will need to perform a full economic

TABLE 6.6

Preliminary Economic Screening Analyses of DSM Options: Benefit-to-Cost Ratios

Cost-Effectiveness Test	For DSM Option 1	For DSM Option 2
1. Participant test	2.71	3.59
2. RIM test	1.07	0.78
3. TRC test	2.69	2.40

analysis of its utility system to determine which DSM option is best for it to add, and whether either DSM option is better than the most economic Supply option (CC Unit A). Although the answer to the question of "which DSM option should be carried forward?" is probably obvious to the astute reader based on the test results, for purposes of discussion our utility system has graciously (again) offered to include both DSM options in its on-going analyses.

In order for it to do this analysis, it is necessary to create two DSM-based resource plans in a manner similar to what was done when the competing Supply options were evaluated. Therefore, we now turn our attention to creating two "With DSM" resource plans.

Creating the Competing "With DSM" Resource Plans

As discussed in Chapter 3, an integrated resource planning (IRP) approach looks at long-term resource plans in order to ensure that all of the impacts that a proposed resource option addition will have on the utility system over an extended period of time are captured. In Chapter 5, we first created, then analyzed, three different resource plans that each featured a different Supply option. We are now at a point where we will create two resource plans to address the two DSM options. Then, in the next section, these resource plans will be analyzed.

Recall that with the Supply options we examined, each Supply option had a different amount of capacity (MW) that would be added in the decision year (Current Year + 5). These different amounts of capacity that would be added in the decision year by each Supply option had a different impact on the timing and magnitude for resource additions in years after the decision year. Therefore, each Supply option required the development of a separate resource plan to address this impact.

This is not the case with our two DSM options. That is because we are assuming that either DSM Option 1 or DSM Option 2 will be implemented to achieve the same amount of MW reduction, 100 MW, by the decision year.* Therefore, in regard to the timing and magnitude of resource additions in years following the decision year, both of the DSM-based resource plans will have identical resource needs. Therefore, for purposes of our discussion, we need only look at a single DSM-based resource plan to examine the impact that 100 MW of DSM will have on the timing and number of the filler units. We will refer to this DSM-based resource plan as the general "With DSM" resource plan.

* We are assuming that our utility has projected that it is possible to sign up at least 100,000 and 200,000 participating customers by the decision year for DSM Options 1 and 2, respectively. In utility parlance, the "achievable potential" for the DSM options are at least 100,000 and 200,000 participants, respectively, in the desired time frame.

However, recall that the two DSM options have different costs, and have different values for the total amount of energy that will be reduced. These different impacts will be addressed by actually evaluating two separate "With DSM" resource plans in which the underlying DSM cost and energy reduction values for each DSM option are accounted for. This evaluation will be discussed in the next section.

We now look at the impact of either of our DSM options—both of which will achieve 100 MW of demand reduction by the decision year—on future resource needs using a similar long-term resource plan format to that previously used in Chapter 5. That format is presented in Table 6.7.

Table 6.7 is almost identical to the format presented in Chapter 5 in Tables 5.4 through 5.9. The only change is that two new columns have now been added in order to address the DSM options.

In the earlier Chapter 5 version of this table, Column (2) presented the "Peak Electrical Demand (MW)" values which represent the forecasted system peak loads assuming no incremental DSM is implemented. We have now labeled the former Column (2) as Column (2a). Then two new columns are added immediately after Column (2a). Column (2b) presents the annual MW reduction values at the system peak hour from the incremental DSM added if either DSM Option 1 or 2 is selected.

As shown by the values in Column (2b), we have assumed that either DSM option will reduce peak load by 20 MW more each year from Current Year + 1 through Current Year + 5 to achieve the desired 100 MW of incremental demand reduction by Current Year + 5. This 100 MW demand reduction is then assumed to be maintained for all subsequent years.

Then Column (2c) computes what we earlier termed the "firm" peak electrical demand that results when subtracting the incremental DSM in Column (2b) from the forecasted peak demand in Column (2a). This firm peak demand value is what the utility will actually have to serve at its peak hour. This firm peak demand value is also used to calculate the utility's reserve margins and resource needs.

As shown in Table 6.7, the incremental 100 MW of incremental demand reduction by Current Year + 5 will allow our hypothetical utility to achieve a 20% reserve margin without adding any additional new generating units in that year. However, the table also shows that additional resources will be needed starting with the next year, Current Year + 6, because our utility's reserve margin is projected to drop to 18.8% in that year. This fact is highlighted by shading the Current Year + 6 row. Additional resources will also be needed in subsequent years as well to continue to meet the 20% reserve margin criterion as the demand for electricity continues to increase. These additional resource needs will be addressed in the same manner as was the case for the Supply options: through the addition of 500 MW filler units starting in Current Year + 6. The number and timing of these filler units is presented in Table 6.8 in Column (1e).

As shown in Column (1e), this general "With DSM" resource plan for either DSM option will have 6 filler units added after the decision year. This is the same number of filler units as in Supply Only Resource Plan 3 (PV) that added 120 MW of capacity.

In fact, the timing and number of filler units in both Supply Only Resource Plan 3 (PV) presented in Chapter 5, and the With DSM resource plan presented earlier, are identical. One would expect this to be the case because the 120 MW of capacity added in Supply Only Resource Plan 3 (PV) and the 100 MW of demand reduction added in the With DSM resource plan will have an identical impact on the utility's reserve margin. Consequently, for all years after the decision year, the utility's resource needs will be the same regardless of whether 120 MW of new generating unit capacity, or 100 MW of demand reduction from DSM is added, in Current Year + 5.

Now that we know what the timing and magnitude of filler units will be for either of the DSM options, we will move forward to an economic analysis of a resource plan that features 100 MW of DSM Option 1 and a resource plan that features 100 MW of DSM Option 2. We will (again, showing great imagination) call these two resource plans the With DSM Resource Plan 1 (which features DSM Option 1), and the With DSM Resource Plan 2 (which features DSM Option 2), respectively. With these two resource plans, we will account for the differences in the two DSM options in regard to DSM total administration and incentive costs and to the amount of energy reduced.

Final (or System) Economic Analysis of DSM Options

Overview

In Chapter 5, our discussion of the Supply options ultimately focused on two tables that summarized the final economic analyses for the Supply Only options. Table 5.16 presented both the CPVRR total cost values and levelized system average electric rate for one of the Supply Only Resource Plans [Supply Only Resource Plan 1 (CC)] in detail. Then Table 5.17 summarized the CPVRR and levelized system average electric rates for all three of the Supply Only resource plans.

We will take the same approach in presenting the results for the With DSM resource plans. And, just as we did in the discussion of the analyses of the Supply options, we will also focus in detail on only one DSM option: DSM Option 1 which is featured in the With DSM Resource Plan 1. As we did in Chapter 5, we will present three different cost components for this resource plan: Fixed Costs, DSM Costs, and Variable Costs. We will then present the total costs for this plan. Once these cost values have been presented, we will discuss how the costs for the With DSM Resource Plan 1 differ from the costs for the Supply Only resource plans.

Finally, the levelized system average electric rate calculation for the With DSM Resource Plan 1 will be presented in detail, followed by a summary of

TABLE 6.7

Long-Term Projection of Reserve Margin for the Hypothetical Utility: With New DSM Added by "Current Year + 5"

	(1a)	(1b)	(1c)	(1d)	(1e)	(1f)	(2a)	(2b)	(2c)	(3)	(4)	(5)
						= (See Formula Below)			= (2a) − (2b)	= (1f) − (2c)	= (3)/(2c)	= [(2c)*1.20] − (1f)
		Cumulative No. of New Unit Additions										
Year	**Previously Projected Generating Capacity (MW)**	**CC Unit A (No. Units)**	**CT (No. Units)**	**PV (No. Units)**	**Filler Units (No. Units)**	**Total Projected Generating Capacity (MW)**	**Previously Forecasted Electrical Demand (MW)**	**Newly Projected DSM (MW)**	**Firm Electrical Demand (MW)**	**Reserves (MW)**	**Reserve Margin (%)**	**Generation Only MW Needed to Meet Reserve Margin (MW)**
Current Year	11,600	0	0	0	0	11,600	9,600	0	9,600	2,000	20.8	(80)
Current Year + 1	12,000	0	0	0	0	12,000	9,700	20	9,680	2,320	24.0	(384)
Current Year + 2	12,000	0	0	0	0	12,000	9,800	40	9,760	2,240	23.0	(288)
Current Year + 3	12,000	0	0	0	0	12,000	9,900	60	9,840	2,160	22.0	(192)
Current Year + 4	12,000	0	0	0	0	12,000	10,000	80	9,920	2,080	21.0	(96)
Current Year + 5	12,000	0	0	0	0	12,000	10,100	100	10,000	2,000	20.0	0
Current Year + 6	12,000	0	0	0	0	12,000	10,200	100	10,100	1,900	18.8	120
Current Year + 7	12,000	0	0	0	0	12,000	10,300	100	10,200	1,800	17.6	240
Current Year + 8	12,000	0	0	0	0	12,000	10,400	100	10,300	1,700	16.5	360
Current Year + 9	12,000	0	0	0	0	12,000	10,500	100	10,400	1,600	15.4	480
Current Year + 10	12,000	0	0	0	0	12,000	10,600	100	10,500	1,500	14.3	600
Current Year + 11	12,000	0	0	0	0	12,000	10,700	100	10,600	1,400	13.2	720
Current Year + 12	12,000	0	0	0	0	12,000	10,800	100	10,700	1,300	12.1	840
Current Year + 13	12,000	0	0	0	0	12,000	10,900	100	10,800	1,200	11.1	960
Current Year + 14	12,000	0	0	0	0	12,000	11,000	100	10,900	1,100	10.1	1,080
Current Year + 15	12,000	0	0	0	0	12,000	11,100	100	11,000	1,000	9.1	1,200
Current Year + 16	12,000	0	0	0	0	12,000	11,200	100	11,100	900	8.1	1,320

Current Year + 17	12,000	0	0	0	0	12,000	11,300	100	11,200	800	7.1	1,440
Current Year + 18	12,000	0	0	0	0	12,000	11,400	100	11,300	700	6.2	1,560
Current Year + 19	12,000	0	0	0	0	12,000	11,500	100	11,400	600	5.3	1,680
Current Year + 20	12,000	0	0	0	0	12,000	11,600	100	11,500	500	4.3	1,800
Current Year + 21	12,000	0	0	0	0	12,000	11,700	100	11,600	400	3.4	1,920
Current Year + 22	12,000	0	0	0	0	12,000	11,800	100	11,700	300	2.6	2,040
Current Year + 23	12,000	0	0	0	0	12,000	11,900	100	11,800	200	1.7	2,160
Current Year + 24	12,000	0	0	0	0	12,000	12,000	100	11,900	100	0.8	2,280
Current Year + 25	12,000	0	0	0	0	12,000	12,100	100	12,000	0	0.0	2,400
Current Year + 26	12,000	0	0	0	0	12,000	12,200	100	12,100	(100)	−0.8	2,520
Current Year + 27	12,000	0	0	0	0	12,000	12,300	100	12,200	(200)	−1.6	2,640
Current Year + 28	12,000	0	0	0	0	12,000	12,400	100	12,300	(300)	−2.4	2,760
Current Year + 29	12,000	0	0	0	0	12,000	12,500	100	12,400	(400)	−3.2	2,880
CC unit A =			500	MW		Formula: (1f) = (1a) + ((1b) × 500) + ((1c) × 160) + ((1d) × 120) + ((1e) × 500)						
CT =			160	MW								
PV =			120	MW								
Filler units =			500	MW								

TABLE 6.8

With DSM Resource Plan 1 With New DSM Added by "Current Year + 5" Plus Filler Units

	(1a)	(1b)	(1c)	(1d)	(1e)	(1f)	(2a)	(2b)	(2c)	(3)	(4)	(5)
						= (See Formula Below)			= (2a) − (2b)	= (1f) − (2c)	= (3)/ (2c)	= [(2c) × 1.20] − (1f)
		Cumulative No. of New Unit Additions										
Year	Previously Projected Generating Capacity (MW)	CC (No. Units)	CT (No. Units)	PV (No. Units)	Filler Units (No. Units)	Total Projected Generating Capacity (MW)	Previously Forecasted Electrical Demand (MW)	Newly Projected DSM (MW)	Firm Electrical Demand (MW)	(Reserves (MW)	Reserve Margin (%)	Generation Only MW Needed to Meet Reserve Margin (MW)
Current Year	11,600	0	0	0	0	11,600	9,600	0	9,600	2,000	20.8%	(80)
Current Year + 1	12,000	0	0	0	0	12,000	9,700	20	9,680	2,320	24.0%	(384)
Current Year + 2	12,000	0	0	0	0	12,000	9,800	40	9,760	2,240	23.0%	(288)
Current Year + 3	12,000	0	0	0	0	12,000	9,900	60	9,840	2,160	22.0%	(192)
Current Year + 4	12,000	0	0	0	0	12,000	10,000	80	9,920	2,080	21.0%	(96)
Current Year + 5	12,000	0	0	0	0	12,000	10,100	100	10,000	2,000	20.0%	0
Current Year + 6	12,000	0	0	0	1	12,500	10,200	100	10,100	2,400	23.8%	(380)
Current Year + 7	12,000	0	0	0	1	12,500	10,300	100	10,200	2,300	22.5%	(260)
Current Year + 8	12,000	0	0	0	1	12,500	10,400	100	10,300	2,200	21.4%	(140)
Current Year + 9	12,000	0	0	0	1	12,500	10,500	100	10,400	2,100	20.2%	(20)
Current Year + 10	12,000	0	0	0	2	13,000	10,600	100	10,500	2,500	23.8%	(400)
Current Year + 11	12,000	0	0	0	2	13,000	10,700	100	10,600	2,400	22.6%	(280)
Current Year + 12	12,000	0	0	0	2	13,000	10,800	100	10,700	2,300	21.5%	(160)
Current Year + 13	12,000	0	0	0	2	13,000	10,900	100	10,800	2,200	20.4%	(40)
Current Year + 14	12,000	0	0	0	3	13,500	11,000	100	10,900	2,600	23.9%	(420)
Current Year + 15	12,000	0	0	0	3	13,500	11,100	100	11,000	2,500	22.7%	(300)

Current Year + 16	12,000	0	0	0	3	13,500	11,200	100	11,100	2,400	21.6%	(180)
Current Year + 17	12,000	0	0	0	3	13,500	11,300	100	11,200	2,300	20.5%	(60)
Current Year + 18	12,000	0	0	0	3	13,500	11,400	100	11,300	2,200	19.5%	60
Current Year + 19	12,000	0	0	0	4	14,000	11,500	100	11,400	2,600	22.8%	(320)
Current Year + 20	12,000	0	0	0	4	14,000	11,600	100	11,500	2,500	21.7%	(200)
Current Year + 21	12,000	0	0	0	4	14,000	11,700	100	11,600	2,400	20.7%	(80)
Current Year + 22	12,000	0	0	0	4	14,000	11,800	100	11,700	2,300	19.7%	40
Current Year + 23	12,000	0	0	0	5	14,500	11,900	100	11,800	2,700	22.9%	(340)
Current Year + 24	12,000	0	0	0	5	14,500	12,000	100	11,900	2,600	21.8%	(220)
Current Year + 25	12,000	0	0	0	5	14,500	12,100	100	12,000	2,500	20.8%	(100)
Current Year + 26	12,000	0	0	0	5	14,500	12,200	100	12,100	2,400	19.8%	20
Current Year + 27	12,000	0	0	0	6	15,000	12,300	100	12,200	2,800	23.0%	(360)
Current Year + 28	12,000	0	0	0	6	15,000	12,400	100	12,300	2,700	22.0%	(240)
Current Year + 29	12,000	0	0	0	6	15,000	12,500	100	12,400	2,600	21.0%	(120)
CC units =			500	MW		Formula: (1f) = (1a) + ((1b) × 500) + ((1c) × 160) + ((1d) × 120) + ((1e) × 500)						
CT units =			160	MW								
PV units =			120	MW								
Filler units =			500	MW								

the CPVRR costs and levelized system average electric rate values for both resource plans featuring the DSM options.

The calculation methodology leading to the results shown are identical to those presented in detail in Chapter 5 for Supply Only Resource Plan 1 (CC). However, as we shall see, the DSM Costs columns, and the column showing the DSM energy reductions for the levelized system average rate calculation, will no longer show zero values as they did in Chapter 5 for Supply Only Resource Plan 1 (CC).

Results for the With DSM Resource Plan 1

We begin by looking at the Fixed Costs for the With DSM Resource Plan 1. These values are presented in Table 6.9.
At this point we will only briefly pause to mention that the total Fixed Costs value for With DSM Resource Plan 1 of $3,257 million CPVRR is lower than the total Fixed Cost value for Supply Only Resource Plan 1 (CC) of $3,582 million CPVRR. We will return to see why this is the case after we have first introduced the DSM and Variable costs for this With DSM resource plan.

Now, as we did in Chapter 5, we carry over Column (6) from this table which provides the total Fixed Costs, then add the DSM Costs. This information is presented in Table 6.10.

We again pause only briefly to note two aspects of the DSM costs. First, they are no longer zero as was the case with the Supply Only resource plans. Second, the net DSM cost for this resource plan featuring DSM Option 1 is actually a negative net cost value of ($10) million CPVRR. We will soon return to discuss these DSM net costs. However, we now carry over the total Fixed Costs [Column (6)] and total DSM Costs [Column (10)], then add in the Variable Costs. This information is presented in Table 6.11.

Finally, we summarize the total costs for the With DSM Resource Plan 1 in Table 6.12.

It is now time to examine more closely the costs for the With DSM Resource Plan 1 to see what the effects are of the introduction of DSM Option 1, and the accompanying avoidance of the most economic Supply option (CC Unit A) in Current Year+5. By looking at the CPVRR cost information on the bottom row of Table 6.12, we immediately see that the With DSM Resource Plan 1 differs from the most economical Supply option resource plan, Supply Only Resource Plan 1 (CC), in two ways.

First, as previously mentioned, the DSM Cost values presented in Column (10) are no longer zero as was the case with any of the Supply Only resource plans. Second, the total CPVRR cost for the With DSM Resource Plan 1, $47,569 million, is lower than the CPVRR cost value for Supply Only Resource Plan 1 of $47,828 million.

The lower CPVRR total cost for With DSM Resource Plan 1 can best be explained by looking at differences in the Fixed Costs, DSM Costs, and Variable Costs sections of the table in comparison to the previously presented

TABLE 6.9

(See color insert.) Fixed Costs Calculation for With DSM Resource Plan 1

	(1)	(2)	(3)	(4)	(5)	(6)
						= Sum of Cols. (2) through (5)
Year	Annual Discount Factor 8.000%	Generation Capital (Nominal $, Millions)	Generation Fixed O&M (Nominal $, Millions)	Generation Capital Replacement (Nominal $, Millions)	Firm Gas Transportation (Nominal $, Millions)	Total Generation Fixed Costs (Nominal $, Millions)
Current Year	1.000	0	0	0	0	**0**
1	0.926	0	0	0	0	**0**
2	0.857	0	0	0	0	**0**
3	0.794	0	0	0	0	**0**
4	0.735	0	0	0	0	**0**
5	0.681	0	0	0	0	**0**
6	0.630	102	3	5	82	**192**
7	0.583	98	3	5	82	**188**
8	0.540	94	3	5	82	**184**
9	0.500	90	3	5	82	**181**
10	0.463	196	7	11	164	**378**
11	0.429	188	7	11	164	**370**
12	0.397	179	7	11	164	**362**
13	0.368	171	7	12	164	**354**
14	0.340	282	11	18	246	**557**
15	0.315	268	11	18	246	**544**
16	0.292	255	11	19	246	**531**
17	0.270	242	11	19	246	**519**
18	0.250	229	12	19	246	**506**
19	0.232	347	16	26	329	**718**
20	0.215	329	16	27	329	**700**
21	0.199	310	16	27	329	**682**
22	0.184	292	17	28	329	**665**
23	0.170	416	21	36	411	**884**
24	0.158	392	22	36	411	**860**
25	0.146	367	22	37	411	**837**
26	0.135	343	23	38	411	**814**
27	0.125	473	28	46	493	**1,040**
28	0.116	443	28	47	493	**1,011**
29	0.107	412	29	48	493	**982**
Total CPVRR =		**$1,553**	**$72**	**$119**	**$1,513**	**$3,257**

TABLE 6.10

(See color insert.) DSM Costs Calculation for With DSM Resource Plan 1

Year	(1) Annual Discount Factor 8.000%	(6) = Sum of Cols. (2) through (5) Total Generation Fixed Costs (Nominal $, Millions)	(7) DSM Administrative Costs (Nominal $, Millions)	(8) DSM Incentive Payments (Nominal $, Millions)	9) (T&D Costs Avoided by DSM (Nominal $, Millions)	(10) = Col (7) + Col (8) − Col (9) DSM Net Costs (Nominal $, Millions)
Current Year	1.000	**0**	0	0	0	**0**
1	0.926	**0**	2	4	0	**6**
2	0.857	**0**	2	4	1	**5**
3	0.794	**0**	2	4	3	**3**
4	0.735	**0**	2	4	4	**2**
5	0.681	**0**	2	4	5	**1**
6	0.630	**192**	0	0	7	**(7)**
7	0.583	**188**	0	0	6	**(6)**
8	0.540	**184**	0	0	6	**(6)**
9	0.500	**181**	0	0	6	**(6)**
10	0.463	**378**	0	0	6	**(6)**
11	0.429	**370**	0	0	5	**(5)**
12	0.397	**362**	0	0	5	**(5)**
13	0.368	**354**	0	0	5	**(5)**
14	0.340	**557**	0	0	5	**(5)**
15	0.315	**544**	0	0	4	**(4)**
16	0.292	**531**	3	4	4	**3**
17	0.270	**519**	3	4	4	**3**
18	0.250	**506**	3	4	4	**3**
19	0.232	**718**	3	4	3	**4**
20	0.215	**700**	3	4	3	**4**
21	0.199	**682**	0	0	3	**(3)**
22	0.184	**665**	0	0	3	**(3)**
23	0.170	**884**	0	0	2	**(2)**
24	0.158	**860**	0	0	2	**(2)**
25	0.146	**837**	0	0	2	**(2)**
26	0.135	**814**	0	0	2	**(2)**
27	0.125	**1,040**	0	0	1	**(1)**
28	0.116	**1,011**	0	0	1	**(1)**
29	0.107	**982**	0	0	1	**(1)**
Total CPVRR =		**$3,257**	**$12**	**$21**	**$43**	**($10)**

TABLE 6.11

(See color insert.) Variable Costs Calculation for With DSM Resource Plan 1

	(1)	(6)	(10)	(11)	(12)	(13)	(14)
		= Sum of Cols. (2) through (5)	= Col (7) + Col (8) − Col (9)				= Sum of Cols. (11) through (13)
Year	Annual Discount Factor 8.000%	Total Generation Fixed Costs (Nominal $, Millions)	DSM Net Costs (Nominal $, Millions)	Generation Variable O&M (Nominal $, Millions)	System Net Fuel (Nominal $, Millions)	System Environmental Compliance (Nominal $, Millions)	Total Variable Costs (Nominal $, Millions)
Current Year	1.000	0	0	0	1,660	275	1,936
1	0.926	0	6	0	1,725	390	2,115
2	0.857	0	5	0	1,790	506	2,296
3	0.794	0	3	0	1,857	624	2,481
4	0.735	0	2	0	1,926	744	2,670
5	0.681	0	1	0	1,997	866	2,863
6	0.630	192	(7)	7	1,989	965	2,961
7	0.583	188	(6)	7	2,065	1,089	3,161
8	0.540	184	(6)	7	2,144	1,215	3,366
9	0.500	181	(6)	8	2,224	1,343	3,576
10	0.463	378	(6)	15	2,219	1,439	3,673
11	0.429	370	(5)	16	2,302	1,570	3,888
12	0.397	362	(5)	16	2,389	1,702	4,107
13	0.368	354	(5)	16	2,478	1,836	4,331
14	0.340	557	(5)	25	2,473	1,929	4,427
15	0.315	544	(4)	26	2,565	2,065	4,656
16	0.292	531	3	26	2,660	2,204	4,890
17	0.270	519	3	27	2,758	2,345	5,129
18	0.250	506	3	27	2,858	2,488	5,374
19	0.232	718	4	37	2,855	2,577	5,469
20	0.215	700	4	38	2,960	2,722	5,720
21	0.199	682	(3)	38	3,067	2,870	5,975
22	0.184	665	(3)	39	3,178	3,020	6,237
23	0.170	884	(2)	50	3,176	3,105	6,330
24	0.158	860	(2)	51	3,290	3,257	6,598
25	0.146	837	(2)	52	3,409	3,411	6,872
26	0.135	814	(2)	53	3,530	3,568	7,151
27	0.125	1,040	(1)	65	3,530	3,649	7,244
28	0.116	1,011	(1)	66	3,656	3,808	7,530
29	0.107	982	(1)	68	3,785	3,970	7,823
Total CPVRR =		$3,257	($10)	$167	$27,328	$16,826	$44,322

TABLE 6.12

(See color insert.) Total Costs Calculation for With DSM Resource Plan 1

	(1)	(6)	(10)	(14)	(15)	(16)	(17)
		= Sum of Cols. (2) through (5)	= Col (7) + Col (8) − Col(9)	= Sum of Cols. (11) through (13)	= (6) + (10) + (14)	= (1) × (15)	
Year	Annual Discount Factor 8.000%	Total Generation Fixed Costs (Nominal $, Millions)	DSM Net Costs (Nominal $, Millions)	Total Variable Costs (Nominal $, Millions)	Total Annual Costs (Nominal $, Millions)	Total Annual NPV Costs (NPV $, Millions)	Cumulative Total NPV Costs (NPV $, Millions)
Current Year	1.000	0	0	1,936	1,936	1,936	1,936
1	0.926	0	6	2,115	2,121	1,963	3,899
2	0.857	0	5	2,296	2,301	1,972	5,871
3	0.794	0	3	2,481	2,484	1,972	7,844
4	0.735	0	2	2,670	2,672	1,964	9,808
5	0.681	0	1	2,863	2,864	1,949	11,757
6	0.630	192	(7)	2,961	3,146	1,983	13,739
7	0.583	188	(6)	3,161	3,343	1,951	15,690
8	0.540	184	(6)	3,366	3,545	1,915	17,605
9	0.500	181	(6)	3,576	3,750	1,876	19,481
10	0.463	378	(6)	3,673	4,046	1,874	21,355
11	0.429	370	(5)	3,888	4,252	1,824	23,179
12	0.397	362	(5)	4,107	4,464	1,773	24,951
13	0.368	354	(5)	4,331	4,679	1,721	26,672
14	0.340	557	(5)	4,427	4,979	1,695	28,367
15	0.315	544	(4)	4,656	5,196	1,638	30,005
16	0.292	531	3	4,890	5,424	1,583	31,588
17	0.270	519	3	5,129	5,651	1,527	33,116
18	0.250	506	3	5,374	5,883	1,472	34,588
19	0.232	718	4	5,469	6,191	1,434	36,022
20	0.215	700	4	5,720	6,424	1,378	37,400
21	0.199	682	(3)	5,975	6,655	1,322	38,722
22	0.184	665	(3)	6,237	6,899	1,269	39,991
23	0.170	884	(2)	6,330	7,212	1,228	41,220
24	0.158	860	(2)	6,598	7,456	1,176	42,396
25	0.146	837	(2)	6,872	7,707	1,125	43,521
26	0.135	814	(2)	7,151	7,964	1,077	44,598
27	0.125	1,040	(1)	7,244	8,283	1,037	45,634
28	0.116	1,011	(1)	7,530	8,540	990	46,624
29	0.107	982	(1)	7,823	8,804	945	47,569
Total CPVRR =		$3,257	($10)	$44,322	$47,569	$47,569	

(Table 5.14) values for the most economic Supply Only resource plan: Supply Only Resource Plan 1 (CC).

In regard to Fixed Costs, the With DSM Resource Plan 1 has lower costs ($3,257 million CPVRR) than the Supply Only Resource Plan 1 (CC) ($3,582 million CPVRR). This $325 million CPVRR cost savings in regard to Fixed Costs is primarily due to the fact that the 100 MW DSM option avoids the need to build a 500 MW CC unit in the decision year. Avoiding the 500 MW CC unit saves a significant amount of Fixed Costs in that year.

However, recall that the With DSM Resource Plan 1 does require the addition of six CC filler units in later years compared to only five filler units with Supply Only Resource Plan 1. This lowers the Fixed Cost savings value from what it would have been if there had been no difference between the two resource plans in the number of filler units. However, the resulting net Fixed Cost savings value of $325 million CPVRR still represents a significant fixed cost savings for the With DSM Resource Plan 1.

We next turn our attention to DSM Costs. We have seen in Table 6.10 that the sum of the DSM administration costs ($12 million CPVRR) and incentive costs ($21 million CPVRR) equals $33 million CPVRR. This cost value represents the direct costs of implementing the DSM option. However, we have also seen that the reduced electrical load that our utility system must meet will also result in a combined capital and fixed O&M savings of $43 million CPVRR in transmission and distribution facilities that would otherwise have been needed. Consequently, the *net* DSM Costs are actually a *savings* of $10 million as shown by the calculation: $33 million CPVRR for administration and incentive costs–$43 million CPVRR in transmission and distribution savings = ($10) million CPVRR.*

In regard to Variable Costs, we see that the With DSM Resource Plan 1 results in higher Variable Costs of $76 million CPVRR as shown by the calculation: $44,322 million CPVRR for the With DSM Resource Plan 1–$44,246 million CPVRR for the Supply Only Resource Plan 1 (CC). There are two reasons for this result. First, the With DSM Resource Plan 1 will result in the utility system serving less energy each year due to the kWh reduction from DSM. This lowers fuel and environmental compliance costs for our utility system. (If there were no other differences in the two resource plans, this impact would have resulted in the With DSM Resource Plan 1 having lower Variable Costs than the Supply Only Resource Plan 1 (CC). However, there are other differences between the resource plans.)

* The combining of direct DSM administrative and incentive costs with savings from avoided transmission and distribution (T&D) facilities is not a universal practice. However, the final result in terms of total utility costs for resource plans is unaffected regardless of whether these costs are combined. (Combining these costs is one way to address the fact that certain computer models do not directly address T&D costs that would be avoided by DSM. One approach for dealing with this is to combine these avoided T&D costs with the direct DSM costs to ensure that the T&D avoided costs are accounted for.)

Second, the With DSM Resource Plan 1, even though it ends up with the same number (6) of new, fuel-efficient CC units (i.e., CC units in the decision year and subsequent CC filler units) as in the Supply Only Resource Plan 1, will have a number of years in which the cumulative number of these new fuel-efficient CC units is less than the cumulative number of these units in the Supply Only Resource Plan 1. Therefore, the utility system's "fleet" of generation units will be less efficient in those particular years, resulting in higher fuel and emission costs in those years compared to the Supply Only Resource Plan 1 (CC).

These two factors serve to counterbalance each other to a degree for our hypothetical utility system, but the second factor is more important in this particular comparison. The net result is a disadvantage in regard to Variable Costs of $76 million CPVRR for the With DSM Resource Plan 1.

Consequently, we see that the total CPVRR cost of the With DSM Resource Plan 1 ($47,569 million CPVRR) is $259 million CPVRR less expensive than the cost of the Supply Only Resource Plan 1 (CC) ($47,828 million CPVRR). As just discussed, virtually all of these savings come from the Fixed Cost category ($325 million CPVRR) with much smaller savings coming from net DSM costs ($10 million CPVRR). Together, these two factors overcome the Variable Cost disadvantage ($76 million CPVRR).

However, before we are tempted to announce that DSM Option 1 is a better choice for our utility system than the best Supply option (CC Unit A), we need to remember that we have not yet completed the final step of an economic evaluation involving both DSM and Supply options. Namely, we have not performed a calculation of the electric rates that will result from the With DSM resource plan.

In fact, to help ensure that we are not tempted to forget this, we will introduce my third Fundamental Principle of Electric Utility Resource Planning:

> **Fundamental Principle #3 of Electric Utility Resource Planning: "Electric Rate Impacts Are the Most Important Consideration When Analyzing DSM and Supply Options; Total Cost Impacts Are Less Important"**
>
> **In regard to economic analyses, projections of total utility system costs for competing Supply options can be used to correctly select the most economic Supply resource option. As previously discussed, this is because the Supply option which results in the lowest total costs will also result in the lowest electric rates. However, when evaluating DSM options versus Supply options, economic analyses must be carried out one step further. Analyses of DSM versus Supply options must account for the fact that DSM options reduce the number of kWh of sales over which a utility's costs (revenue requirements) are recovered. This unique characteristic of DSM options makes it necessary to conduct an electric rate calculation in order to really determine which option, DSM or Supply, is the best economic choice from a customer's perspective. The importance of total costs is lessened in these calculations because total costs are merely one input into the calculation of electric rates.**

One almost always sees the total CPVRR costs for a utility system are lower if a DSM option, rather than a Supply option, is selected. However, the important issue is how the utility's electric rates compare for resource plans with the two types of resource options. In our example, we need to see how the previously calculated levelized system average electric rate of 12.8116 cents/kWh for the Supply Only Resource Plan 1 (CC) compares with the levelized system average electric rate for the With DSM Resource Plan 1.

The levelized electric rate calculation for the With DSM Resource Plan 1, including DSM-driven costs, benefits, and kWh reductions, is presented in Table 6.13.

From Column (25) in this table, we see that the levelized system average electric rate for the With DSM Resource Plan 1 is 12.8095 cents/kWh. Therefore, this DSM-based resource plan featuring DSM Option 1 will result in lower electric rates than the most economical resource plan featuring a Supply Option: Supply Only Resource Plan 1 (CC) which had a levelized system average electric rate of 12.8116 cents/kWh.

From an economic perspective, DSM Option 1 is clearly superior to the best Supply option. This is what we would expect by examining the RIM preliminary screening test analysis results comparing DSM Option 1 versus the CC unit. DSM Option 1 passed the RIM screening test; i.e., it was potentially more cost-effective than the CC unit for our utility system. And, very importantly, this outcome was essentially guaranteed by not adding more of DSM Option 1 than was called for by our utility's resource needs of 100 MW.*

But what about DSM Option 2 versus the same CC unit? DSM Option 2 passed the TRC test, but failed the RIM test. What will the final economic evaluation show for this DSM option? We will now find out.

Results for the With DSM Resource Plan 2

We now add in the economic information for the resource plan that features DSM Option 2. A summary form of both the cost and electric rate information for both With DSM resource plans is presented in Table 6.14.

Table 6.14 shows us several things about the two With DSM resource plans. First, the Fixed Cost value of $3,257 million CPVRR is identical for the two resource plans. This is to be expected because both resource plans have added an identical amount of DSM (100 MW) that avoided a new generating unit in the decision year and they both result in the identical number, and timing, of filler units that are added after the decision year. Therefore, the Fixed Costs for our utility system will be identical with either DSM Option 1 or 2.

Second, the two plans differ considerably in regard to net DSM Costs. The difference is $112 million CPVRR between the two plans; i.e., a $10 million

---

* Adding more DSM than is needed to meet projected near-term resource needs will increase the present value costs of DSM and can result in an otherwise economically preferred DSM option (or portfolio of DSM options) no longer being the economic choice.

TABLE 6.13

(See color insert.) Levelized Electric Rate Calculation for With DSM Resource Plan 1

	(1)	(15)	(18)	(19)	(20)	(21)	(22)	(23)	(24)	(25)	(26)
		= (6) + (10) + (14)		= (15) + (18)			= (20) − (21)	= ((19) × 100)/(22)	= (1) × (23)		= (1) × (25)
Year	Annual Discount Factor 8.000%	Total Annual Costs (Nominal $, Millions)	Other System Costs Not Affected by Plan (Nominal $, Millions)	Total Utility Costs (Nominal $, Millions)	Forecasted Utility Sales/NEL (GWh)	DSM Energy Reduction (GWh)	Net Utility Sales/ NEL (GWh)	System Avg. Electric Rate Nominal (cents/kWh)	System Avg. Electric Rate NPV (cents/kWh)	Levelized System Avg. Electric Rate (cents/kWh)	System Avg. Electric Rate NPV (cents/kWh)
Current Year	1.000	1,936	2,700	4,636	50,496	0	50,496	9.1800	9.1800	12.8095	12.8095
1	0.926	2,121	2,754	4,875	51,022	20	51,002	9.5576	8.8496	12.8095	11.8607
2	0.857	2,301	2,809	5,110	51,548	60	51,488	9.9240	8.5082	12.8095	10.9821
3	0.794	2,484	2,865	5,350	52,074	100	51,974	10.2930	8.1709	12.8095	10.1686
4	0.735	2,672	2,923	5,595	52,600	140	52,460	10.6646	7.8388	12.8095	9.4154
5	0.681	2,864	2,981	5,845	53,126	180	52,946	11.0390	7.5130	12.8095	8.7179
6	0.630	3,146	3,041	6,187	53,652	200	53,452	11.5747	7.2940	12.8095	8.0722
7	0.583	3,343	3,101	6,445	54,178	200	53,978	11.9395	6.9666	12.8095	7.4742
8	0.540	3,545	3,163	6,708	54,704	200	54,504	12.3074	6.6493	12.8095	6.9206
9	0.500	3,750	3,227	6,977	55,230	200	55,030	12.6786	6.3425	12.8095	6.4079
10	0.463	4,046	3,291	7,337	55,756	200	55,556	13.2065	6.1172	12.8095	5.9333
11	0.429	4,252	3,357	7,609	56,282	200	56,082	13.5684	5.8193	12.8095	5.4938
12	0.397	4,464	3,424	7,888	56,808	200	56,608	13.9340	5.5334	12.8095	5.0868
13	0.368	4,679	3,493	8,172	57,334	200	57,134	14.3035	5.2594	12.8095	4.7100
14	0.340	4,979	3,563	8,541	57,860	200	57,660	14.8135	5.0434	12.8095	4.3611
15	0.315	5,196	3,634	8,830	58,386	200	58,186	15.1747	4.7837	12.8095	4.0381
16	0.292	5,424	3,707	9,131	58,912	200	58,712	15.5517	4.5394	12.8095	3.7390

17	0.270	5,651	3,781	9,432	59,438	200	59,238	15.9216	4.3031	12.8095	3.4620
18	0.250	5,883	3,856	9,739	59,964	200	59,764	16.2962	4.0781	12.8095	3.2056
19	0.232	6,191	3,933	10,124	60,490	200	60,290	16.7922	3.8910	12.8095	2.9681
20	0.215	6,424	4,012	10,436	61,016	200	60,816	17.1593	3.6815	12.8095	2.7483
21	0.199	6,655	4,092	10,747	61,542	200	61,342	17.5201	3.4805	12.8095	2.5447
22	0.184	6,899	4,174	11,073	62,068	200	61,868	17.8976	3.2921	12.8095	2.3562
23	0.170	7,212	4,258	11,469	62,594	200	62,394	18.3819	3.1307	12.8095	2.1817
24	0.158	7,456	4,343	11,799	63,120	200	62,920	18.7528	2.9573	12.8095	2.0201
25	0.146	7,707	4,430	12,137	63,646	200	63,446	19.1293	2.7932	12.8095	1.8704
26	0.135	7,964	4,518	12,482	64,172	200	63,972	19.5117	2.6380	12.8095	1.7319
27	0.125	8,283	4,609	12,891	64,698	200	64,498	19.9869	2.5021	12.8095	1.6036
28	0.116	8,540	4,701	13,241	65,224	200	65,024	20.3633	2.3604	12.8095	1.4848
29	0.107	8,804	4,795	13,599	65,750	200	65,550	20.7459	2.2266	12.8095	1.3748
Total CPVRR =		**$47,569**							**155.7432**		**155.7432**

TABLE 6.14

Economic Evaluation Results of Both With DSM Resource Plans: CPVRR Costs and Levelized System Average Electric Rates

	(1)	(2)	(3)	(4)	(5)	(6)
				= Sum of Cols. (1), (2), and (3)		
Resource Plan	**Fixed Costs (Millions, CPVRR)**	**DSM Costs (Millions, CPVRR)**	**Variable Costs (Millions, CPVRR)**	**Total Costs (Millions, CPVRR)**	**Difference from Lowest Cost With DSM Plan (Millions, CPVRR)**	**Levelized System Average Electric Rate (cents/kWh)**
With DSM resource Plan 1	3,257	(10)	44,322	47,569	262	12.8095
With DSM resource Plan 2	3,257	102	43,948	47,307	0	12.8445

CPVRR net *savings* for the resource plan featuring DSM Option 1 versus a $102 million CPVRR net *cost* for the resource plan featuring DSM Option 2. A logical question is "why does this cost difference exist?"

We just discussed how the various components of DSM costs were calculated for DSM Option 1. In addition, we had previously discussed how the administration and incentive costs would be more than twice as high for DSM Option 2 compared to DSM Option 1. We will now examine how these net DSM costs actually worked out.

The previously discussed CPVRR administration and incentive costs for DSM Option 1, $12 million and $21 million, respectively, add up to an administrative and incentive cost total of $33 million. For DSM Option 2, this total increases initially to $66 million CPVRR just due to the fact that twice as many participants need to be signed up for DSM Option 2 as for DSM Option 1 because each DSM Option 2 participant only provides 0.5 kW reduction compared to 1.0 kW reduction for DSM Option 1 participants.

The $43 million CPVRR savings in avoided T&D expenditures we saw previously for DSM Option 1 are driven solely by the peak demand (kW) reductions achieved by DSM Option 1. Because the identical peak demand reduction, both annually and in total, is also achieved for DSM Option 2, the Resource Plan with DSM Option 2 will also realize the same $43 million CPVRR savings from avoided T&D expenditures. In other words, there is no difference between the two DSM Options in regard to avoided T&D expenditures.

Consequently, when considering just the doubling effect on the administration and incentive costs, plus the identical T&D savings, we realize that the net DSM Costs have (so far) increased from a net *savings* of $10 million CPVRR for DSM Option 1 to a net *cost* of $23 million CPVRR for DSM Option 2 as shown by the calculation: $66 million CPVRR in administration and incentive costs–$43 million CPVRR in T&D savings = $23 million CPVRR of net costs.

However, as we see from the results of Table 6.14, the total net cost for DSM Option 2 is not $23 million CPVRR, but $102 million CPVRR. Where does this additional cost for DSM Option 2 come from? As suggested earlier in this chapter, the answer lies in the difference in the life expectancy of the equipment that is installed for DSM Option 2 versus DSM Option 1.

We previously discussed how the fact that the much shorter (5 years) "life expectancy" for the DSM equipment installed with DSM Option 2, compared to the 15 year life expectancy for DSM Option 1's equipment, would further result in higher DSM administrative and incentive costs for DSM Option 2. This is because the utility must continue to replenish the kW (and accompanying kWh) savings from DSM Option 2 that would otherwise begin to vanish 5 years after a customer is signed up to participate in DSM Option 2. (This replenishment is also needed for DSM Option 1's equipment, but less frequently due to the 15 year life expectancy for that equipment.)

By looking at the final total net DSM Cost for DSM Option 2 of $102 million CPVRR, we see that the life expectancy of a DSM measure can be a truly significant factor in the total DSM CPVRR cost. In our example, the difference in the life expectancy between DSM Option 2 and DSM Option 1 increased the net DSM costs for DSM Option 2 by $79 million CPVRR as shown by the calculation: $102 million CPVRR total net cost–the previously calculated net cost of $23 million CPVRR prior to accounting for life expectancy = $79 million CPVRR of costs due to life expectancy.

Finally, we examine the Variable Costs for our utility system. From the table we see that the Variable Costs for DSM Option 2 ($43,948 million CPVRR) are significantly lower than for DSM Option 1 ($44,322 million CPVRR) as expected. This difference of $374 million CPVRR is driven by the fact that DSM Option 2 will result in triple the energy savings on the utility system compared to DSM Option 1. (Recall that from Row (7) of Table 6.1, the "Energy-to-Demand Reduction Ratio" or, MWh-to-MW reduction ratio, was 6,000 for DSM Option 2 versus 2,000 for DSM Option 1.)

When all three cost categories are combined, the net result is a CPVRR cost for the With DSM Resource Plan 2 of $47,307 or $262 million CPVRR less than for the With DSM Resource Plan 1. However, as we just reminded ourselves, DSM Option 2 will also result in significantly fewer kWh sales over which our utility's costs must be recovered. So we now turn our attention to the more important issue of what effect the With DSM Resource Plan 2 will have on the utility's electric rates.

From the last column in Table 6.14, we see that, despite the lower CPVRR costs for the Resource Plan with DSM Option 2, this resource plan will result in a higher levelized electric rate (12.8445 cents/kWh) than will the With DSM Resource Plan 1 do (12.8095 cents/kWh). Therefore, DSM Option 2 is a worse choice than DSM Option 1 in regard to the electric rates that our utility's customers will be charged. Thus DSM Option 1 is a better economic choice for our utility's customers than DSM Option 2.

We can now close this chapter by summing up the results of both the previous and current chapters. Our utility has now completed the final (or system) economic analyses of all three Supply options: the CC option, the CT option, and the PV option. The utility has also completed the final (or system) economic analyses for both of the DSM Options: DSM Option 1 and DSM Option 2.

Therefore, from an economic perspective, our utility should have the information it needs to determine which one of these five resource options is the best economic choice for its customers. Our utility is also at a point at which it can begin to examine these five resource options from a non-economic perspective. Then, with information from both the economic and non-economic perspectives, a final decision can be made regarding which of the five resource options is the best overall choice of resource option with which to meet its resource needs 5 years in the future.

In Chapter 7, our utility takes these next steps.

7

Final Resource Option Analyses for Our Utility System

Economic Comparison of the Resource Plans

Our hypothetical utility system has now completed its economic analyses of all five resource plans, each resource plan containing one of the five resource options it is considering in order to meet its next resource need. The numeric results of these analyses are summarized in Table 7.1. The resource plans are listed in the order they were discussed in previous chapters.

This table presents the economic results from two perspectives: (i) total costs (presented in Columns (1) and (2)), and (ii) electric rates (presented in Columns (3) and (4)). We will now examine the results from each of these perspectives separately by looking at rankings of the five resource plans from each perspective. This should allow us to understand the results more clearly.

We shall start with a ranking of resource plans from a total cost (CPVRR) perspective. This information is presented in Table 7.2.

This table shows that—from a total cost perspective—the With DSM Resource Plan 2 (that featured DSM Option 2 that failed the RIM test, but passed the TRC test) is projected to result in the lowest total CPVRR costs over the analysis period. The With DSM Resource Plan 1 (that featured DSM Option 1 that passed both the RIM and TRC tests) is projected to result in the second lowest total costs. (Note that both of the resource plans featuring the DSM options are projected to result in lower total costs than any of the three Supply Only resource plans. This is a common outcome in analyses of DSM and Supply resource options.)

From a total cost perspective, Supply Only Resource Plan 1 (CC) is projected as the most economic Supply Only resource plan and as the resource plan that has the third lowest total costs. Supply Only Resource Plan 2 (CT) is projected as the resource plan with the fourth lowest total costs and Supply Only Resource Plan 3 (PV) is projected as having the highest total costs.

However, as we have discussed previously, the total cost perspective ignores a very important economic impact that occurs when DSM resource options are considered: the upward pressure that will be placed on electric

TABLE 7.1

Economic Evaluation Results of All Five Resource Plans: CPVRR Costs and Levelized System Average Electric Rates

Resource Plan	(1) Total Costs (Millions, CPVRR)	(2) Difference from Lowest Cost Resource Plan (Millions, CPVRR)	(3) Levelized System Average Electric Rate (cents/kWh)	(4) Difference from Lowest Levelized Average Electric Rate Resource Plan (cents/kWh)
Supply Only Plan 1 (CC)	47,828	521	12.8116	0.0021
Supply Only Plan 2 (CT)	47,950	643	12.8283	0.0188
Supply Only Plan 3 (PV)	48,225	918	12.8699	0.0604
With DSM Plan 1	47,569	262	12.8095	0
With DSM Plan 2	47,307	0	12.8445	0.0350

TABLE 7.2

Ranking of Resource Plans in Regard to CPVRR Costs

Resource Plan	(1) Total Costs (Millions, CPVRR)	(2) Difference from Lowest Cost Resource Plan (Millions, CPVRR)	(3) Relative Ranking
With DSM Plan 2	47,307	0	1st
With DSM Plan 1	47,569	262	2nd
Supply Only Plan 1 (CC)	47,828	521	3rd
Supply Only Plan 2 (CT)	47,950	643	4th
Supply Only Plan 3 (PV)	48,225	918	5th

rates for all of the utility's customers because the utility's total costs will be spread over fewer kWh of sales due to the kWh reduction characteristic of DSM options. For that reason, a total cost perspective when evaluating both Supply and DSM options can only provide an incomplete economic picture. No final decision regarding resource options should ever be made based solely on projections of total costs when one or more DSM options are being evaluated.

Therefore, we now turn our attention to examining a ranking of the five resource plans from an electric rate perspective. Unlike a total cost perspective, the electric rate perspective does provide a complete economic picture because it *does* account for the kWh reduction characteristic of DSM options. Table 7.3 provides the electric rate perspective.

TABLE 7.3

Ranking of Resource Plans in Regard to Electric Rates

	(1)	(2)	(3)
Resource Plan	**Levelized System Average Electric Rate (cents/kWh)**	**Difference from Lowest Levelized Average Electric Rate Resource Plan (cents/kWh)**	**Relative Ranking**
With DSM Plan 1	12.8095	0	1st
Supply Only Plan 1 (CC)	12.8116	0.0021	2nd
Supply Only Plan 2 (CT)	12.8283	0.0188	3rd
With DSM Plan 2	12.8445	0.0350	4th
Supply Only Plan 3 (PV)	12.8699	0.0604	5th

From the more complete, and meaningful, perspective of electric rates, we see that the ranking of the five resource plans has significantly changed. The With DSM Resource Plan 1 (featuring DSM Option 1 that passed both the RIM and TRC tests) has moved up to first place in the ranking. Therefore, this resource plan not only remains better than all three Supply Only resource plans, but is now seen as significantly better than the With DSM Resource Plan 2 (featuring DSM Option 2 that failed the RIM test, but passed the TRC test).

In regard to the three Supply Only resource plans, the relative ranking of these three plans remains the same. As previously discussed, this is expected because Supply options do not result in changes in the number of kWh over which the utility system's total costs are spread. Therefore, the best Supply Only resource plan from a total cost perspective will also be the best Supply Only resource plan from an electric rate perspective. Consequently, Supply Only Resource Plan 1 (CC) remains more economical than Supply Only Resource Plan 2 (CT). And both of these resource plans remain more economical than Supply Only Resource Plan 3 (PV).

With this information in hand, our utility is able to determine which resource plan is its best plan in regard to economics. Recalling Fundamental Principle #3 (which states that an electric rate perspective is the meaningful economic perspective to take when evaluating both DSM and Supply options), our utility selects DSM Option 1 (i.e., the DSM option that passed both the RIM and TRC tests) as the best resource option from an economic perspective.

However, our utility is not quite finished in its analyses. It decides that it wants to see how the five resource plans compare in regard to three noneconomic considerations: the amount of time until the winning economic resource plan becomes the best economic resource plan (i.e., the cross over time), system fuel use, and system emissions. We will examine these considerations next.

Non-Economic Analyses of the Resource Plans

"Cross Over" Time to Being the Most Economic Resource Plan

The first of the non-economic considerations we will examine is the determination of how long it takes until one resource plan emerges as clearly the most economic resource plan. For this determination, we will use the tabular format for cross over time that was introduced in Chapter 3. Table 7.4 presents the cumulative present value levelized system average electric rate information for each year for all of the five resource plans that our utility has analyzed. This information is presented as a ranking in which "1st" represents the resource plan with the lowest cumulative present value (CPV) electric rates through that year; "2nd" represents the resource plan with the second lowest CPV electric rates through that year, etc.

As shown in this table by the solid line across the page, the economic ranking of these five resource plans has largely stabilized by Current Year + 7. The only changes after Current Year + 7 are in Current Year + 9. In that year, the Supply Only Plan 1 (CC) and Supply Only Plan 2 (CT) "swap" the 2nd and 3rd positions for 1 year (due to differences in the timing of the filler units). Most importantly, the With DSM Resource Plan 1 emerges as the most economic resource plan beginning in Current Year + 6.

Recalling that our utility's resource need was in Current Year + 5, and that the economic ranking of the five resource plans has been virtually established within only 2 years of the resource need date, the generational equity consideration that we discussed in Chapter 3 is not an issue for our utility.

Before we leave this subject, there is one interesting aspect of the two types of resource options, Supply and DSM options, which shows up in this table. That is the fact that DSM options, no matter how economic they are over the full term of the analysis (as is the case with the With DSM Resource Plan 1), typically result in higher electric rates in the years prior to the year of resource need (Current Year + 5) and perhaps for a short time thereafter.

The reason for this is that the DSM options require expenditures (which are often recovered almost immediately from the utility's customers) prior to the year of resource need in order to sign up participating customers.* In those early years, these DSM costs are generally not completely offset by avoiding utility costs for system fuel, system environmental compliance costs, transmission, and distribution. For truly cost-effective DSM options (such as DSM Option 1 that passes the RIM cost-effectiveness test and which results in the

* The construction of new Supply options (i.e., new generating units) also require expenditures prior to the resource need year when the new generating unit will begin to operate. However, a utility will not typically recover these construction expenditures before the unit goes into operation. Exceptions do exist, particularly when the construction cost of the new generating unit is high (as, for example, with a new nuclear unit).

TABLE 7.4

"Cross Over" Table for All Five Resource Plans CPV Electric Rate of Resource Plans by Year

Current Year +	Supply Only Resource Plan 1 (CC)	Supply Only Resource Plan 2 (CT)	Supply Only Resource Plan 3 (PV)	With DSM Resource Plan 1	With DSM Resource Plan 2
1	1st	1st	1st	4th	5th
2	1st	1st	1st	4th	5th
3	1st	1st	1st	4th	5th
4	1st	1st	1st	4th	5th
5	3rd	1st	5th	2nd	3rd
6	3rd	2nd	5th	1st	4th
7	2nd	3rd	5th	1st	4th
8	2nd	3rd	5th	1st	4th
9	3rd	2nd	5th	1st	4th
10	2nd	3rd	5th	1st	4th
11	2nd	3rd	5th	1st	4th
12	2nd	3rd	5th	1st	4th
13	2nd	3rd	5th	1st	4th
14	2nd	3rd	5th	1st	4th
15	2nd	3rd	5th	1st	4th
16	2nd	3rd	5th	1st	4th
17	2nd	3rd	5th	1st	4th
18	2nd	3rd	5th	1st	4th
19	2nd	3rd	5th	1st	4th
20	2nd	3rd	5th	1st	4th
21	2nd	3rd	5th	1st	4th
22	2nd	3rd	5th	1st	4th
23	2nd	3rd	5th	1st	4th
24	2nd	3rd	5th	1st	4th
25	2nd	3rd	5th	1st	4th
26	2nd	3rd	5th	1st	4th
27	2nd	3rd	5th	1st	4th
28	2nd	3rd	5th	1st	4th
29	2nd	3rd	5th	1st	4th

1st = lowest electric rate.

lowest system electric rates), the additional benefits derived from avoiding the generating unit that would otherwise be needed, is enough to result in this type of DSM option beginning to deliver lower electric rates to the utility's customers either in the year of resource need or shortly thereafter.

A further note of caution is also warranted. We have just seen how the selection of a DSM option will typically result in increased electric rates in the near-term compared to the selection of a Supply option. We have also seen that selection of a DSM option that fails the RIM test (such as DSM Option 2) will result in increased electric rates in both the near- and long-terms compared to a DSM option that passes the RIM test (DSM Option 1). There is yet one other way in which electric rates can be further increased if the resource option selected is a DSM option.

This occurs if a *greater amount* of DSM is implemented than is needed to meet the projected resource need. This drives up electric rates for two reasons. First, DSM expenditures are incurred earlier than they need to be to meet any subsequent resource need (i.e., a resource need after Current Year+5 in our example). This drives up the present value of these additional DSM costs compared to what the present value of those costs would have been if the additional DSM had been signed up later in time (i.e., closer to the time when the additional DSM could have met a subsequent resource need). Second, the additional kWh reduction resulting from the additional DSM further reduces the number of kWh upon which electric rates are calculated.

Therefore, the implementation of more DSM than is needed to meet the next projected resource need both increases the numerator (costs), and decreases the denominator (kWh sales), in an electric rate calculation. This "double whammy" will definitely increase electric rates. Thus, it is simply not a good idea from an electric rate perspective to "push" for more DSM than is needed to meet a utility's next resource need. Customers' electric rates are increased needlessly.

System Fuel Use

The next non-economic consideration that our utility takes a look at is the impact of the resource plans on its system fuel use. We start by examining our utility's total system fuel use over all of the years of the analysis period.* Our utility system's fuel use will be expressed in terms of million mmBTU.† The projected total amount of fuel that would be used by our utility over the analysis period from each of the five resource plans is presented in Table 7.5. The order of the resource plans in this table is the same order in which the resource plans were originally discussed earlier in the book.

* The use of all of the years addressed in the analysis when examining total system fuel use allows us to include the longer-term system impacts of differences between the resource options that could be chosen for the decision year. The primary differences are the capacity (MW) of these options (which, as we have seen, affects the timing and number of the filler units and their fuel use) and the options' respective capacity factors or, in the case of the DSM options, equivalent capacity factors.

† This value, a million mmBTU, represents a lot of energy. To put this value in perspective, one million mmBTU represents the equivalent energy content that could be supplied by approximately 156,000 barrels of oil.

TABLE 7.5

Comparison of Total System Fuel Use by Resource Plan

Resource Plan	System Total Fuel Use (Million mmBTU)
Supply Only Plan 1 (CC)	16,889.10
Supply Only Plan 2 (CT)	16,931.38
Supply Only Plan 3 (PV)	16,875.43
With DSM Plan 1	16,875.24
With DSM Plan 2	16,782.33

The projected total fuel use results are not too surprising. One would expect that, all else equal, the resource plans featuring the DSM options, and the PV option, might result in lower system total fuel usage. This is because the DSM options themselves not only consume no fuel, but also reduce the number of kWh that our utility must serve. The PV option also consumes no fuel (but it does not reduce the number of kWh the utility must serve). Conversely, the two resource plans that feature the CC and CT options are based on resource options that do burn fossil fuel. One might expect that the total system fuel usage for these two resource plans would be relatively high compared to the other three resource plans.

Therefore, it is not surprising that the With DSM Resource Plan 2 (that features the DSM option with the greatest kWh reduction) results in the lowest total fuel usage of any of the five resource plans for our utility system.

The resource plan that results in the second lowest total fuel usage is the With DSM Resource Plan 1 that features a DSM option with somewhat less kWh reduction. The Supply Only Resource Plan 3 (PV) results in the third lowest total fuel usage.

The two resource plans that feature the CC and CT options, Supply Only Resource Plans 1 and 2, result, as expected, in the highest total fuel usage with Supply Only Resource Plan 2 featuring the less efficient CT option resulting in the highest projected system fuel use.

However, at this point, the astute reader is likely to ask the following question:

> "This total fuel use information in terms of mmBTU might be interesting, but how important is it really? After all, the cost of the fuel for each resource plan has already been accounted for in the economic analyses. Is there another reason that it is important to look at system fuel use information?"

This is a very good question. (You really are becoming quite proficient at this.) It is true that the actual fuel costs have been accounted for in the economic analyses. Furthermore, utilities do not often look at their system fuel use in terms of total mmBTU of fuel consumed. The reason they do not often do is actually evident if one reexamines the previous table; such a look tells us nothing about the *types of fuels* that are being consumed with each resource plan.

As mentioned in the overview discussion at the end of Chapter 3, utilities frequently examine their projected system fuel usage to ensure that they are not becoming overly dependent upon any one type of fuel. There are at least three reasons for this concern.

First, utilities want to see if a resource option makes them overly dependent upon a particular type of fuel in regard to the risk of sudden cost increases in that type of fuel. The costs of certain fuels are prone to significant cost changes in a relatively short time. This has historically been the case for oil and/or natural gas at various times. For this reason, utilities strive over the long-term to avoid becoming too reliant on such fuels because rapid fuel cost increases must be paid by the utility's customers and can lead to unexpected shocks in utility bills. In most regulatory jurisdictions, fuel costs are recovered quickly; i.e., without the lag time inherent in a formal rate case. Therefore, customers can feel the impact of fuel cost increases relatively quickly after the increases occur.

Second, too much reliance on any one type of fuel is also a system reliability issue. If the available supply of any type of fuel is reduced, even temporarily, the utility's ability to produce electricity from plants that rely on that type of fuel will be diminished. This can also result in system fuel cost increases that must be passed on to customers if the utility has to switch to higher cost (but more available) fuels. Furthermore, if the availability of the fuel supply is either significantly reduced and/or reduced for a long period of time, the reliability of the utility system as a whole (i.e., the ability of the utility to meet its customers' demand for electricity) may come into question.

Taken together, these two considerations are often referred to under a general heading as the "fuel diversity" profile of a utility system. And, before we introduce a third reason why there one might be interested in examining utility system fuel usage, we will take a look at how our utility system's fuel diversity profile is affected by each of the five resource plans.

Table 7.6 presents this information. It shows the percentage of the total energy served (GWh) by our utility system that is generated by the following types of fuel: nuclear, coal, natural gas, oil, and renewables. This information is first shown for the Current Year (i.e., the year that the utility will be making its decision as to how to address its resource needs 5 years in the future). The table also shows the projected values for Current Year + 5 (i.e., the decision year). The values for the Current Year provide a point of comparison that enables our utility to see how the selection of any of the five resource plans will affect or alter the utility's fuel use profile in 5 years. (Subsequent decisions about resource options that would be added in later years would impact our utility's fuel use profile in those years. However, because our utility is not currently making a resource decision for any year after Current Year + 5, projections of our utility's fuel mix for years after Current Year + 5 are of relatively little importance.)

TABLE 7.6

Fuel Diversity Profile of the Five Resource Plans (% of System GWh Served by Type of Fuel)

Year	Resource Plan	Nuclear (%)	Coal (%)	Natural Gas (%)	Oil (%)	Renewables (%)	Total (%)
Current Year	Not Applicable	14.7	48.6	26.0	10.6	0.0	100.0
Current Year+5	Supply Only Plan 1 (CC)	14.0	46.2	30.2	9.6	0.0	100.0
Current Year+5	Supply Only Plan 2 (CT)	14.0	46.2	28.3	11.5	0.0	100.0
Current Year+5	Supply Only Plan 3 (PV)	14.0	46.2	28.0	11.4	0.4	100.0
Current Year+5	With DSM Plan 1	14.1	46.3	28.1	11.5	0.0	100.0
Current Year+5	With DSM Plan 2	14.2	46.6	27.8	11.4	0.0	100.0

The first row of Table 7.6 shows us that the energy generated by our utility system to serve its customers in the Current Year is derived 14.7% from nuclear fuel, 48.6% from coal, 26.0% from natural gas, and 10.6% from oil. None of our utility's energy is currently derived from renewable sources.

The remaining rows of the table then show us the projected percentages of total energy served from each of these fuel types 5 years in the future for each of the resource plans under consideration.

A review of this information reveals several things. Most importantly for the two considerations we are discussing under the "fuel diversity" heading, there is not a significant difference in these percentage values between the five resource plans. All five resource plans will generally "move" our utility system's fuel diversity profile in the same direction: slightly lower percentage contributions from nuclear and coal (the utility's non-marginal fuels), and slightly higher percentage contributions from natural gas and oil (the utility's primary marginal fuels). (Note that the addition of the new CC unit in Supply Only Resource Plan 1 is an exception to the last part of this statement because it decreases system oil usage. We will return to that fact in a moment.)

This result, that the fuel diversity profiles vary relatively little between the five resource plans, means that our utility will not reject the results of its economic analyses due solely to fuel diversity considerations. This is important information for our utility to have.

But before we leave this table, let us further our education a bit more and look at two more items that are revealed by the table's results. First, we note that the percentage values 5 years in the future for nuclear and coal decline for each resource plan. The nuclear and coal units are projected to provide just as much energy (GWh) in 5 years as they do in the Current Year. However, the number of total GWh that will be served by the utility has increased due to our assumption of continuing load growth for our utility. Because no new nuclear or coal capacity is being added by our utility, the percentage of GWh served by nuclear and coal, compared to the total GWh served, will shrink. (Note that this "shrinkage" is slightly less for the two With DSM resource plans. This is because the DSM options will result in somewhat less total GWh being served by our utility.)

The other item that is revealed by the table is that the percentage of energy that is served by oil increases for all resource plans except the Supply Only Resource Plan 1 (CC). We recall that this new CC unit will have an annual capacity factor of 80%. This enables the new CC unit to replace a considerable amount of energy that would have otherwise been generated by existing oil and gas generating units in Current Year + 5.

Finally, we now return briefly to discuss the third reason that utilities may wish to examine their projected fuel use patterns. An analysis of a utility's fuel use patterns can, indirectly, shed considerable light on the utility system's projected air emissions. However, for the sake of expediency, we will move directly to examine our utility's system air emissions. We turn to that non-economic consideration next.

System Air Emissions

As previously mentioned, we will be examining three types of air emissions for the entire utility system: sulfur dioxide (SO_2), nitrogen oxides (NO_X), and carbon dioxide (CO_2). Although there are a number of other emissions that utilities are concerned with (mercury, particulates, etc.), our discussion of air emissions will focus solely on the three types of air emissions listed earlier. This will enable us to simplify our discussion while, at the same time, bring out certain key aspects of how resource options may impact system air emissions for a given utility system.

In addition, it allows us to look at two types of air emissions (SO_2 and NO_X) that are currently regulated due to the direct effects of these emissions on human health. It also allows us to look at one type of air emission (CO_2) that, at the time this book is written, is not currently regulated in the U.S., but is of interest largely due to concerns not directly tied to human health.*

We will start our discussion by presenting Table 7.7. This table provides a comparison of total system emissions for our utility system for all of the years addressed in our analysis. The total emissions (presented in terms of total tons of emissions) for each of the three types of emissions are presented separately. The values represent the total projected system emissions over the entire period of our analyses.

* Just as a point of interest, electric utilities typically have very accurate data regarding air emissions through the use of equipment that continually monitors air emissions. This data is recorded and reported regularly to various governmental agencies.

TABLE 7.7

Comparison of System Air Emissions by Resource Plan

Resource Plan	SO_2 (tons)	NO_X (tons)	CO_2 (tons)
Supply Only Plan 1 (CC)	1,702,426	474,525	1,140,009,500
Supply Only Plan 2 (CT)	1,735,379	487,861	1,144,298,834
Supply Only Plan 3 (PV)	1,725,122	483,416	1,140,817,610
With DSM Plan 1	1,725,057	483,390	1,140,794,059
With DSM Plan 2	1,709,520	476,946	1,135,120,409

An initial glance at this table shows that, regardless of which resource plan one may be looking at, there are significant differences in the magnitude of the values between the SO_2, NO_X, and CO_2 columns. The CO_2 values are almost three orders of magnitude (i.e., about 1,000 times) greater than the SO_2 values, and the SO_2 values, while not an order of magnitude higher, are more than three times larger than the NO_X values.

Upon reflection, the differences in the magnitude of the values shown in this table for the three types of air emissions are expected. Recall that in Table 5.1, the emission rates for each of the three emission types (in pounds of emission per mmBTU) for each of the Supply options was shown. By examining the emission rates for SO_2 and CO_2 for only the new CC unit, we see the following:

$$SO_2 = 0.006\ \text{lb/mmBTU}$$

$$CO_2 = 119\ \text{lb/mmBTU}$$

From looking only at these emission rates for this one type of generating unit (a CC unit) that essentially burns one type of fuel (natural gas), we see that there are significant differences in the various magnitudes of this one generating unit's emission rates. For example, the CO_2 emission rate is more than 1,000 times greater than the SO_2 emission rate so it is not surprising that the differences in system total emissions for these two types of emissions follows a similar pattern. Other significant differences exist (and often in reverse direction in the case of SO_2 and NO_X emissions) for other types of generating units and different fuels. This look at the emission rates for one type of generating unit burning one type of fuel (natural gas) is provided for illustrative purposes to help indicate why there are significant differences in the utility system emission totals shown in Table 7.7 above.

Now we turn our attention to a result in the table that is even more interesting. A closer examination of the emission total presented in Table 7.7 reveals two results that may be counterintuitive at first glance. First, there is no one resource plan that results in the lowest system emissions for all three types of emissions. Second, the resource plan that results in the lowest system emissions for two of the three types of emissions, SO_2 and NO_X, is

TABLE 7.8

Ranking of System Air Emissions by Resource Plan

Resource Plan	SO_2	NO_X	CO_2
Supply Only Plan 1 (CC)	1st	1st	2nd
Supply Only Plan 2 (CT)	5th	5th	5th
Supply Only Plan 3 (PV)	4th	4th	4th
With DSM Plan 1	3rd	3rd	3rd
With DSM Plan 2	2nd	2nd	1st

1st = lowest emissions; 5th = highest emissions.

Supply Only Resource Plan 1 (CC) which features a new generating unit (a CC unit) that will operate 80% of the hours per year.

In order to see these two points more clearly, we turn our attention to Table 7.8 that transforms the numerical emission ton values in Table 7.7 into a ranking of total system emissions by type of air emission. The designation of "1st" indicates the resource plan with the lowest system emissions for that type of emission, continuing to the designation of "5th" which indicates the resource plan with the highest system emissions.

When faced with these results, some readers may be inclined to offer up their best Gary Coleman (the actor) character impression: "Willis, what you talking about?"

This is not an uncommon reaction. The idea that a resource plan (Supply Only Resource Plan 1 (CC)) that features a large new generating unit that operates most of the hours in a year, can result in lower system air emissions than a resource plan that features either a renewable energy generator (Supply Only Resource Plan 3 (PV)) or DSM (in the two With DSM resource plans), is a result that may seem counterintuitive to some readers. In fact, this possibility may never have occurred to those readers when they hear about energy efficiency and/or renewable energy options being "clean" or "green" resource options. The conclusion that one is tempted to leap to is that anything touted as "clean" or "green" must result in a lowering of all types of air emissions. As we can see from the analyses of our utility system example, this conclusion may well be wrong.

From experiences gained in previous discussions regarding this topic, I believe that some readers will find these results counterintuitive, while other readers will readily understand how the results presented in Table 7.8 could have occurred after accounting for a number of factors such as: the greater efficiency of the new generating units, the lower emission rates of the new generating units compared to the emission rates of the utility's existing generating units, the relative capacity factors of the resource options, etc.

(For those readers who do understand why these results make sense, no explanation is needed. For the other group of readers, an explanation is necessary. In either case, an explanation of why results such as these may occur on a utility system takes a bit of time. Rather than provide that explanation here and take a chance of disrupting the flow of the narrative, I refer readers who need/wish to read an explanation to refer to Appendix G. This appendix provides an explanation of why results such as these can occur using two different examples.)

Our examination of air emissions for our utility system has revealed insights into how resource options will impact a utility system's total air emissions. However, the importance of those insights goes far beyond the issue of air emissions which is being discussed at the moment. The importance of those insights is summarized in my Fundamental Principle #4 which applies to *all* aspects of resource planning, not just in regard to system emissions:

> **Fundamental Principle #4 of Electric Utility Resource Planning: "Always Ask Yourself: 'Compared to What?'"**
>
> **In all aspects of resource planning, one must always ask the question "compared to what" when one is analyzing a particular resource option.**

For example, the fact that resource option A is projected to either lower the number of kWh served by a utility (such as DSM options), or to produce kWh without burning fossil fuel (such as renewable energy options), is no guarantee that resource option A will actually result in lower total system air emissions (or lower system fuel usage, etc.) for a specific utility. The key point is "what is resource option A being compared to?" Although resource option A may reduce air emissions from what the emissions otherwise would be *if no other resource option were added or considered*, there may be a resource option B that would result in even lower system air emissions.* (In such a case, the selection of resource option A would actually *increase* system emissions compared to the selection of resource option B.) Therefore, in order to know which resource option will really result in lower system air emissions, all applicable resource options should be evaluated. In regard to system air emissions, the answer may vary from one type of air emission to another.

Our utility system has performed just such an analysis that examined all resource options available to it. The results from the analyses of system air emissions, combined with the economic analysis results and the results of the analysis of economic cross over times and system fuel diversity profiles, provides comprehensive information that our utility system can now use to make its final resource decision.

---

* In addition, there are many circumstances in which "doing nothing"; i.e., considering no other resource option, is not a viable option. Examples of such circumstances include: changes needed to address new environmental regulations, resource additions needed to meet growing load, etc.

Summary of Results from the Resource Option Analyses for Our Utility System

In its decision-making process, our utility first summarizes the results of the analyses it has conducted. Its summary consists of the following points:

1. From a system reliability perspective, all five resource plans will enable the utility system to meet its reliability (i.e., reserve margin) criterion by the decision year. Consequently, all five resource plans are judged to be acceptable from a system reliability perspective.
2. In regard to economics, the With DSM Resource Plan 1 is projected to result in the lowest electric rates for our utility's customers. By comparison, all other resource plans will result in electric rates higher than would be experienced with the With DSM Resource Plan 1 (i.e., these other four resource plans would result in electric rate increases for the utility's customers compared to the selection of the With DSM Resource Plan 1). In addition, the "cross over" time to when the With DSM Resource Plan 1 becomes the best economic plan is very short: Current Year+6, which is only 1 year after our utility needs to add a new resource option. Consequently, the cross over time is acceptable.
3. From a system fuel perspective, two points are evident. First, the costs of system fuel use have already been accounted for in the economic analyses. Second, the fuel use profile analysis for each of the five resource plans does not show significant differences in the relative percentages of energy by fuel type. In particular, none of the resource plans will result in an overdependence on one type of fuel. Therefore, there are no significant differences between the five resource plans in regard to system fuel-based issues regarding the volatility of fuel costs or the reliability of fuel supply. Consequently, all five resource plans are acceptable in this regard.
4. Two points are also evident in regard to a system air emission's (SO_2, NO_X, and CO_2) perspective. First, the costs of system air emissions have already been accounted for in the economic analyses. Second, no one resource plan emerged as the plan with the lowest emissions for all three types of emissions. The Supply Only Resource Plan 1 (CC) is projected to be the resource plan with the lowest SO_2 and NO_X system emissions, while the With DSM Resource Plan 2 is projected to be the resource plan with the lowest CO_2 system emissions.

Our utility further summarizes this information in tabular form in Table 7.9.

Based on this information, our utility system selects the With DSM Resource Plan 1 as the best resource option with which it will meet its resource needs

TABLE 7.9

Summary of Rankings of Resource Plans

Resource Plan	System Reliability (Meeting Reserve Margin Criterion by the Decision Year)	System Economics (Levelized Electric Rate)	System Fuel Diversity (Fuel Use Profile)	System Air Emissions		
				SO_2	NO_X	CO_2
Supply Only Plan 1 (CC)	Acceptable	2nd	Acceptable	1st	1st	2nd
Supply Only Plan 2 (CT)	Acceptable	3rd	Acceptable	5th	5th	5th
Supply Only Plan 3 (PV)	Acceptable	5th	Acceptable	4th	4th	4th
With DSM Plan 1	Acceptable	1st[a]	Acceptable	3rd	3rd	3rd
With DSM Plan 2	Acceptable	4th	Acceptable	2nd	2nd	1st

1st = best in category; 5th = worst in category.

[a] Time for With DSM Plan 1 to become the most economic plan is acceptable.

5 years in the future. This resource plan is acceptable in regard to both system reliability and system fuel use profile, places the utility in reasonable position (i.e., 3rd ranking in regard to system emissions) in case environmental regulations were to unexpectedly tighten, and, most importantly of all, will accomplish all of the aforementioned items while serving our utility's customers with the lowest levelized system average electric rates.

As a backup resource plan, our utility would select the Supply Only Resource Plan 1 (CC). It provides a bit more protection in regard to the possibility of a tightening in environmental regulations (i.e., it has rankings of 1st, 1st, and 2nd in regard to system air emissions), but does so with higher levelized system average electric rates for the utility system's customers than does the With DSM Resource Plan 1.*

The remaining three resource plans; the Supply Only Resource Plan 2 (CT), Supply Only Resource Plan 3 (PV), and the With DSM Resource Plan 2, are clearly less desirable choices from an economics-only perspective, or from a combined economic and system emission perspective. The utility system clearly has two better overall choices than any of these three remaining resource plans.

Therefore, our utility system has completed its resource planning work. It has conducted a comprehensive integrated resource planning (IRP) analysis

* If such a tightening of environmental regulations were to occur, it is possible that the utility would have to take certain actions (install more pollution control equipment, perform fuel switching, etc.) for any of the five plans. The economic advantage of the With DSM Resource Plan 1 allows our utility more "room" or "cushion" economically to make such changes because it would be starting from the position of the lowest electric rates for its customers.

and it has made its selection of how it will meet its future resource needs. You may well be asking: "What is left for us to discuss?"

I'm glad you asked. We will next discuss in qualitative terms (yes, that's correct—no more numbers for awhile; you are welcome) various issues that could have complicated the resource planning analyses our hypothetical utility system has just completed, and which will complicate the resource planning efforts of many utilities in the future.

8

Are We Done Yet? Other Factors That Can (and Will) Complicate Resource Planning Analyses

Constraints on Solutions: Six Examples

We ended the previous chapter by stating that we would now discuss various issues that could have further complicated the resource planning analyses our hypothetical utility system has just completed, and which will complicate the resource planning efforts of many utilities in the future. In fact, the issues we will discuss are routinely addressed by many utilities today in their resource planning. Not all issues apply to every utility, but at least some of these issues are faced, to one degree or another, by virtually all electric utilities.

From the perspective of a utility resource planner, these issues can be thought of as "constraints" that must be dealt with as the resource planner attempts to determine in his or her analyses a solution to the question of which resource option(s) is best for his or her specific utility. There are a number of such constraints, or types of constraints, that a particular utility may face. These may include: geographic constraints, regulatory/legislative constraints, self-imposed constraints, financial constraints, etc. For purposes of our discussion, we will focus on six specific constraints or types of constraints. We shall first place each of these six constraints into one of three general categories.

These three general categories of constraints are ones we label as: (i) "absolute" constraints (i.e., types of constraints that have existed for many years and over which the utility has little/no direct control); (ii) legislative/regulatory-imposed constraints; and (iii) utility-imposed constraints. We shall first list two examples each of constraints that could be placed into the three categories of constraints. Then we shall examine each type of constraint in more depth.

The six constraints that we shall discuss, and the three categories into which we shall put these constraints, are as follows:

1. "Absolute" constraints:
    a. Siting/geographic constraints
    b. Tightening of environmental regulations

2. Legislative/regulatory-imposed constraints:
   a. Imposition of "standards"/quotas for specific resource options
   b. Prohibition of specific resource options
3. Utility-imposed constraints:
   a. System reliability constraints on specific resource options
   b. Load shape constraints

Examples of "Absolute" Constraints

Siting/Geographic Constraints

This first type of constraint, siting/geographic constraints, almost exclusively affects Supply options. This type of constraint has been a factor in utility resource planning almost since the first day that the electric utility industry emerged and the industry began to build new generating units.

A number of considerations come into play regarding where new generating units of various types could be sited in a utility's service territory. A partial list of the considerations that come into play as a utility examines a potential site for a new generating unit includes the following:

- Proximity of the site to existing transmission lines;
- Proximity of the site to the utility's "load center";
- Proximity of the site to adequate water resources for operation of the generating unit;
- Proximity of the site to existing natural gas pipelines, railroad lines, navigable waters/ports, etc., in regard to fuel transportation; and,
- Proximity of the site to non-attainment air emission zones.

The first consideration, the proximity of the site to existing transmission lines, takes into account that the energy produced by any new generating unit must be carried over electrical lines away from the plant site before it can be delivered to customers. All else equal, a potential site is much more attractive from an economic perspective if the site is close to existing transmission lines.

For example, assume that a utility has two potential sites. Site A is 10 miles from an existing transmission line and Site B is 25 miles from the same existing transmission line. In either case, a new transmission line will need to be built that connects the new generating site with the existing transmission line. Assuming a rough cost of $1 million to $2 million per mile for building a new transmission line that will connect with the existing transmission line, the utility's costs to build this new transmission line will be $10 million to $20 million if Site A is chosen, or will be $25 million to $50 million if Site B is chosen. Assuming all else equal, the utility will likely choose Site A.

This consideration is one that is common to all types of generating units. However, it is one that can be especially relevant in regard to certain renewable energy generators. Wind energy generators are a good example because the geographic region of higher winds is often remote from the utility's load center (perhaps because people found it less desirable to live in, much less play golf in, areas with frequent high winds). Thus, new transmission lines that cover a great distance may need to be built in order to access the wind resource and bring the wind-generated electrical energy to existing transmission lines.

The second consideration, the proximity of the site to the center of the utility's load center, addresses how far the site being considered for a new generating unit is from where the majority of the utility's customers are (i.e., the utility's "load center"). The greater the distance is between the site and the load center, the further the energy produced by the new plant has to travel before it can be delivered to customers.

This distance to the load center may affect the cost to build the new transmission lines (as just discussed), and will definitely affect the magnitude of electrical losses that occur as electricity travels along the electrical lines from the generating plant to the customers.* The greater these losses are, the less energy the generating unit is able to actually deliver to the utility's customers.

The third consideration is the proximity of the site to adequate water resources. Sufficient quantity and quality of water must be available for operating the chosen generating unit. These requirements may vary considerably depending upon the type of generating unit in question.

The fourth consideration, the proximity of the site to existing natural gas pipelines, railroad lines, navigable waters/ports, etc., addresses the fact that most types of new generating units (except for renewable energy facilities such as solar and wind) will typically require delivery of fuel to the plant site. Depending on the type of generating unit being considered, the fuel may be delivered by a number of delivery "avenues" such as pipeline, rail, truck, or barge/ship.

The lack of existing delivery systems or avenues near a potential site may eliminate one or more type of generating units for that site because the fuel which that type of plant will require cannot be delivered by existing pipelines, rail lines, etc., and because the cost to develop new delivery systems to serve the site is prohibitive. In certain cases, the potential plant site

* These losses are typically referred to as "line losses." In a very rough sense, one can think of them almost as a type of electrical "friction" loss as electricity travels a long distance through electrical wires. The amount of the line losses will vary from one utility to another, but losses in the range of 6% to 9% are fairly typical. In general, the longer that electricity must travel, the greater the line losses are. Obviously, the lower the line loss value, the more efficient the utility system is and the lower its costs (assuming all else equal).

may be completely eliminated because no type of fuel can be delivered at a reasonable cost.

The fifth consideration we will address, the proximity of the site to non-attainment air emission zones, recognizes that certain federal and/or state regulations have designated certain geographic areas as zones in which air emission standards either are not being met or are in danger of not being met. Consequently, a particular type of new generating unit may not be able to be permitted for a potential site if the projected air emission output from that unit's operation will result in a failure to meet air emission standards in the zone. (However, other types of new generating units may still be acceptable for the site.)

How do utility resource planners attempt to address these considerations? For illustrative purposes, think of the general approach as one in which a utility starts with computer-generated maps of their service territory. One map may show the general geography of the area including rivers, roads, etc. The next map may show the electrical load or population centers. A subsequent map may show the location of existing transmission lines. Another map may show the location of existing natural gas pipelines, etc. This process continues until maps that present all of the relevant considerations for the utility in regard to siting new generating units have been obtained.

Then these maps are (figuratively) laid on top of each other to develop a single composite map that includes all of this information. From this composite map, the utility can begin to locate promising potential sites for new generating units of various types. Once these potential sites are identified, the utility can begin to develop estimates for each site of the costs to build and operate types of generating units. In addition, the utility will develop estimates of the time it will take to obtain all of the necessary permits that will be needed for the new units at each site. In this way, the utility identifies the types of new generating units that can be built at specific sites. The utility can then estimate the costs associated with building a specific type of generating unit at each site.

Potential Tightening of Environmental Regulations*

At the end of Chapter 7, our utility system considered the projected levels of system air emissions for each resource plan in the course of making a decision regarding which resource option was the best selection for meeting its resource needs. This examination of system air emissions is, in part, a

* This constraint could have been placed in the grouping that will be discussed next: the "legislative/regulatory-imposed" constraint grouping. I have chosen not to do so because I view the tightening of environmental regulations constraint as being primarily based on attempts to reduce environmental impacts from utility operation of existing generating units. Conversely, the two constraints that will be discussed in the "legislative/regulatory-imposed" grouping are ones that I view as being primarily based on forcing utility resource planning into resource options that are "favored" by the legislature or regulators.

way to judge the utility's risk level in light of a potential tightening of environmental regulations, the second constraint that we shall discuss. Such a tightening of environmental regulations will likely alter the operation of the utility system as a whole. The costs of that altered operation are passed on to the utility's customers through higher electric rates (just as are all other prudently imposed costs of utility system operation).

By "tightening" of environmental regulations, we refer both to the imposition of more restrictive standards for air emissions currently being regulated, and to new regulations of emissions (or other environmental concerns) not currently being regulated.

The issue of potential tightening of environmental regulations is typically considered by utilities when they make a resource option decision. Resource options that the utility is now considering to meet a future resource option will not be placed into service immediately, but will become operational in a future year of resource need. (Recall that our utility system is placing a resource option in service in Current Year + 5, or 5 years in the future from its Current Year.) In addition, most resource options will have an operating life of many years. For example, a new generating unit will typically have an expected operational life of at least 25 years.

Therefore, a resource decision that is made today may not begin to impact the utility system for a number of years due to the length of time it takes to construct or acquire the resource option, but the decision will certainly impact the utility system for a number of years or decades after the resource is constructed or acquired.

Utilities must make decisions about resource options that will operate in future years, and they must do so recognizing that there is uncertainty about potential environmental regulations that the utility, and its resource options, will face. Utilities are also mindful that the overall trend of environmental regulations is generally one of continually tightening regulations.

Of particular concern to utility resource planning is the potential emergence of new environmental regulations that address an entirely new type of emission that has not been regulated in the past. Relatively speaking, utility resource planners find that an incremental tightening of an air emission (for example, SO_2 or NO_X) that is already being regulated is fairly easy to address. This is true, in part, because manufacturers of generating equipment, pollution-control equipment, etc., anticipate that such tightening of currently regulated emissions will occur and they guide their development work accordingly.*

---

* This is almost a "chicken or the egg" situation. Environmental regulations frequently become tighter as soon as it becomes clear that emission control technology has advanced to the point where it is now technologically possible to lower emissions. Thus technology advancements and tighter regulations often move in lockstep fashion. One may wonder if decisions on whether environmental regulations should be tightened are based less on analyses of whether the regulations *should* be tightened, than on whether it is technologically *possible* to meet tighter standards.

However, the emergence of new, or even of potential new, regulations for air emissions (or other concerns) that are not currently regulated is more problematic in utility resource planning. This situation greatly increases the uncertainty that utility resource planners must consider or address in two ways.

First, there is uncertainty as to whether there will even be regulation of this type and, if so, when the regulations will be in place. Second, there is additional uncertainty regarding the eventual costs of complying with any such regulation. Only when regulations are actually in place will this uncertainty be entirely removed or minimized.

A good example of this is the potential regulation of so-called greenhouse gas (GHG) emissions, including CO_2. At the time this book is written, there is no federal regulation of GHG emissions in the United States, despite the fact that the possibility of GHG emission regulation has been seriously discussed for at least two decades.*

During this time, utility resource planners and utility regulators have wrestled with whether, and how, to address potential GHG emission regulation as decisions are made about resource options. A frequently used approach is to consider various scenarios of compliance costs with GHG emissions (often presented as CO_2 allowance costs).

Such a scenario approach can certainly utilize a wide range of GHG compliance costs in the analyses of resource options. If the utility finds that one resource option emerges from its analyses as the best resource option under all, or most, of the scenarios, there should be little debate regarding the selection of this resource option as the best selection for that utility.

However, a utility may find that Resource Option A is the best solution for a scenario with high GHG compliance cost projections, Resource Option B is the best solution for other scenarios with medium GHG compliance cost projections, and Resource Option C is the best solution for still other scenarios with lower GHG compliance cost projections. In such a situation, the utility is often faced with a dilemma in regards to which scenario does the utility believe is the "most likely" one to occur as it must select a single resource option as the best solution for its future resource needs?

Then the question becomes whether the GHG compliance cost scenario the utility selects as the most likely scenario is also perceived by the utility's regulators as the most likely scenario. The two parties can reach different conclusions. In such a case, there will likely be disagreement regarding which resource option is in the best interest of the utility's customers. And, unfortunately, the answer as to whether the best overall resource option was chosen is often not known for many years.

* To further point out the problems which utility resource planners face regarding uncertainty with potential environmental regulation, recall that during the 1970s there was considerable concern over global *cooling*. The possibility of new environmental regulations to address this began to be discussed. In this case, increased CO_2 emissions were considered as one possible way to address this perceived problem.

In future years, with the benefit of hindsight, once the question of potential environmental regulation for a specific type of emission is finally settled, a review of that particular resource option decision may show that the utility's customers were harmed by the decision. The customers may be harmed in several ways. Customers can be harmed if a resource option was chosen based on a scenario of passage of a specific environmental regulation, or of a specific projected level of environmental regulation compliance costs, but the regulation never occurred or the actual level of environmental compliance costs is considerably different than what was projected. Such an outcome, realized only with the benefit of hindsight, could show that the selected resource option was not the best economic choice. As a consequence, the selected resource option is likely to have unnecessarily increased electric rates for the utility's customers.

Another way the utility's customers may be harmed is by selecting a resource option that results in greater reduction of one type of emission that *might* be regulated in the future, but increases system emissions of other types of emissions that are currently being regulated. (In fact, we saw such an example in the analyses for our utility system in which one resource plan resulted in the lowest CO_2 system emissions, but resulted in higher SO_2 and NO_X system emissions than another resource plan.)

Therefore, the potential for tightening of environmental regulations is truly problematic for utility resource planners. Unfortunately, this problem has existed for many years, and it is unlikely to disappear from resource planners' radar screens at any time in the foreseeable future.

Examples of Legislative/Regulatory-Imposed Constraints

"Standards"/(Quotas) for Specific Types of Resource Options

The third constraint is the use of so-called "standards" for specific types of resource options. These standards are actually quotas for selected types of resource options.* At the time this book is written, such "standards"/quotas are currently being directed at two types of resource options: (i) renewable energy generation options and (ii) DSM options. We will first take a look at "standards"/quotas for renewable energy generation options.

The use of "standards"/quotas for renewable energy options in utility resource planning is commonly referred to at the time this book is written as "renewable portfolio standards" or RPS. The basic rationale behind the introduction of RPS is that renewable energy-based resource options are not

* Because these "standards" are actually quotas, I will refer to them as "standards"/quotas to remind the readers that we are discussing quotas imposed by governmental actions.

typically being selected in great volume by utilities as they determine what the best resource options are for their individual utility systems.* Proponents of renewable energy resource options offer a couple of reasons why these options should be utilized more. Essentially, they argue that greater implementation of renewable energy options will reduce the United States' use of fossil fuels to produce electricity. This, in turn, is supposed to reduce utility system air emissions and the United States' dependence on imported fuel, particularly imported oil.

However, as our earlier evaluation of resource options for our utility system showed, this argument may not hold up when one asks the meaningful question of "lower air emissions and oil usage *compared to what other resource options*?"

Nevertheless, as this book is written, roughly half of the states in the United States have some form of RPS regulation in effect (and RPS regulation is also being considered by the federal government). A wave of state RPS regulations appeared in the mid-2000s after a national conference of state governors was held that focused on energy issues as a primary topic. Almost immediately, numerous states began to issue RPS regulations. Many of these RPS regulations have been imposed without the benefit of IRP-type analyses to determine what the full range of projected impacts for the utilities (and their customers) will really be. Therefore, the potential benefits (and, to some extent, costs of complying with the regulations) that accompanied the introduction of RPS regulations were often speculative, at best, when the initial RPS regulations were issued.

The individual state RPS regulations largely took a form that suggests that the regulations were designed mostly with a catchy marketing slogan in mind. For example, a number of states issued regulations that touted "15% by 15" (i.e., 15% of the annual energy produced by electric utilities will be produced from renewable energy sources by 2015) or "20% by 20" (i.e., 20% of the annual energy produced by electric utilities will be produced from renewable energy sources by 2020—you get the picture). Such slogans are indeed catchy and looked good in the press.

In addition, the RPS "standards"/quotas frequently have more than one layer of quotas. For example, an RPS quota may call for 15% of the utility's annual energy production to be produced by renewable energy sources by 2015. In addition, the RPS regulation may also require that, of the 15% amount, one-third (or 5% of the annual energy production of the utility) is to be produced by a specific type (or combination) of renewable energy source such as wind energy, solar thermal, and/or PV. Thus, apparently operating

* The results of the resource option evaluation work for our utility system in earlier chapters provides some insight into why certain renewable energy resource options are not typically being selected by utilities in their resource planning work. The resource plan featuring the PV option ranked very low in regard to both economics and system air emissions when competing head-to-head with the other resource options.

on the theory that "more is better," there are often quotas-within-quotas in RPS regulations.

However, just as with the case regarding whether an RPS regulation should even be implemented, relatively little rigorous analyses may be performed to determine the "best" value to set as a quotas-within-quotas for specific renewable energy sources' contribution to annual energy production. Such analyses may have proved problematic in regard to actually justifying setting any quota when viewed from the perspective of a utility's customers and the utility's electric rates. Again, recall the analyses we conducted for our utility system. Our analyses of a specific PV option for our utility system, even with the use of an unrealistically optimistic assumption regarding what portions of this PV option's output could be considered as firm capacity, and how much energy could be expected to be produced from year-to-year, showed that adding this renewable energy option to our utility system was not attractive. The results of our analyses showed that higher electric rates, and higher system air emissions, for our utility's customers would be the outcome if this PV option was selected.

The legislators and regulators across the United States who have imposed RPS regulations in various states recognized this reality of higher electric rates. Consequently, they have generally sought to protect electric customers, to a degree, by including in the RPS regulation a form of a "safeguard" against electric rates that are "too much higher." This safeguard usually takes the form of a "cap" on annual expenditures for renewable energy generation. Such a cap can take the form of an edict that utilities can (or should) stop spending for renewable energy generation in a given year once these expenditures equal a set percentage (such as 2% or 5%) of the utility's annual revenue requirements.*

The theory is that the use of an annual cap on expenditures, with an RPS "standard"/quota, will allow the implementation of renewable energy sources with an "acceptable level" of economic harm to electric utility customers (i.e., electric rates will definitely increase, but will not increase beyond a certain level set by the legislators/regulators). However, the reality of combining an expenditure cap with an RPS "standard"/quota is that one of two things will likely happen.

First, if the expenditure cap does a reasonably adequate job of protecting utility customers from higher electric rates, the annual amount of energy produced from renewable energy sources may be less (perhaps much less) than the amount called for by the RPS. In other words, the RPS "standard"/ quota may not be met. Second, the only way that the amount of energy called for by the quota value will be met may be to relax the expenditure cap safeguard. This would result in greater costs being passed on to utility

* Although not often stated in these terms (probably due to public relations concerns), this edict to spend money on renewable energy options up to the expenditure "cap" is an edict to raise electric rates for utility customers by the same percentage as the cap percentage.

customers through electric rates that are higher than called for by the cap. Stated another way, the cap is relaxed or ignored.

In either case, at least one of the two objectives of the RPS "standard"/quota and expenditure cap approach may not be met. One either implements less renewable energy than is called for by the RPS "standard"/quota (but the utility's electric rates still increase up to the cap level), or one meets the renewable energy "standard"/quota and the utility's electric rates increase beyond the amount called for in the original expenditure cap.

When RPS "standards"/quotas are looked at in this way, such "standards"/quotas do seem to be a strange departure from the concept of integrated resource planning (IRP) for electric utilities. As previously mentioned in Chapter 3, a key IRP principle is to ensure that all resource options *compete* on a level playing field in order to determine which resource option(s) are best for an individual utility. By comparison, the use of RPS "standards"/quotas *seeks to avoid competition* and force the selection of one type of resource option: renewable energy options. This is clearly a departure from, or an abandonment of, IRP principles.

We earlier mentioned that another type of resource option, DSM, has been the recipient of a "standard"/quota approach imposed by legislators and/or regulators. This approach was first introduced in the early 1990s as a way to force certain utilities to seriously consider DSM in their resource planning work. This approach was sidelined to an extent when a number of states and utilities moved away from a traditional, regulated electric utility structure to a competitive, "unregulated" market structure. As part of this move, DSM became less of a factor for utilities which were now in this new market structure.

Other states and utilities retained the traditional, regulated utility market structure. If such a state or utility continued with an IRP approach to utility planning, then DSM was automatically considered on a level playing field with Supply options. Most importantly, if the state adopted DSM "standards"/quotas, but set those DSM "standards"/quotas based solely on the outcome of the utility's IRP work, the introduction of DSM "standards"/quotas caused little (if any) harm to the utility's customers. In other words, this logical type of DSM "standard"/quota basically mandates the utility to use its IRP process to (i) first determine how much DSM might be needed to fully meet the utility's projected future resource needs, and (ii) then determine how much of that amount was actually cost-effective to utilize.

Such an approach ensures that the utility utilizes only the amount of resource options (DSM and/or Supply) that is actually needed to meet future resource needs and is cost-effective. Therefore, the analyses we conducted in previous chapters for our utility system would have been the same regardless of whether our utility system was facing a state-imposed DSM "standard"/quota or not, as long as the DSM "standard"/quota was based on the utility's IRP process.

For example, our utility system's projected resource needs were first identified (100 MW if the resource need is met solely by DSM), then DSM options (along with Supply options) were analyzed to determine how much DSM was cost-effective to add. In the analyses, 100 MW of DSM Option 1 were selected as the most cost-effective resource option for the utility's customers. As long as our utility's regulators set a DSM "standard"/quota based on our utility's IRP analyses, the utility's customers will not be harmed. However, if another DSM option had been mandated, and/or a greater amount of DSM had been mandated, by a different type of DSM "standard"/quota, then our utility's customers would have been harmed due to the resulting higher electric rates.

At the time this book is written, there are concerted efforts in a number of states to alter an IRP-based approach to DSM "standards"/quotas. These efforts seek to replace this IRP- or competition-based approach to DSM regulation with a set of largely arbitrary percentage values similar to an RPS "standard"/quota approach. As discussed earlier, in the RPS "standard"/quota approach, an arbitrary percentage of the utility's total annual energy production must be produced by renewable energy sources.

The efforts to change the DSM "standard"/quota approach frequently involve an attempt to move in an opposite direction.* This type of DSM "standard"/quota approach typically requires that the utility *reduce* its annual energy production by an arbitrary percentage (such as 2% to 5%) instead of *producing* an arbitrary percentage of energy as with RPS regulations. This would result in the amount of energy the utility sold to be lowered, year after year, through the implementation of DSM.

Consequently, the type of DSM options that would be most favored under this type of DSM quota would be those DSM options that have the highest kWh reduction characteristics. In other words, such a DSM "standard"/quota system would move a utility system to implement DSM options that are more similar to DSM Option 2 that was analyzed in Chapters 6 and 7 for our utility system than DSM Option 1. However, as you may recall, DSM Option 2 was demonstrated *not* to be one of the better resource options that our utility system could have selected. In fact, this resource option was among the worst resource options in regard to how much it would increase electric rates for our utility's customers.

Furthermore, such a DSM "standard"/quota system could require that a much greater amount of DSM options similar to DSM Option 2 would be implemented *than what is actually needed to meet the utility's projected future resource needs*. This is because this type of DSM "standard"/quota is focused solely on energy (MWh) reduction, not on the demand (MW) reduction aspect of DSM that actually avoids the need for new power plants. This type of DSM "standard"/quota that focuses on energy (MWh) reduction can end up with an associated amount of demand (MW) reduction that is greater than what is needed to meet the utility's resource needs. This greater-than-needed amount of a less desirable type of DSM options results in *even greater increases* in the utility electric rates.

* Such DSM-based quotas are often referred to as DSM "goals" or "energy efficiency resource standards (EERS)."

This prescriptive approach to DSM options is, like a prescriptive RPS approach to renewable energy options, also a departure from, or abandonment of, IRP's key principle of requiring all resource options to compete with each other to earn a role in the utility's resource plan. Instead, this prescriptive approach ignores this underlying IRP principle and mandates that a certain (usually arbitrarily selected) amount of DSM is to be included.

Later, at the end of this chapter, we will return to these RPS and DSM "standards"/quota constraints, and other constraints that we will soon address, to further discuss the impacts that these constraints typically have on a utility system and its customers. In the meantime, we will now move on to several other constraints that affect utility resource planning.

Prohibition of Specific Resource Options

The fourth constraint involves either a direct or de facto ban on specific types of resource options. State legislators or regulators may impose an outright ban on building a certain type of generating unit. Another form of effective prohibition can occur absent such direct legislation/regulations. In this case, utilities may find that one or more permitting agencies in their state consistently rule against applications to authorize the building of a new generating unit of a particular type.

The types of resource options in question are typically Supply options that would otherwise operate with high capacity factors (i.e., baseload generating units). At the time this book is written, the type of generating unit that is most affected by some form of a ban is coal units.

The state of Florida's recent experience in the middle of the 2001–2010 decade is a good example of this. Growing concerns over the increasing dependence of the state on natural gas as a utility fuel, caused by the introduction of a number of new natural gas-fired generating units in the late 1990s and early 2000s, resulted in a clear message being sent from the Florida Public Service Commission (FPSC) to the state's utilities to begin to add new non-gas-fired generating units. The utilities had also recognized that they were becoming quite dependent upon natural gas and they were already evaluating non-gas-fired resource options on their own.

Because new nuclear units would take at least a decade to permit and build, and facing steady demand growth that resulted in projections of significant future resource needs in much less time than the 10 years or more it would take to add new nuclear units, Florida utilities turned their attention to advanced coal generation technologies. By 2006–2007, several electric utilities in Florida petitioned the FPSC for authorization to build advanced technology coal units.

However, by that time, Florida had a new governor. The new governor was concerned about GHG emissions to the point that he issued an Executive Order that laid out goals for reducing state GHG emissions to certain levels by 2017. Also about that time, a number of analyses emerged from various sources around the country that showed a wide range of compliance costs for GHG emissions that might occur if federal GHG emission regulation were to occur. A projection of high potential compliance costs for GHG emissions would be detrimental to coal units because the CO_2 emission rate for coal units is higher than the CO_2 emission rate for any other type of new Supply option.

Therefore, despite the previous concerns over Florida's high dependence on natural gas, by 2006–2007 the political winds had changed regarding the desirability of adding new coal-fired generating units. As a consequence, a couple of Florida utilities' petitions to the FPSC for authorization to build new coal units were denied due, in large part, to uncertainty over potential future GHG compliance costs.* In addition, another utility's petition for a new coal unit that had previously been approved by the FPSC was effectively blocked in the environmental permitting process. These actions constituted a de facto ban on new coal units in Florida that continues as this book is written.†

Examples of direct prohibition of specific types of generating units, including non-coal units, also exist. One example occurred several decades ago via a federal ban on the use of natural gas in all new generating units (other than peaking units). This ban was put in place primarily due to concerns over the projected availability of domestic natural gas supplies. (This federal ban was subsequently lifted as the projections of natural gas availability changed to show that significant quantities of natural gas were available.) Other examples of direct prohibition of specific types of generating units have focused on new nuclear units. As this book is written, the states of Minnesota and Wisconsin, for example, effectively prohibit serious consideration of new nuclear units.

We shall return to this prohibition type of constraint at the end of this chapter when we shall further discuss the impacts of various constraints on utility resource planning and utility customers.

Examples of Utility-Imposed Constraints

System Reliability Constraints

We just looked at a pair of constraints that are imposed by governmental actions and which require that a utility react to the constraints. We now turn our attention to a pair of constraints that may be self-imposed by utilities themselves.

* In fairness to the FPSC's decision, the utility analyses presented scenarios that showed that if GHG regulations were instituted, and if the costs of complying with these regulations were high, then the advanced technology coal units were not projected to be cost-effective.

† As a consequence of this, more gas-fired generating units have been added and the state's dependence upon natural gas has increased.

The fifth constraint we examine is a system reliability constraint that the utility may impose upon itself in its resource planning work. This system reliability constraint also affects specific types of resource options. However, while the prohibition constraint, for example, typically affects baseload Supply options, this self-imposed reliability constraint typically affects DSM and renewable options.

Recall from the discussion in Chapter 3 that utilities perform reliability analyses to ensure that they are able to continue to supply their customers with reliable electric service. As part of these reliability analyses, DSM and renewable energy options deserve special attention because DSM and renewable energy options have certain characteristics that differ from the characteristics of more conventional Supply options.

From a system reliability perspective, these key characteristics are: (i) the voluntary nature of most DSM options, and (ii) the intermittent nature of a number of renewable energy sources (such as wind and solar).

The Voluntary Nature of DSM Options

All DSM options are "voluntary" in the sense that customers must voluntarily sign up to participate in the utility's DSM program. In addition, there are at least two other voluntary aspects to specific DSM options.

First, certain DSM options require a participating customer to voluntarily sign up for either a new time-differentiated (or peak load-differentiated) electric rate or to give the utility the ability to temporarily control the operation of certain electrical equipment on the customer's premises. These types of DSM programs, referred to as load management DSM programs, were discussed earlier in Chapter 6.

For purposes of this discussion, the key point is that these DSM load management programs require the *continued* voluntary participation of utility customers that have signed up for these programs. Participating customers can usually choose to drop out of these programs (particularly for programs that address residential customers) with very little notice to the utilities. Therefore, there is always a risk that a significant number of participating customers may choose to cease participating in the utility's DSM program and leave the utility with diminished resources with which to meet electrical load.

As we also discussed in Chapter 6, the other basic type of DSM program is an energy conservation (or energy efficiency) program. For this type of DSM program, a participating customer is typically paid an incentive to select a more efficient appliance, a higher level of insulation, etc. The more efficient appliance, or higher level of insulation, will then be in place for a number of years. Therefore, in contrast to a load management program in which a customer may choose at any time to cease participating in the program, once a more efficient appliance, etc., is installed in a building, it will likely be there for many years without the participating customer having to make a choice about whether to participate in the same utility program.

However, even with this type of DSM program, there is another aspect of the program to consider that can, in a sense, be considered a voluntary aspect. This voluntary aspect is that customers can choose to alter their electrical usage patterns in ways that can result in thwarting the projected kW and kWh reductions that were projected for the utility DSM program. For example, customers can decide to: lower their air conditioning thermostat in summer, raise their thermostat in winter, raise their water heater thermostat settings, etc. Such actions may be the result of these customers using some/all of the savings in their monthly bills from participating in the DSM program to "purchase more comfort."

Customers could also use the savings in their electric bill resulting from participating in the DSM program to purchase other goods/services that require electricity (such as new computers, televisions, etc.). Both of these examples are typically discussed as "rebound" effects. While there is little question that rebound exists, the magnitude of impacts of rebound are less well known.

The key point is that, because the projected kW and kWh reductions from conservation/energy efficiency types of DSM programs are subject to decisions and actions by customers over which the utility has no control, the projected impacts of DSM options on the utility system are inherently less certain than are the projected impacts of Supply options. The same is true for load management type DSM options because participating customers can choose to drop out of these programs with little notice given.*

These voluntary aspects of both types of DSM programs can lead to problems if a utility places too much reliance on DSM options to meet its system reliability criterion.

For an example of this, we again return to Florida, but this time we take a look at what happened in the late-1990s. The second largest electrical utility in the state, Florida Power Corporation (now known as Progress Energy Florida), was enjoying a very high level of success in regard to signing up its residential customers for a load control program. This very successful program offered monthly billing credits to participants for allowing the utility to remotely control the operation of certain home equipment

* In addition to the voluntary aspect of DSM programs, there is another aspect that makes their projected impacts less certain, at least in the short term, than the impacts of Supply options. That is measurement and verification (M&V) of the impact of the two types of resource options. M&V of the output of Supply options is continuous and virtually instantaneous. Conversely, M&V of most DSM options requires collecting and evaluating monitored data from relatively large numbers of utility customers. This process can take months or years to perform, particularly when monitoring impacts at the summer and winter peak hours. More than one year may be needed to gather data on seasonal peaks to ensure the data is accurate. Thus there is a time lag in a utility knowing what it is actually getting in terms of output from DSM options compared to Supply options.

such as air conditioners, space heaters, and water heaters. Florida Power Corporation signed up approximately 40% of its residential customers for this program.

At that time, the utility utilized a 15% reserve margin as one of its reliability criteria. The huge success of its residential load control program resulted in the utility deferring/avoiding the need to construct new power plants. As a consequence, this DSM program accounted for a substantial portion of the utility's ability to meet its 15% reserve margin criterion.

However, during one prolonged summer hot spell in 1998, the utility needed to control the operation of participating customers' equipment much more frequently than had been the case in prior years. In particular, the customers participating in this program had their air conditioners controlled several times a week, for a number of weeks in a row.

Load control programs that involve utility control of air conditioners and/or space heating systems frequently result in participating customers facing a type of "balancing" decision. Is the lower bill (due to monthly billing credits) worth brief periods of a slightly warmer (in summer) or cooler (in winter) home? Florida Power Corporation's residential load control program in the late-1990s was no exception.

Unfortunately, during that very hot 1998 summer, the tolerance level for warmer homes diminished considerably as the frequency of Florida Power Corporation's load control events became more frequent. As a result, large numbers of participating customers (approximately 70,000) choose to drop out of the program over a very short time. The loss of so many participating customers significantly diminished the utility's ability to lower its peak load through the load control program.* Consequently, the utility's actual reserve margin dropped almost overnight.

The result was that the utility had to quickly add new generating units to increase its reserve margin. As for the residential load control program, the utility essentially stopped signing up new participants for a number of years to ensure that its system settled into a better balance between DSM and Supply options in regard to meeting its reserve margin criterion.

The key point of this discussion of the voluntary nature of DSM options is just that: the balance between DSM and Supply options. Due to the voluntary nature of DSM programs, there is inherently more uncertainty regarding the future contribution toward utility system reliability of DSM options than there is of most Supply options. Consequently, utilities may wish to impose some form of limit on how much of their system reliability is really based on DSM. We will return to this topic after taking a look at renewable energy resource options.

---

* In addition, the utility was reluctant to continue to control the load of equipment at the remaining participants' homes due to the concern that even greater numbers of participants would drop out of the program.

*Intermittent Nature of Renewable Energy Resource Options**

Now let's turn our attention to renewable energy options. We have previously discussed the intermittent nature of a number of types of renewable energy options such as wind and solar, particularly in Chapter 5 in regard to the PV resource option that our hypothetical utility system was evaluating. At that time, for the sake of simplicity, the intermittent nature of the PV option was ignored (thus making the PV option more attractive in our analyses than would actually be the case by assuming that 100% of the PV option's output could be counted on as firm capacity at our utility's system peak hours).

However, in real life, the intermittent nature of certain renewable energy options must be addressed in a more realistic manner.† The question is: "how to deal with the intermittent aspect of those renewable energy options to which it applies, particularly from a system reliability perspective?"

Let us first refresh our memory about what we mean by the intermittent aspect of the output from certain renewable energy resource options. One cannot simply look at the "nameplate" rating (for example, 2MW) for a wind energy generator or a PV installation and assume that the full 2MW output will consistently be available at the utility's system peak hour. This is because the wind is not always blowing at the speed used to calculate the nameplate rating of the wind energy generator. Nor is the sun always shining at the intensity at which the nameplate rating for a PV module/array was calculated for.

Therefore, for reserve margin analyses, the utility cannot include the full 2MW nameplate rating value. The question then becomes: "what less-than-100% value should the utility assume for the output of these renewable energy sources in reserve margin calculations?" This is not an easy question to accurately answer. It may literally take years before such an answer is possible.

When considering the placement of renewable energy resource options such as wind and PV options at a specific site, one can begin to answer this question by gathering and analyzing data regarding either the energy source itself (i.e., the speed/duration of the wind or the amount of solar energy) at the site. Or, one could proceed with the installation, then proceed to measure the actual output of the renewable energy option.

In the first case, unless one is fortunate enough to have years of reliable energy source data for that specific site, it may take years before sufficient data can be gathered to provide accurate projections of the output of the

* Because this book is focused on resource planning, only resource planning issues with intermittent resource options are addressed. However, there are also significant operational issues associated with intermittent resource options.

† Other renewable energy options, such as biomass facilities, do not have such an intermittent aspect to the output of their facilities. They are capable of generating electricity as long as there is sufficient fuel which can be stored on-site much as oil or coal are stored on-site for oil- and coal-fired generating units.

facility on an hour-by-hour basis.* In the second case, it will likely also take a number of years before the actual output of the renewable energy facility on an hour-by-hour basis can be gathered.

For the sake of discussion, let's assume that it takes 5 years to gather this site-specific data so that one feels comfortable in knowing the range of values that can be expected at that site at the utility's summer or winter peak hour. The key point here is that there will almost certainly be *a range of values.* Furthermore, this range is likely to be a broad one, conceivably from 0% to 100% for a given technology and site. The utility will still need to select a single capacity (MW) output value (or, equivalently, a percentage of the nameplate rating) for each renewable energy installation or option for its reserve margin calculations. In reliability analyses, this can be problematic, particularly if the utility has added, or is planning to add, a significant amount of renewable energy facilities to its system.†

Now that we have introduced the voluntary and intermittent considerations for DSM and certain renewable energy options, respectively, what should a utility do about these considerations? Because the objective of reliability analyses is to ensure that the utility system can continue to be able to reliably deliver energy to its customers during all hours of the year, including peak load hours, certain utilities have developed their own constraints that are used in their resource planning work to ensure that DSM and/or renewable energy installations are not relied upon too heavily to meet peak loads.

In regard to DSM, one can apply a constraint that calculates reserve margins with and without the contribution of DSM. The first reserve margin calculation, the calculation with DSM, is the utility's official reserve margin projection. Let's assume it shows a reserve margin of 20% for a given year.

The second reserve margin calculation, the calculation without DSM, will naturally show a lower reserve margin value. Let's assume the result of this second calculation is a projected "generation-only" reserve margin of 12% assuming no DSM contribution. Therefore, DSM's contribution to the utility's reserve margin is 8% (or 40% of the original 20% value).‡ This second

* Databases and computer models for renewable energy sources exist that can provide approximate energy source values for general geographic areas (and these are continually improving). However, at the time this book is being written, the databases/models fall short—at least in the author's opinion—of being able to provide accurate enough information with which to confidently be able to predict the output of renewable facilities at a specific hour of a specific day; i.e., the information needed for reserve margin calculations.

† The reader should note that this resource planning problem is compounded if an RPS quota is instituted for the utility. The quota may force the utility to add a significant amount of renewable energy installations in a relatively short time at various sites. However, the utility must attempt to account for the RPS-driven greater magnitude of renewable energy options at multiple sites in its system reliability analyses. The accuracy level in the system reliability analyses is certain to suffer as a result.

‡ Note that these percentage values in this example are completely arbitrary and are used solely for illustrative purposes. The example does assume a significant DSM contribution for a utility.

calculation shows what the utility's reserve margin would be if all of the voluntary customer actions assumed for the projected DSM contribution to the official reserve margin projection somehow failed to occur.

Because it is extremely unlikely that 100% of the utility's projected DSM contribution would fail to materialize at the utility's system peak hour, this second calculation is useful only as a first step, or in setting a "floor," for examining the inherent uncertainty in DSM's contribution. Other calculations that used various higher-than-zero-contributions for DSM would provide additional insight.

These other calculations should be based on the type of DSM options (load control, time-differentiated electric rates, etc.) the utility utilizes and the relative certainty the utility has in the contribution of each type of DSM option. As the utility works through this exercise, the original extreme "assume zero DSM contribution" reserve margin value of 12% in our example would increase to a higher value. Let us say that, after accounting for the voluntary aspect of utility's specific DSM programs, the revised reserve margin calculation rises from 12% to 16%.

The utility sees that it will have at least a 16% reserve margin even if the voluntary nature of DSM options takes a turn for the worse, and will have a 20% reserve margin if the negative aspect of the voluntary nature of DSM options plays no role at all. The utility could then better judge the wisdom of deciding how much of its projected reserve margin in the future it is comfortable with being supplied by DSM.*

Once this information is in hand, the utility could use it to set its own DSM-related constraint. This constraint might state that: "the utility will not implement DSM that would result in a system reserve margin lower than X% assuming no or limited DSM contribution." The use of such a constraint would ensure that the utility does not become too reliant upon DSM resources that inherently have a voluntary aspect that can result in less-than-projected demand and energy reductions being achieved, particularly at the utility's peak load hour.†

Another DSM consideration exists for utilities whose DSM portfolio of programs consists of a considerable amount of load management. In such a case, the frequency with which the utility will have to implement load management will almost certainly increase as it grows more dependent upon these DSM resources to meet its system reliability criterion. The example of the former Florida Power Corporation (now Progress Energy Florida) experience with heavy reliance upon load control in the late 1990s pointed out

* Stated another way, the utility would be considering what an acceptable minimum reserve margin might be based solely on generation-only resources.

† Conversely, a utility might decide to increase its current reserve margin criterion by some level (e.g., from 20% to 22%) if it believed it was becoming too dependent upon DSM resources for system reliability. This approach might work, but the utility would have to take care that it did not then allow all of the extra reserves (2% in this example) to be provided by additional DSM.

the inherent risk of this approach. As load control frequency increases to a point where participating customers are inconvenienced too much, significant numbers of those participating customers may choose to leave the program.

A similar renewable energy-related constraint could also be utilized by a utility. Such a constraint would select an appropriate value from within the range of projected outputs for a renewable energy installation at the utility's peak load hours. A separate value would be selected for each type of renewable energy resource. This could even extend to a separate value for each combination of renewable energy resource type and geographic region or site.

Each such value would likely start in the lower portion of the projected range (possibly including a value of 0%) to ensure that the utility was not placing too much reliance on these renewable energy resources for system reliability. These values would then be reviewed annually as additional data for the type of renewable energy resource and site became available. In this way, this type of constraint for renewable energy resources would be refined annually.

We now focus on another type of constraint that a utility may impose upon itself in regard to resource options.

Load Shape Constraint

The sixth constraint is one that we will refer to as a "load shape" constraint. This constraint applies only to DSM options, particularly load management (or demand response) type DSM options. Before we discuss this constraint, we will first return to the concept of load shape that was discussed in Chapter 2.

By load shape, we are referring here to the utility's peak day load shape. For the discussion that follows, we will focus solely on the load shape for the summer peak day. You may recall that we previously presented the summer peak day load shape for our hypothetical utility system in Figure 2.1. We will now take that same load shape and focus only the hours from about noon to 10 p.m. using a different scale for the values. This information is presented in Figure 8.1.

This portion of our utility's summer peak day load shape shows the portion of the summer peak day in which our utility's load is increasing from late morning to its late afternoon peak of 10,000 MW. The load then decreases as the evening hours unfold. This noon-to-10 p.m. portion of a summer peak day is the portion of the peak day that DSM options typically seek to address in terms of reducing system summer peak demand, thus assisting the utility to meet its summer reserve margin criterion.

Virtually any DSM option that seeks to reduce system peak demand will affect the utility's peak day load shape during these hours. There are two basic impacts that DSM options will have and these are tied to two different

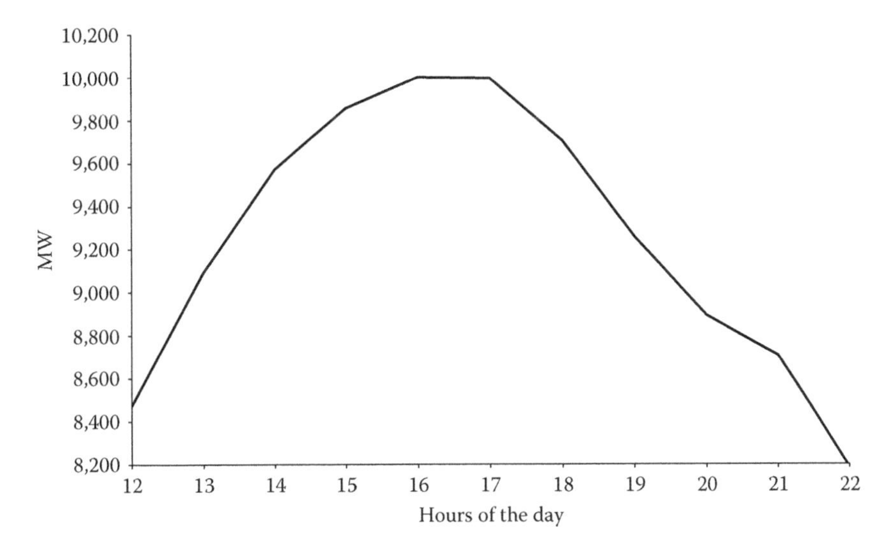

FIGURE 8.1
Summer peak day load shape (noon to 10 p.m.).

types of DSM options. First, certain DSM options will tend to lower the utility's load, to some degree, over most/all of these late morning-to-evening hours on the summer peak day. The energy conservation type DSM options, such as higher efficiency air conditioning systems, are these types of DSM options that tend to lower the load curve in all/many of the hours shown in Figure 8.1.

The other type of DSM option, a load management option, will impact the summer peak day load shape in a more selective manner. We will discuss this by using a load control program as our example. Load control is implemented by the utility literally "pushing a button" that either causes electrical power to be temporarily shut off to some electrical equipment (such as a water heater or swimming pool pump), or causes air conditioning/space heating equipment to cycle on and off in a manner different than it normally would operate when only the thermostat was controlling the equipment. (The basic effect for an individual air conditioning unit, when the utility implements load control, is that the length of time that the air conditioner would be "off" is slightly longer than would be the case if the utility were not controlling the equipment.)

The combined effect of this, when applied to tens of thousands, or hundreds of thousands, of participating load control customers, is to significantly reduce the system demand when the utility hits the load control "button." This reduction in peak load occurs very quickly on the utility system. A somewhat simplified view of what the resulting change in our utility system's peak day load shape may be is represented graphically in Figure 8.2.

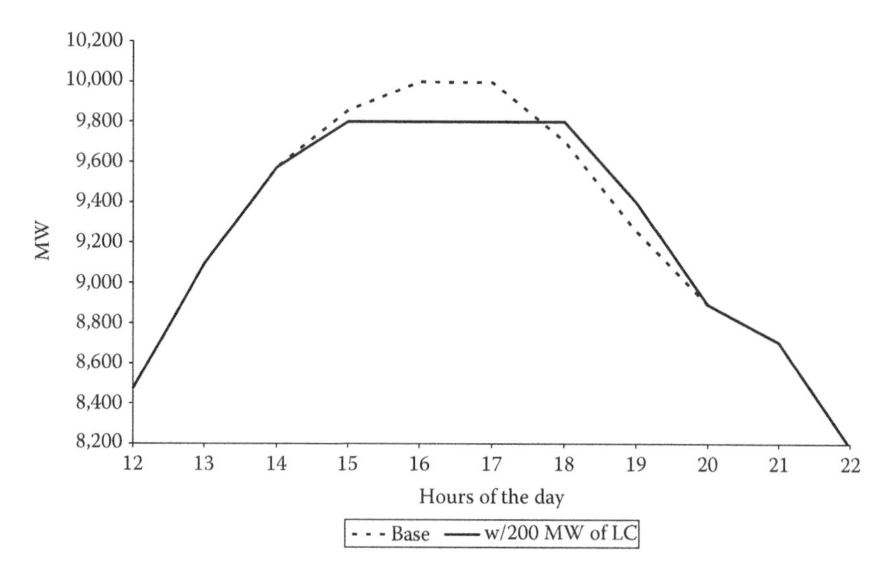

FIGURE 8.2
Summer peak day load shape after implementing 200 MW of load control.

In Figure 8.2, the dotted line now represents the original load shape on the peak day assuming no load control is implemented; i.e., the same load shape that was presented in Figure 8.1. The solid line now represents the altered load shape after load control is implemented. We are assuming that a relatively modest amount of load control capability, 200 MW, has been signed up by the utility and that all of it has been implemented on this summer peak day. We see that the former peak load of 10,000 MW has now been lowered by 200MW to 9,800 MW, as we would expect.

However, perhaps unexpectedly, we also see that while the former peak was only 1 hour in duration, the new, lower peak after load control has been implemented is a peak that now lasts for 3 hours (i.e., from hours 15 to 18) in order to lower the peak by 200 MW. In other words, the load management program not only lowers the peak demand, it also "flattens" the load shape.

This flattening effect occurs for two reasons. First, the objective is to lower the former peak hour's load. In order to ensure that this is accomplished, the load control button is typically pushed before the former peak hour occurs. Second, when the utility removes its finger from the button, an increase in load occurs as water heater and air conditioning thermostats switch their respective equipment back on in order to bring either the water temperature in a water heater tank, or the room temperature in a home or business, back to the temperatures settings on the thermostats for both types of equipment. This thermostat-driven increase in load when a utility releases load control is often referred to as a "payback" effect (which we will return to shortly).

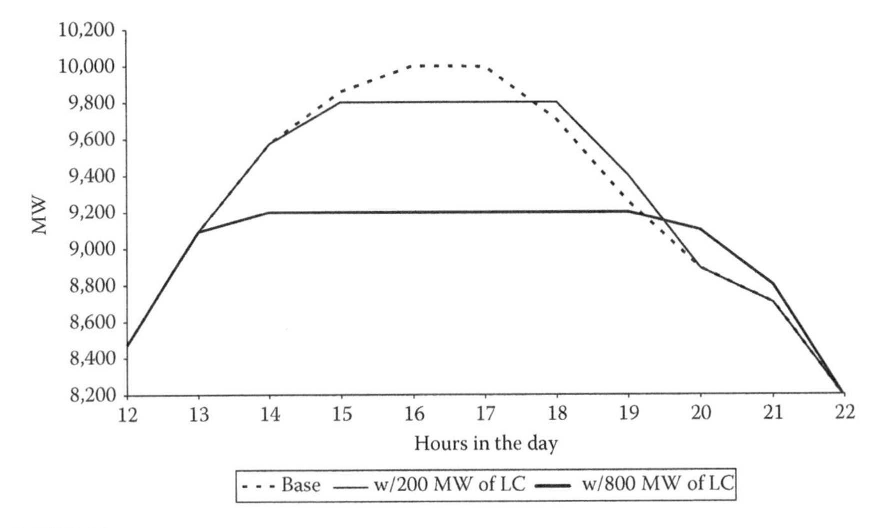

FIGURE 8.3
Summer peak day load shape after implementing 200 or 800 MW of load control.

Now suppose that our utility system is considering adding 600 MW *more* of load control capability. The resulting impact when the new total of 800 MW (= 200 MW originally + 600 MW more) of load control are implemented on the summer peak day is shown graphically in Figure 8.3. For comparison purposes, this figure also shows both the original load shape (without any load control) and the projected load shape when 200 MW of load control is implemented; i.e., the same information conveyed in Figure 8.2.

We see that when the greater amount of load control is signed up and implemented, our utility now experiences a peak that has been extended so that it lasts for 5 hours (i.e., from hours 14 to 19). From these examples, it is clear that, the greater the amount of load control capability the utility has, the longer the duration of the utility's peak when the utility implements all of its load control capability. In other words, the more a utility tries to drive the peak load down by implementing increasing amounts of load control, the more hours that the control of the participating customers' equipment must be maintained. As a result, the utility's peak day load shape becomes even more flattened.

This points out that there are a couple of practical issues to consider when considering the actual operation of load control. We will start with a consideration of the participating customer. Participating customers will tolerate control of their air conditioning, etc., equipment for some number of hours, but will not readily tolerate longer durations of control. In other words, if the duration of load control becomes too lengthy, participating customers may find that their homes are too hot in summer (or too cold in winter) and/or their water is not warm enough.

Recognizing this, utilities typically limit the duration of load control in a tariff.* This limit on the duration of load control is usually based on time periods that ensure that the tolerance levels of participating customers will not be exceeded. Absent an outright emergency on the utility system, the duration of load control for any one participating customer cannot exceed the maximum duration of control set forth in the applicable load-control tariff.

Now recall what the Figures 8.1 through 8.3 have shown us: the more load control capability that is added, the longer the duration of the new flattened peak load is when this load control capability is implemented. What all of this leads to is the conclusion that a utility system cannot indefinitely increase the amount of load control on its system by signing up ever increasing numbers of load control participants. As a utility signs up more load-control participants, the peak day load shape becomes flatter for even more hours.

At some point, adding one more MW of potential load control capability would require that the utility's control of customers' load would be for periods of time which are longer than the control durations allowed by the tariff. This point can be referred to as a "physical limit" to load control for the utility system. A utility *could* attempt to get around this problem by operating load control as a form of "relay race." Such an attempt would result in: (i) one participating customer being controlled for all or part of his or her allowable control duration, (ii) control of that customer then ceasing, and (iii) control of a second participating customer immediately beginning. In this way, control could be implemented for longer durations than allowed by the tariff's restrictions for any one participating customer.

However, there is a significant problem with this approach. The utility is now signing up and paying credits on the monthly electric bill to *two* customers in order to get one the same amount of demand reduction through load control that was previously supplied by signing up and paying monthly credits to just one customer. Once this point has been reached, the cost-effectiveness of adding one more participating customer in the load control program has been cut in half from what it was before.

One can look at this same phenomenon in a different way. Even if one were to disregard the additional cost aspect of such an approach, and focus solely on the demand reduction perspective, the impact of the physical limit would show in a different way. One would reach the point at which signing up additional participating customers provided diminishing MW reductions.

In the utility example just discussed, our utility first signed up 200MW of load management and was able to obtain 200MW of peak load reduction. Then, our utility signed up another 600MW of load management and was able to obtain a total of 800MW of peak load reduction. Now let us suppose

* An electric utility's tariff is basically a type of rule sheet or contract for the utility and its customers. In part, it states the conditions under which a specific type of electrical service (such as a load control offering) may be offered and operated.

TABLE 8.1

Impacts of Increasing Amounts of Load Management for Our Utility System

(1)	(2)	(3)	(4)	(5)
				= (4)/(1)
Incremental Amount of Load Management Signups (MW)	**Cumulative Amount of Load Management Signups (MW)**	**Cumulative Amount of Projected Peak Load Reduction (MW)**	**Incremental Amount of Projected Peak Load Reduction (MW)**	**Percentage of Projected Peak Load Reduction Compared to Signups (%)**
200	200	200	200	100
600	800	800	600	100
100	900	867	67	67
100	1,000	917	50	50

that 800 MW was the point at which the flattening of the peak day load shape would result in diminishing MW reductions for any additional load management signups.

What would happen if our utility then attempted to sign up another 100 MW of load management (in an attempt to have a total of 900 MW of load management), then attempted to sign up yet another 100 of load management (in an attempt to have 1,000 MW of load management)? An example of a likely outcome is presented in Table 8.1.

In Table 8.1, the first two rows provide in tabular format the same results presented earlier in Figures 8.2 and 8.3. Both the 200 and 600 MW increments of load management signups resulted in a projected 200 and 600 MW, respectively, of peak load reduction. However, an additional 100 MW of signups would now be projected to show diminishing returns of only 67 MW more of peak demand reduction. If our utility persisted and added yet another 100 MW of load management, the diminishing returns impact would be even more pronounced as only 50 MW more of peak demand reduction would be expected.

Therefore, regardless of whether one thinks of this diminishing returns impact from increased signups in terms of significantly reduced cost-effectiveness, or diminishing incremental peak load reductions, it is a characteristic of load management programs that needs to be accounted for in utility resource planning.

This characteristic of load management programs' impact on a utility system peak day load shape can be described by stating that load management programs have a "physical limit" to the programs that is driven by the utility's peak day load shape. The physical limit is reached when the amount of load management capability on the system reaches a point where one more participating customer cannot provide the same demand reduction as each previous participating customer provided. Another way to describe this is to refer to the amount of load management up to that physical limit as being the "usable" amount of load management for the utility system.

A utility should address this by including a load shape-based form of analysis in its IRP work. Such an analysis identifies the physical limit for, or the usable amount of, load management for that utility's system. Once the amount of usable load management is identified, the utility uses that as the upper limit for how much load management should be signed up by the utility. In other words, the utility imposes a "load shape constraint" on itself in regard to its load management programs.*

The use of a load shape constraint in a utility's resource planning eliminates a couple of potential problems. First, it prevents the utility from planning to meet its future resource needs with a load management program whose incremental signups, after the utility's physical limit for such programs has been reached, will be unable to provide what would otherwise be the projected demand reduction capability for a new participating customer. This eliminates a potential system reliability problem by ensuring that the utility will not overestimate the contributions from the load management programs.

Second, the use of a load shape constraint also prevents the utility from spending monies on load management programs that, once the utility's physical limit for load management has been reached, will result in greatly reduced benefits for this load management program (due to the reduced kW reduction from each incremental participating customer who is signed up after the physical limit has been reached). The use of a load shape constraint can allow a utility to increase the economic efficiency of its portfolio of various DSM programs, and other resource options, because the use of this constraint guides the utility to know when to "back off" DSM expenditures for additional load control and focus on other, potentially more cost-effective resource options.

Before we leave the discussion of the load shape constraint, a couple of additional points should be made. First, the need to use a load shape constraint may vary widely from one utility to another based on their peak day load shape.† Our hypothetical utility system has a quite broad summer peak day in terms of the number of hours of relatively high load. Another utility may have a peak day load shape that is more of a "spike" peak of very short duration. This is often seen with utilities that are more concerned with a relatively short-lived winter peak load than a broader summer peak load. A load shape constraint may not be needed for such a utility.

Second, let's assume that two utilities have the exact same system peak load shape. The values for the physical limits to load control for each system may vary considerably based on the type of load control program that is implemented. All else equal, a utility whose load control primarily addresses

* Although we have used a load control type program in this explanation, the same situation occurs with other load management/demand response programs including special time-differentiated electric rate-based programs.

† A reminder of one of the many reasons we have introduced Fundamental Principle of Electric Utility Resource Planning #1.

large business customers will have a higher physical limit (and can sign up more MW of incremental load control) than can a utility whose load control primarily addresses residential and small business customers. This is because the majority of the demand reduction achieved with residential and small business load control involves control of equipment that utilizes a thermostat (such as air conditioners and water heaters).

As these devices are controlled, water in the water heater tank cools and the air in the home heats up in summer or cools off in winter. When the utility ceases its control of this equipment, the equipments' thermostats cause the appliance to use electricity continually for an extended period of time to heat the water in the tank and to cool or heat the home. As previously mentioned, this particular amount of energy is commonly referred to as "payback" energy.

This payback energy must be accounted for in the planning and operation of residential and small business load control so that a new, artificial peak load is not created when control is released. This serves to lower that physical limit for residential and small business load control options compared to large business load control options.

In large business load control options, thermostatically driven equipment is less likely to be controlled by the utility. Instead, industrial processes, motors, etc., are more commonly controlled. Particularly for larger business customers, a participating customer may also have a backup electrical generator that allows these business functions to continue even though the utility has temporarily cut off the electrical supply to this equipment. Therefore, there is generally much less (and perhaps zero) payback energy associated with large business load control options. Therefore, there may be a higher physical limit for large business load control programs compared to residential and small business load control programs.

Finally, one may ask how one actually calculates what the load management physical limit is for a specific utility. One method that has proven successful is to use non-linear programming (NLP) techniques involving optimization software. The basic approach is to start with a projected peak day load shape for a future year. The projected load is laid out in 15 minute interval data. Then the projected demand reduction capability impacts, and the associated payback energy impacts, are developed for each 15 minute interval if load management is implemented for that interval. (This is done separately for each type of load management program the utility offers its customers.)

The NLP optimization program then applies all possible permutations of load management operation to achieve an objective of lowering the utility's peak day load to the lowest possible peak-load value.* In regard to the load

* The lowest peak day load value is typically calculated on an hourly basis because utilities generally perform reserve margin calculations on an hourly basis. The 15 minute data is used in the optimization model's calculations in order to more accurately capture the payback impacts that can vary considerably from one 15 minute interval to the next.

management operation, the optimization program considers not only which load management programs may be implemented, but what the possible starting and stopping times can be for each aspect of a load management program (i.e., residential water heaters, small business air conditioners, etc.).

This type of analytical approach has proven very useful in determining the physical limits of various types of load management programs/rates. It can also be useful in an operational sense by providing another tool for utility system control center personnel, i.e., the folks who actually coordinate the running of all of a utility's generating units and the dispatching of load management programs, to use.

We have now completed an overview discussion of a number of potential constraints to utility resource planning. Our attention now turns to the impacts these constraints will have, or will likely have, on utility systems and their customers.

What Are the Impacts of Addressing These Constraints?

Now that we have introduced the concept of constraints on utility resource planning analyses, and have discussed six constraint examples, the natural question is: "what impacts will these constraints likely have on a utility's resource planning work?" After all, the title of this chapter includes the phrase "…can (and will) complicate resource planning analyses."

We shall begin to address this question by stating what should be an obvious point in regard to any situation in which one is trying to find the best solution to a complex problem (such as utility resource planning). This point is that:

> All else equal, the less flexibility one is allowed to have when finding a solution, the less likely it is that the best possible solution will be found.

However, equally obvious is the fact that no one has absolute flexibility in regard to finding solutions for certain problems. For example, when considering how best to paint a vaulted ceiling in one's home, it is not possible (except perhaps for Michael Jordan) to leap into the air and seemingly hover for an extended period of time while you paint the ceiling. The law of gravity removes this as a possible solution.

In regard to this ceiling painting example, the law of gravity can be considered as an "absolute" constraint on the problem of how the ceiling might best be painted. One then moves on to more feasible or practical potential solutions.

Similarly, a utility will typically accept the first constraint that we discussed (siting/geographic constraints) as an absolute constraint, consider what resource options remain as feasible after recognizing this constraint, and then move on with their analyses. A utility will also typically accept the second constraint (tightening of environmental regulations) as an absolute constraint because a utility typically has little or no direct control over such changes in regulations, especially in the short term. The utility will make its best projections of what future environmental regulations may be, then examine projected system emissions in its resource planning work to give it insight into how potential tightening of regulations may impact its resource decisions.

A key point is that not only utilities, but legislatures and regulators as well, recognize these types of constraints as "givens" when considering utility resource planning. Therefore, the resulting loss of flexibility, in regard to potential resource option decisions, that emanates from these two constraints is simply not much of an issue. In other words, there is really no decision to be made regarding whether these two constraints should be used in utility resource planning (although there will likely be decisions or judgments involved in regard to how the second type of constraint is accounted for in specific resource planning work).

The same cannot be said for the remaining four constraints that we have discussed. In each case, a decision has been made by the legislature, the regulatory agency, or the utility to impose one or more of these four constraints. Let's examine these four constraints to see what their impacts are likely to be on utility resource planning and the utility's customers.

We will start by jumping to the two utility-imposed constraints that we previously discussed. The fifth and sixth constraints (that address system reliability and peak day load shapes) are constraints that may be imposed by a utility after consideration of the specific characteristics of its individual utility system. Keep in mind that the utility not only knows its particular system better than any other party can, but the utility also has the responsibility to serve its customers in a reliable and cost-effective manner.

Therefore, a utility would only impose the fifth and sixth constraints if it believed that it needed to do so to ensure that system reliability is maintained and that various resource options are only signed up to a level that is usable and cost-effective. In other words, the utility imposes these constraints upon itself to help it identify and select the best resource option for its customers.

Consequently, any impact that the fifth and sixth constraints will have on a utility system and the utility system's customers are likely to be beneficial for the utility's customers. This leaves us with the two legislative/regulatory-imposed constraints (the third and fourth constraints) to consider in regard to what the impacts are to utility resource planning and a utility's customers.

It is obvious that both of these two constraints (i.e., the imposition of "standards"/quotas for certain resource options and the prohibition of other resource options) restrict the flexibility that a utility has in its resource planning. In effect, they limit or "squeeze" the number of resource options that can be considered in two ways. The third constraint ("standards"/quotas for specific types of resource options) dictates that a certain amount of electricity (or energy services) that a utility delivers to its customers must come from certain sources, generally renewable energy and/or DSM sources. Conversely, the fourth constraint (prohibition of specific resource options) dictates that a utility cannot add any new resources of a certain type (such as new coal units).

Together, these two types of constraints remove a significant amount of flexibility that a resource planner would otherwise have when determining the best resource option to select for his or her utility. Certain resource options are prohibited outright while the resource planner is forced to select other types of resource options to at least partially "fill out" his or her resource plan.

This reduced flexibility will, as stated earlier, make it less likely that the best possible solution for the utility's customers will be found. This is likely to lead to a non-optimum selection of resource options which, in turn, will lead to higher electric rates for the utility's customers. In addition, as was demonstrated in the analyses we did for our hypothetical utility system, the selection of a non-optimum resource option may have other downsides in regard to non-economic considerations as well such as higher system air emissions for certain types of emissions.

Consequently, the use of the legislative/regulatory-imposed constraints ("standards"/quotas for specific resource options and prohibition of other resource options) can definitely result in negative impacts to a utility's customers.

We will return to this possibility of negative impacts to utility customers from these legislative/regulatory constraints in the next chapter as we conclude our discussion of utility resource planning. In addition, a few opinions will be offered regarding several of the topics/issues that we have touched on so far.

9

Final Thoughts (Including Some Opinions)

We have covered a lot of ground in our earlier discussions of electric utility systems and how they are operated, and in our later discussions of utility resource planning. It is now time to begin wrapping up these discussions. We do so first by presenting summaries of these two topics. After these two summaries are completed, we will switch gears a bit. Up to this point, I have endeavored to focus the discussion on the presentation of fundamental concepts, analytical approaches, and principles in regard to utility systems and resource planning. In so doing, I have tried to minimize presenting personal opinions. However, later in this chapter, I shall offer some personal opinions regarding some of these issues we have discussed. We will then close this chapter with a brief discussion of what may lie ahead for utilities in regard to their future resource planning efforts.

We now head for the two summaries regarding utility systems and utility resource planning. You may surprise yourself with how much you now know about those topics.

Summary of the Key Points We Have Learned about Utility Systems in General

For many of you, it is likely that before you started reading this book, the extent of your knowledge regarding electric utility systems was "drive by" knowledge; i.e., you could recognize a power plant when you drove by it.* Hopefully, your knowledge of utility systems has expanded considerably. In regard to electric utility systems, together we have actually created a hypothetical electric utility system and have examined how it operates.

This has enabled us to learn a number of things about utility systems in general. A list of the key points we have learned about utility systems includes the following:

* Or at least you could identify the facility as a power plant when you drove close enough to the plant property so that you could read the sign which provided the name of the plant.

- The demand for electricity that a utility system's customers will have varies considerably from hour to hour, and from season to season. We saw that from examining both peak day load shapes and annual load duration curves.
- A utility system will have a variety of different types of generating units with which it will meet the hourly demand for electricity. These generating units will be operated (or "dispatched") according to their operating costs. All else equal, the least expensive-to-operate generating units will be dispatched first, followed in turn by increasingly expensive-to-operate generating units.
- Each utility will have one or more different types of generating units that regularly operate "at the margin" of the hourly demand for electricity (i.e., their operation ramps up or down as the demand for electricity increases or decreases).
- The type of fuels these "at the margin" generating units burn as they ramp up or down are termed the utility's "marginal" fuels. These are typically fossil fuels; natural gas, oil, and, for some utility systems, coal. These "at the margin" generating units are important not only in respect to the day-to-day (and hour-to-hour) costs and air emissions of the current utility system, but they also play a pivotal role in the utility's analyses regarding which resource option(s) is the best choice when new resources are needed. This is because, when a utility needs to add new resources, each of the resource options being considered will be impacting the future operation of these "at the margin" generating units. Also, each resource option will impact the future operation of these "at the margin" generating units differently.
- Finally, and perhaps most importantly, we have learned that, because of inherent differences in electrical demand patterns and the number and type of generating units, each utility system is unique. This fact is so important that it has been expressed as my Fundamental Principle #1 of Electric Utility Resource Planning.

We then turned our attention to how a utility actually performs its resource planning function to determine what type of resources it should add to meet the future needs of its customers.

Summary of the Key Lessons We Have Learned Regarding Utility Resource Planning

We utilized our hypothetical utility system to demonstrate how a utility may proceed to make decisions regarding future resource options. We used an integrated resource planning (IRP) approach to perform economic and non-economic analyses of a variety of types of resource options. In so doing, we learned the following:

- There are three questions that must be answered in every resource planning analysis: (i) When does the utility need to add new resources; (ii) What is the magnitude (MW) of the new resources that are needed; and (iii) What is the best resource option with which to meet this need?
- There are three distinct types of analyses that utilities typically undertake in an IRP approach: (i) reliability analyses; (ii) economic analyses; and (iii) non-economic analyses. All of these three types of analyses in an IRP approach are designed to place all resource options on a level playing field so that the options can compete fairly against each other; i.e., to perform analyses of resource options with no bias and/or predetermined outcome for any type of resource option.
- Reliability analyses are used to determine the answers to the first two questions listed earlier: when the utility needs to add new resources (in what year), and what is the magnitude (MW) of the new resources needed, in order to maintain the reliability of the utility system.
- Economic and non-economic analyses are then used to determine the answer to the third question: which resource option(s) available to the utility would be the best choice to meet the next resource need?
- Economic analyses may be conducted in two stages. In the first stage, preliminary economic screening analyses may be used to screen out the least economically attractive resource options.
- These preliminary analyses are often performed in an individual-option-versus-individual-option approach. While such preliminary screening analyses *may* be useful in very limited circumstances, these preliminary analyses cannot, and do not, account for all of the system impacts a resource option will have on the entire utility system. Therefore, the results of these preliminary economic screening analyses cannot be used to make a meaningful final decision about resource option selection. The importance of this fact is so high that it has been expressed as my Fundamental Principle #2 of Electric Utility Resource Planning.
- In the second stage of economic analyses, final (or system) economic analyses are conducted to determine which resource option is in the best economic interests of the utility's customers. The final economic analyses are designed to fully account for all of the system impacts a resource option will have if it is selected. These final (or system) economic analyses are typically conducted using different resource plans in which each resource plan includes at least one of the competing resource options.

- Furthermore, in order to ensure that Supply and DSM options that are competing for a role in a utility's resource plan are evaluated on a level playing field, a final (or system) economic analysis involving both of these types of resource options must be performed using an electric rate perspective. If only a total cost perspective is used, then important impacts associated with DSM options are overlooked. Utilizing an electric rate perspective is so important that it is expressed as my Fundamental Principle #3 of Electric Utility Resource Planning.
- Non-economic analyses generally address non-cost attributes associated with resource options' impacts on the utility system as a whole. (However, the costs of those impacts are typically addressed in IRP final economic analyses.) Examples of these non-cost attributes include, but are not limited to: (i) the time it takes for a resource option to emerge as the best option; (ii) the types of fuel the utility system uses; and (iii) the amount and types of system air emissions for the utility. Such attributes are meaningful, particularly in regard to issues involving utility system reliability and risk.
- Non-economic analyses are particularly useful in regard to avoiding making decisions about certain types of resource options based on commonly held (but often incorrect) beliefs about whether those types of resource options really reduce system fuel usage and system air emissions. The non-economic portion of an IRP analysis ensures that decisions are made only after a thorough comparison of these impacts has been made for all of the resource options being considered. In other words, IRP-based non-economic analyses ensure that one answers the question "compared to what?" regarding any preconceived beliefs that might exist regarding certain types of resource options. The importance of this fact is so high that it forms the basis of my Fundamental Principle #4 of Electric Utility Resource Planning which applies to both economic and non-economic analyses.

Finally, a utility will typically use the results from both the economic and non-economic analyses to ensure that the resource options have been examined from a total system perspective. This helps ensure that the utility is making a fully informed decision as to which resource option is the best selection for its customers.

A Few Opinions on Various Topics

It is now time for me to switch hats. To this point, I have attempted to focus the discussion on fundamental concepts, analytical approaches, and principles regarding utilities and utility resource planning. These concepts, analytical approaches, and principles could readily be demonstrated and explained, as we have done using our hypothetical utility system.

In doing so, I have tried to keep my opinions to a minimum.* Stated another way, I have been wearing an "instructor" hat. But now, I will remove the instructor hat and will put on a "yes-I-have-an-opinion" hat. With this new hat in place, I will discuss at least a few issues using a question and answer format.† I hope that this change of pace proves interesting.

Question (1): Do you believe that an IRP approach is the best way to select resource options to meet a utility's resource needs?

Answer: Yes. Before I explain the reason for this belief, let's review the definition of IRP that we introduced in Chapter 3. That definition is:

IRP is an analytical approach in which both types of resource options, Supply and DSM, are analyzed on a level playing field. For each resource option, an IRP analysis accounts for all known cost impacts on the utility system that are passed on to its customers through the utility's electric rates. In addition, non-economic impacts to the utility system from the resource options are also evaluated. In this way, IRP analyses result in a comprehensive competition among resource options.

From this definition, we see that an IRP approach is designed to foster a competition between different types of resource options in which all costs to the utility system from each resource option that are passed on the utility's customers are accounted for. Other non-economic impacts to the utility system from each resource option are also accounted for.

An IRP approach requires that resource options compete to earn a place in a utility's resource plan in a competition that is specific to the individual utility in question. Therefore, the competition uses information specific to that particular utility including: electric load patterns, fuel costs, types of generating units, etc. In other words, an IRP approach allows one to make a fully informed decision, both from an economic and a non-economic perspective, regarding which resource option is the best selection for a specific utility. Therefore, an IRP approach is the best way to select resource options to meet a specific utility's future resource needs.

Question (2): In regard to electric utility regulatory policy, is there a trend toward using a true IRP approach?

Answer: Unfortunately, at the time this book is written, I believe the answer to this question is "no." A number of utilities and states often

* You may have noticed a few opinions that somehow sneaked into our earlier discussions. I will speak sternly to Security about this.

† Let me emphasize that I am providing my own personal opinions in this section. My opinions may, or may not, be in agreement with the opinion or position of FPL.

claim to use an IRP approach. However, at the same time, a number of regulatory policies are being set that are clearly diametrically opposed to a true IRP approach; i.e., these regulatory policies are actually "anti-IRP" policies. The reason this is true is because the regulatory policies seek to limit the competition between resource options that an IRP approach is designed to provide.

One example of anti-IRP regulatory policies is the implementation of prescriptive constraints that we discussed in the previous chapter. Both the third constraint ("standards"/quotas for specific resource options) and the fourth constraint (prohibition of other specific resource options) do not allow a true IRP approach to be used. The "standards"/quotas constraint mandates that all utilities in the regulatory body's jurisdiction select some level of the "favored" resource options while the prohibition of specific resource options constraint prohibits these utilities from even considering certain "unfavored" resource options. These two constraints work to prohibit a true competition among resource options (the objective of an IRP approach) from taking place.

Another example of anti-IRP regulatory policies is when decisions about resource options are made by a regulatory body based on incomplete information regarding the resource options. Various organizations that regularly intervene in utility resource option hearings before state regulatory bodies often argue for a particular type of resource option using incomplete information. For example, the incomplete information may come from an intervening party using only the results of a preliminary economic screening analysis such as a screening curve analysis ($/MWh) or a particular DSM cost-effectiveness screening test.*

If a utility or a regulatory body chooses to make a resource option decision based largely, or solely, on incomplete information from preliminary economic screening analyses, then the utility or regulatory body has decided to use an anti-IRP approach in its decision making and to forego a true competition between resource options. It has also chosen to make a decision based on incomplete information. This choice virtually guarantees that a less-than-optimum decision for the utility's customers will be made.

Question (3): Are electric utility customers harmed if a state or the federal government decides to use an anti-IRP approach when making resource option decisions or when setting electric utility regulations or policy?

* To make the situation even worse, these organizations frequently use preliminary economic screening analysis results from states other than the state the utility in question is located in. Therefore, this incorrect approach violates, at a minimum, both Fundamental Principles of Electric Utility Resource Planning #1 and #2.

Answer: Yes. Electric utility customers will almost certainly be harmed by taking an anti-IRP approach. Common sense tells us that decisions made on incomplete information are likely to be the incorrect decision.

By definition, an anti-IRP approach that does not allow a full competition between resource options will only use an incomplete analysis that can, by definition, only provide incomplete information. Therefore, any decision regarding resource options that is made with incomplete information is far more likely to harm a utility's customers compared to a decision that is based on a complete set of information that would be provided by an IRP approach.

Question (4): Why have certain states and the federal government moved in the direction of anti-IRP policies?

Answer: I believe that the primary driving force behind this movement toward anti-IRP-based policy making is largely due to the interest in GHG emissions and potential legislation and/or regulation that may address these emissions. This high level of interest, sustained over a number of years (regardless of the fact that no such federal legislation has yet been enacted at the time this book is written), has allowed what might be called a "creeping" movement away from IRP principles.*

I believe that this movement away from IRP principles is also driven by frustration on the part of certain interests in the fact that federal GHG legislation in the United States has not yet been enacted. Furthermore, this movement away from IRP principles may have gone largely unnoticed by legislators and regulators because it has happened gradually over the years. In addition, the interest in GHG emissions has provided a significant amount of political "cover" under which this movement away from true competition fostered by an IRP analysis approach could take place.†

Question (5): Would you further explain your belief that interest in GHG emissions is a primary driver in a creeping movement away from IRP principles?

Answer: Yes. In my explanation I will start with the third and fourth constraints as examples of the result of a movement away from IRP principles due to interest in, and frustration with, GHG emission regulation.

* A case can be made that referring to this as a "creeping" movement away from IRP principles understates the situation and that a more correct description would be to call this a "galloping" movement away from IRP principles.

† In making these statements, I make no judgment regarding the need for GHG legislation/regulation. The point I am making is that one of the side effects, whether intentional or unintentional, of interest in GHG emissions is a move away from analyzing resource options completely through the use of IRP principles.

In my opinion, a connection exists between these two constraints ("standards"/quotas for specific resource options and prohibition of other specific resource options) and the second constraint (potential tightening of environmental regulations). Consideration of new or tighter environmental regulations invariably is accompanied by consideration of how utilities can meet the regulations. In recent years, the primary concern in environmental circles has been to reduce GHG emissions, particularly CO_2.

One consideration that comes to mind when contemplating GHG regulation is that the use of all fossil fuels (oil, natural gas, and coal) will result in some level of CO_2 emissions. Among generating units that use fossil fuels, coal units emit the most CO_2 per unit of energy (approximately 210 pounds of CO_2/mmBTU) and natural gas units emit the least CO_2 per unit of energy (approximately 110 pounds of CO_2/mmBTU). Therefore, new coal units are frowned upon in comparison to new CC units by parties highly concerned with CO_2 emissions.

In addition, another consideration in regard to GHG regulation is what types of Supply options can deliver energy without CO_2 emissions. The answer is nuclear and renewable energy options. Therefore, both nuclear and renewable energy generating units should logically be favored in regard to CO_2 emission legislation/regulations by parties highly concerned with CO_2 emissions.

Therefore, it is not surprising that parties who strongly favor GHG regulations would want to immediately begin to "push" for non-CO_2 emitting generation options (such as nuclear and renewable energy options) and to push against high CO_2 emitting generating options (such as coal units).

It is probably safe to state that the majority of individuals and organizations that are pushing GHG environmental regulations favor renewable energy options and DSM, and are not in favor of new coal units.* These "favored" resource options are generally perceived by these parties (and by much of the general public) as "greener."†

Consequently, it is not a surprise that as soon as momentum began to build for GHG/CO_2 emission legislation/regulation, the inclination to prohibit new (and, in some cases, existing) coal units, and to push renewable energy and DSM options, would arise and be championed by certain parties. These parties have been successful in certain states and that inclination has often

* Interestingly, a number of parties pushing GHG regulation are not in favor of new nuclear generating units even though the operation of nuclear units produces no GHG (or other) air emissions and operate at very high capacity factors.

† However, as the analyses of our hypothetical utility showed, the lower capacity factors of many renewable energy and DSM options may result in greater utilization of a utility system's existing marginal generating units; i.e., fossil fueled units that are older and less efficient than newer units. Thus, while certain system emissions may be lowered, other air emissions may be increased by the selection of these lower capacity factor options. So the issues become: "green" in regard to what type of emission, and "green" compared to what other resource option? In other words, Fundamental Principle of Electric Utility Planning #4 comes into play again.

manifested itself there into prescriptive "standards"/quotas for renewable energy options and/or DSM, and the prescriptive prohibition of new coal units.

In other words, one could say that the third and fourth constraints ("standards"/quotas for specific resource options and prohibition of other specific resource options) have been applied to accompany and to assist the second constraint (the potential tightening of environmental regulations) in regard to CO_2 emission regulation. However, one could also take a more cynical view and make a case that the third and fourth constraints are a frustrated reaction to the fact that federal legislation/regulation of GHG emissions does not yet exist (at the time this book is written).

The case that the implementation of the third and fourth constraints is a frustrated reaction to the lack of GHG legislation/regulation is based on the opinion that believers in the potential dangers of global warming/climate change, frustrated by the many years spent trying to get GHG legislation/regulation, have decided to do an "end run" around the stalled legislation/regulation. These parties have decided to simply jump to what they perceive as the solutions to global warming/climate change by enacting other, easier-to-get regulations that mandate "solutions" (i.e., "standards"/quotas for specific resource options and prohibiting other specific resource options) that they believe would be chosen if only the "correct" legislation/regulation would be enacted. In other words, they perceive the problem of global warming/climate change as too serious to wait for the passage of legislation/regulation, so efforts to run around this delay are warranted.

These parties are likely aided by other parties who see this as an opportunity to realize other objectives. These opportunists realize that the long battle over GHG legislation/regulation provides "cover" to attempt to force utilities to utilize more renewable energy and/or DSM options than what otherwise could be justified through rigorous analyses. Whether driven more by strictly business interests (such as the ability to increase sales of renewable energy equipment or to otherwise improve the bottom line of a company), or for other reasons, the level of attention that has been given to GHG emissions has certainly offered an opportunity to force constraints on utility resource planning.*

In these ways, I believe that the interest in GHG emissions has led to a movement away from IRP principles and to anti-IRP policies such as the use of the third and fourth constraints. I also believe that the enormous amount of media coverage over the years regarding the GHG emission issue has further aided this movement away from the IRP principles that foster true competition.

---

* Reasons other than global warming/climate change have been given for the use of the third constraint ("standards"/quotas for specific resource options). An example is the national security aspect of less imported oil and natural gas use. However, these other reasons strike me as secondary to the global warming/climate change argument that typically is advanced first in any discussion of the issue.

Question (6): Do you see evidence of an attempt to retain at least some IRP principles as the anti-IRP movement takes place? For example, are the impacts on other system air emissions, particularly SO_2 and NO_X, or the impacts on electric rates, considered before decisions to implement prescriptive constraints were enacted?

Answer: I cannot definitively answer for each of the states that have decided to enact constraints such as the third and fourth constraints we have discussed, because I do not have in-depth knowledge of the decision-making processes in each of these states. However, based on the knowledge that I do have I believe the general answer to this question is "no."

In regard to a consideration of system air emissions, it appears that the general belief may be that "anything that lowers CO_2 emissions will automatically lower other air emissions." However, from our previous analyses of our hypothetical utility system, we see that this is not necessarily true.* Therefore, it is entirely possible that the use of the third and fourth constraints will increase system emissions of SO_2 and/or NO_x, two types of emissions that are directly linked to human health concerns, for a number of utility systems.

In regard to whether the impact of these constraints on electric rates was considered, I believe the answer is probably "yes" in part, and "no" in part.

Let's discuss the "yes" part first. As previously mentioned in regard to the third constraint ("standards"/quotas for specific resource options), the states that have adopted this type of prescriptive constraint generally recognize that mandating the use of renewable energy options will increase total system costs for a utility, thus increasing electric rates for the utility's customers. As we have previously discussed, in order to provide some level of protection for the utility's customers, these states have usually adopted a type of "price cap" as part of the prescriptive constraint. Therefore, states that have adopted the third constraint have acknowledged that there will be increases in electric rates, and have attempted to minimize this (to a degree) by instituting a "price cap" provision.

In regard to the fourth constraint (the prohibition of other specific types of resource options), the impact on electric rates of the prescriptive constraint is usually addressed, but in an even less satisfying manner. Let's use a prohibition of new coal units as an example and let's assume that the prohibition is largely (if not solely) based on that state's regulatory view of potential future compliance costs that may be imposed on CO_2 emissions. The state regulatory body may look at the economics of the new coal unit versus other

* We are reminded once again of Fundamental Principle of Electric Utility Resource Planning #4 that points out that, when it comes to system air emissions, particularly different types of air emissions, one much carefully analyze all resource options for each utility system before one can be sure of resource option impacts on specific types of emissions.

resource options using a range of projected CO_2 compliance cost scenarios. All else equal, the new coal unit's projected economics will be more favorable with lower CO_2 cost scenarios and less favorable with higher CO_2 cost scenarios.

Because there will be some higher CO_2 cost scenarios that would show that the new coal unit would be more expensive than other resource options, these scenarios would also show that the utility's electric rates would be higher if the new coal unit were chosen instead of other resource options. The regulatory body may decide to prohibit new coal units because of the possibility that electric rates would be higher if high CO_2 compliance costs were imposed in the future. One can view that as evidence that the fourth constraint was imposed while considering the impact of the constraint on electric rates.

However, a decision to prohibit the coal unit also ignores or disregards the fact that the lower CO_2 cost scenarios showed that the new coal option was the most economic option. In those scenarios, electric rates for the utility's customers would have been lower than with the other resource options. Therefore, the regulatory body's prohibition can be seen as having considered electric rate impacts, but the consideration may only have been in one direction.

Because the central premise for the prohibition (that not only will CO_2 compliance costs be imposed, but that they will be imposed such that the costs are consistent with a high compliance cost scenario) is based on a future event that may not occur, one can question whether the prohibition is prudent. After all, if the regulatory body's central premise is wrong, then the electric rates for the utility's customers will have been unnecessarily increased. If the prohibition is primarily driven by possible CO_2 compliance costs, other beneficial aspects of the new coal unit may be given little consideration.*

Let us now discuss examples when the answer to the question of "whether electric rates were really considered when applying prescriptive constraints" is "no."

We do so by returning to the third constraint and looking at what may happen in regard to electric rates if "standard"/quotas are set for DSM options. If the "standards"/quotas are set for DSM using two criteria, there should be no adverse impact on electric rates that are charged to the utility's customers. In fact, if these two criteria are used in setting DSM "standards"/quotas, the utility's customers will be served with the lowest possible electric rates when considering all of the resource options that could have been chosen.

These two criteria are (i) using only DSM options that pass the RIM preliminary economic screening test and (ii) using only the amount of DSM that is needed to meet the utility's near-term projected resource needs. We

* One could also note that increased use of domestic (i.e., U.S.) coal could also reduce the use of imported oil (and natural gas) for electricity production, thus having positive national security implications.

have previously discussed the fact that DSM options that pass the RIM preliminary economic screening test are DSM options that will likely result in lower electric rates than a competing Supply option, or a competing DSM option that fails the RIM screening test (but passes the TRC screening test). Therefore, this criterion is no surprise.

In addition, if more DSM than is needed to meet the utility's projected resource needs is selected, this additional amount of DSM is usually not cost-effective compared to the amount of DSM that is just needed to meet the projected near-term resource needs. This is because this additional DSM is no longer competing against the same Supply option in the decision year that the DSM option was originally evaluated against in the preliminary economic screening analyses. The amount of DSM that met the utility's projected resource needs has already avoided the need to add a Supply option in the decision year.

Therefore, this additional amount of DSM would have to compete in cost-effectiveness screening analyses versus a Supply option that is further out in the future than the decision year. (Relating this back to the analyses of our hypothetical utility system, in such an analysis any additional DSM would now have to compete with a Supply option that would be added further out in the future than Current Year + 5.)

However, it has been my experience that such "second step" preliminary economic screening analyses in which additional DSM is evaluated versus a later Supply option are seldom performed or considered. Instead, the results of the original preliminary economic screening analyses versus the decision year Supply option are all that is considered. A regulatory decision may be rendered that essentially states that all DSM that passes the preliminary economic screening analyses should be implemented regardless if that amount of DSM exceeds, perhaps far exceeds, the utility's resource need for the decision year. Such a decision is not in the best interests of the utility's customers because it unnecessarily increases electric rates for a utility's customers.

Because the additional DSM is actually competing with a Supply option further out in time than what the DSM was compared to in the original preliminary economic screening analyses, several things occur that are detrimental to the cost-effectiveness of the additional DSM.* First, in regard to the benefits of DSM (i.e., the costs that would be avoided by not building the Supply option that would now be built further out in the future), the net present value of those benefits are reduced. This is because there are simply more years over which those benefits would be present valued back to the current year.

Second, if the additional DSM is added in the same time frame (such as in the time frame of Current Year + 1 through Current Year +5 as in our utility system example), the costs of this additional DSM will *not* be reduced by being present valued back over more years as the benefits will be. Therefore,

* A couple of these items have been previously discussed, but they bear repeating.

this amount of additional DSM faces the hurdle of diminished present valued benefits, but undiminished present valued costs. Therefore, diminished cost-effectiveness must occur for the additional DSM.

Third, the "additional" DSM serves to further lower the number of kWh over which the utility recovers its costs. From an electric rate calculation perspective, this additional DSM not only adds costs earlier than needed (i.e., it increases the numerator), it reduces kWh sales (i.e., reduces the denominator), thus unnecessarily providing a double negative hit on electric rates.

This is one way in which the answer of "no" may apply to the original question of whether the parties that imposed the third constraint really considered the electric rate impacts of their decision. Let us now look at a second example of how the answer to this question in regard to the third constraint may also be "no."

This second example involves a request from an interested third party (or a regulatory body) to a utility to spend an amount of money (usually an arbitrary amount) annually to promote a type of resource option (for example, a particular type of DSM option) that has not been shown to be cost-effective for that utility. The third party or regulatory body may attempt to roughly gauge the impact on electric rates, but may do so in an incomplete manner.

For example, let's assume that the request is that the utility spend $1 million annually for such a resource option.* The third party or regulatory body might attempt to see what this will do to the utility's electric rates. It may start with the current electric rates, factor in the $1 million of additional costs, and recalculate the electric rate to gauge the resulting increase. However, that is an incomplete and, therefore, incorrect way to gauge the electric rate impact.

Recall that the calculation of an electric rate value utilizes two basic components: (i) the utility's total costs that appear in the numerator, and (ii) the utility's total kWh sales that appear in the denominator. The calculation approach just described would account for the $1 million increased costs that would be included in the revised (higher) numerator value. However, the described approach would not account for the reduction in total kWh sales that would result from the implementation of the specific resource option in question. This kWh reduction impact should be accounted for by using a revised (i.e., lower) denominator.

The lower denominator, by itself, will result in an increased electric rate, but this impact is simply not accounted for in the calculation approach just described. A complete, and correct, calculation approach to gauging the impacts of this directive on electric rates must account for both the $1 million

* In my opinion, a request (or prescriptive edict) to spend any amount of money on resource options that do not even pass preliminary economic screening analyses for a given utility is not a wise thing to do. Such a request or edict is usually being performed largely, if not solely, for public relations, political, and/or business advantage purposes.

of higher costs and the reduction in total kWh sales. This correct calculation approach will show that the directive to spend money on this type of resource option will have a higher (and perhaps significantly higher) impact on electric rates than will an incomplete, "cost only" calculation approach.

Question (7): You seem to believe that the use of prescriptive constraints, such as the third and fourth constraints, is a bad thing. Would you please summarize your opinions regarding such prescriptive constraints?

Answer: Yes. I believe that prescriptive constraints such as these two types of constraints are generally unnecessary, harmful to a utility's customers when enacted, and have the potential to become even more harmful over time.

These types of constraints are only "necessary" if one's objective is to avoid competition and guarantee an outcome in which one's "favored" resource options are chosen. However, if one's objective is to have all feasible resource options compete with each other in order to determine which option(s) is the best for a specific utility, the last thing you want are prescriptive constraints such as these. Only an IRP approach to resource planning can meet this objective.

Because prescriptive constraints such as these can only result in resource option decisions being made without complete information regarding the system costs, electric rates, fuel, and emission impacts of each feasible resource option, it is virtually certain that utility customers can only be harmed by the imposition of these constraints. The only thing that is not known with such constraints is the degree by which customers will be harmed.

Furthermore, if the trend of utilizing prescriptive constraints were to continue, the potential looms large for utility customers to be harmed to an ever increasing degree over time. And, as in the discussion just completed, the degree to which the utility customers are being harmed may not even be known.

Let's discuss this last issue a bit further using an example. We will assume that a state has imposed the third constraint ("standards"/quotas for specific resource options) for both DSM and renewable energy options. If the imposition of these two "standards"/quotas, one for DSM options and one for renewable energy options, does not occur at the same time with a single regulatory decision, it is certainly possible (and probably likely) that the full electric rate impact of each individual "standard"/quota will not be known.

If the imposition of the two "standards"/quotas does not occur at the same time, then one must occur before the other. For example, let's say that a regulatory (or legislative) body imposes a "standard"/quota that a certain percentage of the utility's energy production must come from renewable energy sources. Then, sometime later in a different regulatory docket or legislative decision, the regulatory (or legislative) body imposes a second "standard"/quota that the utility must implement a certain amount of DSM.

Let's also assume that, in both cases, the regulatory body believes that it has correctly calculated the complete electric rate impact of each "standard"/quota at the time it imposes the "standard"/quota. In each regulatory docket, the regulatory body decides that the rate impact of each "standard"/quota is acceptable (by whatever criterion it uses to judge this).

However, in this case, the reduction in total kWh sales that occurs from the imposition of the DSM-based second "standard"/quota was not factored in when the renewable energy-based first "standard"/quota decision was made. If the kWh reduction from the second (DSM) "standard"/quota had been factored in, then the electric rate impact of the first (renewable energy) "standard"/quota would have been higher. Would the regulatory body have made the same decision to impose the first "standard"/quota?

There is no way to know. All we know is that when they made the decision to proceed with the first "standard"/quota, they did not account for the impacts that would occur from the imposition of a second "standard"/quota. Had they done so, they would have gotten a different (higher) projection of electric rate impact.* One can make a case that the same basic situation exists if the order in which the two "standards"/quotas were imposed had been reversed. This is because the additional costs of the second (renewable energy) "standard"/quota were not accounted for when the kWh reduction aspect of the first (DSM) "standard"/quota was addressed. Although it may be possible to correctly calculate the electric rate impact of the combined effects of the two "standards"/quotas, the original calculation of the electric rate impact of the first "standard"/quota is no longer valid once the second "standard"/quota is introduced.

Therefore, the same question still applies: would the regulatory body have made the decision to impose the first "standard"/quota if it knew that other "standards"/quotas would later be applied?

This question takes on added importance when considering two items. The first item is that the regulatory or legislative body may impose even more "standards"/quotas than just the two we are discussing. This is not unlikely. RPS "standards"/quotas require that a certain percentage of the total energy produced by the utility must come from renewable energy sources. This is similar to the example we have just discussed.

However, as we have previously discussed, RPS "standards"/quotas often have smaller quotas within the overall renewable energy quota. For example, some definitive portion of the overall renewable energy percentage may have to come from solar thermal or PV, and another percentage may have to

* One may recall that the usual accompaniment of a cost-not-to-exceed provision to the imposition of a percentage-of-energy-produced-must-be-from-renewable-energy "standard"/quota provides some measure of protection to the utility's customers. However, the issue we are discussing is whether the regulatory body would have been as likely to have imposed the constraint if it had accounted for the kWh reduction impact of the second (DSM) "standard"/quota. All else equal, the logical answer is that the regulatory body would have been less likely to have imposed the first constraint.

come from wind or biomass energy. Therefore, some regulatory bodies have already imposed two or more "standards"/quotas while appearing to have imposed only one.

The second item to consider is that there is no evidence to-date of regulatory bodies deciding to lower or eliminate these "standards"/quotas (or to even review the need for them after a certain period of time has passed). To the contrary, the evidence points to greater use of "standards"/quotas both in terms of more states imposing them and by the federal government considering the same type of approach.

The key point is that "standards"/quotas will likely be with us for some time and may increase in number. And, when these are imposed on a piecemeal basis, without going back to see if previous "standards"/quotas should now be modified or eliminated, it is likely that an accurate view of the full impact of each of the individual "standards"/quotas on electric rates will never even be asked for, much less considered.

Thus, prescriptive constraints such as the third and fourth constraints are unnecessary, harmful to a utility's customers, and have the potential to become even more harmful over time. From that perspective, they make no sense unless one's objective is to avoid a full analysis and true competition in order to "fix" the selection process of resource options so that favored options are chosen.

The pitfalls of imposing prescriptive constraints such as the third and fourth constraints include the following:

- Inability for the utility, the utility's customers, the regulatory body, and other interested parties to be able to look at a complete list of resource options and to judge how each of those options, if selected, would impact that particular utility system and its customers;
- Higher electric rates for the utility's customers; and,
- Often unexpected, and sometimes counterintuitive, results regarding certain system air emissions, system reliability problems, and system fuel diversity.

In short, the imposition of prescriptive constraints that call for bypassing a true IRP approach in order to force some party's desired outcome for resource option selection is simply a bad idea.

Question (8): What should a legislative or regulatory body do if it is genuinely interested in finding out what role certain types of resource options, such as renewable energy options, should be in a utility's resource plan?

Answer: The answer is actually quite simple. The legislative or regulatory body should request that the utility perform a full IRP analysis in which a number of resource plans that meet the utility's reliability criterion are fully developed, including resource plans that feature specific levels of the resource options of particular interest (such as renewable energy options). The request should ensure that the following information, at a minimum, is provided for each resource plan:

- A projection of annual and cumulative present value of electric rates;
- A projection of annual and cumulative system SO_2, NO_X, and CO_2 emissions (plus other types of emissions that may be of interest);
- A projection of how much of the utility's reserve margin is dependent upon DSM and Supply options; and,
- A projection of the annual reliance upon each fuel type (natural gas, oil, coal, etc.).

Such a request would allow the utility, the utility's customers, the legislative or regulatory body, and all other interested parties to judge the numerous impacts, both pro and con, of selecting any of a wide variety of possible resource options for that specific utility. One could then easily see the inherent tradeoffs that come with each type of resource option. Examples of these types of tradeoffs will vary from one utility system to another, but may include: (i) higher electric rates versus a lower system air emission for one type of emission; (ii) lower system air emissions for one type of emission, but higher system air emissions for other types of emissions; and, (iii) increased reliance on certain types of fuel versus lower system air emissions, etc.

When it comes time to make a decision regarding resource option selection, a decision can then be made with a full set of information. The ultimate decision that is reached *might* be to select the same amount of (for example) renewable energy options that would have been imposed with a "standard"/quota constraint, but at least the decision would now be made with "eyes wide open," knowing what all of the tradeoffs really were. In other words, the utilization of a true IRP approach would allow a fully informed decision to be made.

In addition, the use of a true IRP approach is an approach that can be (and probably should be) repeated at regular intervals (for example, perhaps every few years), to ensure that the best resource option decisions continue to be made as new information regarding fuel costs, environmental compliance costs, costs of the resource options themselves, etc., becomes available. This is certainly a preferable approach to the imposition of arbitrary and prescriptive constraints that, once imposed, are very likely to take on a life of their own with little/no ongoing scrutiny.

The utilization of a true IRP approach, instead of the imposition of arbitrary and prescriptive constraints, is an approach that can readily be updated on a regular and ongoing basis. It is, therefore, a far more logical and prudent course of action to take than is the imposition of arbitrary and prescriptive constraints.*

What Lies Ahead for Electric Utilities and Utility Resource Planning?

The only correct answer to this question is "no one knows." Indeed, this has been the correct answer to this question every since I first began my work in electric utility resource planning analyses years ago.

There are simply too many factors that can quickly, and significantly, change what one may have thought was a safe-to-predict direction for the electric utility industry as a whole, or for a specific electric utility. A listing of such "game changing" factors would certainly include, but not be limited to, the following (with this listing in no particular order regarding either the likelihood or importance of occurrence):

- *Changes in fuel costs and the availabilities of fuels.* Perhaps more so than any other aspect of utility planning and operation, the cost and availability of different fuels is greatly affected not only by national events, but by international events as well. This ensures that fuel-related decisions must be made knowing that directions can change quickly and significantly.
- *Federal and state legislation/regulation.* Enactment of far reaching legislation/regulation (such as regulating GHG emissions) could certainly be a game changing event. This is especially true because such legislation/regulation would almost certainly have unexpected consequences.
- *New nuclear generating units.* How successful the first wave of new nuclear generating units are that are now moving toward construction and eventual commercial operation for various electric utilities across the United States will be closely scrutinized. Whether these outcomes are seen as successful or unsuccessful, these outcomes will result in significant impacts to utility resource planning for a number of utilities.
- *The continuing trend of improvement in existing technologies.* A few examples include: the continued fuel efficiency gains in already very

* As previously stated, the only parties who would not want to utilize a true IRP approach to resource option selection are parties who do not favor a head-to-head competition in which their preferred option might lose. For these parties, it is far preferable to convince legislative and/or regulatory bodies to "rig the game" by mandating that their preferred option be selected (with little or no scrutiny) and to prohibit other "unfavored" options.

efficient CC generating units; the projected fuel efficiency gains in combustion turbines, the declining capital cost and increasing efficiency of solar generating technologies such as PV; and the ongoing emergence of increasingly efficient electrical equipment used in homes and offices. Two key questions are how far, and how fast, these trends will go.

- *The emergence of new electricity supply technologies.* New technologies that bring such items as CO_2 capture and removal equipment for generating units, home/office-sized electricity generation and storage equipment, plus entirely new types of electricity generating sources to commercial operation could have widespread (and, once again, unexpected) impacts to utility planning and operation.
- *The emergence of new electricity demand technologies.* New technologies that bring such items as long-range electric cars, "must have" electronic devices for the home and office, and "smarter" appliances into widespread use will result in large and long-term changes in the demand for electricity.
- *Changing economic conditions.* The timing and duration of economic cycles will affect all aspects of electric utility planning and operation from capital investment in new/more efficient technologies to the levels of demand for electricity.
- *Unforeseen events.* There will almost certainly be future events that affect the electric utility industry that were not predicted, or if they were predicted, were seen to have a very low probability of occurring. An example of this is the recent (at the time this book is written) earthquake- and tsunami-induced damage to several Japanese nuclear reactors. As unforeseen events such as these occur, certain resource options will be looked upon more favorably while other resource options will be looked upon less favorably. This will impact utility resource planning, at least until the next unforeseen event occurs.

Therefore, in regard to electric utility planning and operation, the only certainty is that there will be an enormous amount of *uncertainty* that utility resource planners will face. Both the challenge and the complexity of electric utility resource planning will likely continue to increase due to this uncertainty.

With this uncertainty in mind, I will close this book with a two final points. These points represent hopes that I have in regard to electric utility planning in the future.

First, the inherent uncertainty in electric utility planning and operation is the best reason to undertake utility resource planning using an approach that allows one to examine all of the important impacts to the utility system as a whole. A true IRP approach, such as the one discussed at length throughout this book, is the best utility resource planning approach with which to address this future uncertainty.

An IRP approach which provides decision makers with the most complete set of information possible is infinitely preferable to any other approach, especially the imposition of arbitrary and prescriptive constraints.

It is my opinion that the movement away from IRP principles that we discussed earlier in this chapter is a mistake. This "trend" is both unnecessary and harmful to utility customers. Therefore, I hope that this trend is short-lived and that future generations of legislators, regulators, and utility resource planners will continue to recognize the inherent advantages of an IRP approach in making fully informed decisions.

Second, and in conclusion, I hope that I have been able to convey through this book both the continual challenges that face anyone whose job is to conduct resource planning for electric utilities, and the enjoyment I have had, and continue to have, in trying to meet those challenges.

Appendix A: Fundamental Principles of Electric Utility Resource Planning

Throughout the book we have introduced a number of concepts that are important to utilities in their ongoing work of planning for future resources that are needed to maintain a reliable and cost-effective utility system. Of these concepts, I believe there are a handful of concepts that can be viewed as truly "fundamental principles" of electric utility integrated resource planning.

These fundamental principles, discussed previously in the book, are listed again in this appendix in order to provide a handy reference. In addition, a few comments about how to "apply" these fundamental principles are provided.

Fundamental Principle #1 of Electric Utility Resource Planning: "All Electric Utilities Are Different"

Each electric utility is different in regard to (at least) its electrical load characteristics and its existing generating units. Therefore, when faced with a particular problem or issue, such as "which resource option is the best selection?", the correct answer for one electric utility may not be the correct answer for another electric utility.

Because each utility system is different, any assumption (for example) that what is the best resource option selection for utility A will automatically be the best choice for another utility (or, even worse, for all other utilities) is very likely to be wrong, and, in some cases, spectacularly wrong.

Understanding this fundamental principle is of particular importance to legislative and/or regulatory bodies that may be tempted to impose certain resource option selections, or certain levels of resource options, on all utilities within its jurisdiction. Although such an action may be tempting due to expediency, political and otherwise, a "one-size-fits-all" approach will almost certainly guarantee that the best choices for individual utilities are not made.

In regard to how best to apply this fundamental principle, the answer should be obvious to one who has read through the book to this point. Each utility needs to perform a complete set of analyses (as is the case when an IRP resource planning approach is used), utilizing utility-specific assumptions and inputs, to ensure that the best choices for the individual utility are made.

Fundamental Principle #2 of Electric Utility Resource Planning: "System Cost Impacts of Producing or Conserving Electricity Are of Upmost Importance. Individual Resource Option Costs of Producing or Conserving Electricity Are of Little or No Importance When Considered Separate from the Utility System as a Whole"

The projected costs of producing/conserving electricity for any individual resource option by itself, often expressed in terms of cents/kWh or $/MWh, whether a Supply or DSM option, is of little/no consequence when performing economic evaluations whose objective is to select the most economic resource option for the utility system as a whole. When selecting a resource option for a particular utility, the objective is to identify the resource option that results in the lowest electric rates that are charged to customers. Only an analysis that accounts for all of the impacts that a resource option will have on the entire utility system can determine the most economic resource option.

The application of this fundamental principle is most important when one runs across an argument that resource option A should be selected because "resource option A can produce/save energy at a cost of only 'X' cents/kWh." Alarm bells should go off immediately when one reads or hears such a statement because the party issuing such a statement either: (i) doesn't understand the significant limitations of such cents/kWh comparisons of individual resource options, or (ii) understands these significant limitations, but is trying to pull a fast one on his/her audience (who may not realize these limitations).

In order to obtain a complete answer to what the cost of a particular resource option would be for a specific utility, analyses of the utility system as a whole must be conducted that evaluate different resource plans that address not only resource option A, but other possible resource options as well. Only in this way can one obtain a complete and, therefore, meaningful picture of the system cost impacts of competing resource options.

Fundamental Principle #3 of Electric Utility Resource Planning: "Electric Rate Impacts Are the Most Important Consideration When Analyzing DSM and Supply Options; Total Cost Impacts Are Less Important"

In regard to economic analyses, projections of total utility system costs for competing Supply options can be used to correctly select the most economic Supply resource option. As previously discussed, this is because the Supply option which results in the lowest total costs will also result in the lowest electric rates. However, when evaluating DSM options versus Supply options, economic analyses must be carried out one step further. Analyses of DSM versus Supply options must account for the fact that DSM options reduce the number of kWh of sales over which a utility's costs (revenue requirements) are recovered. This unique characteristic of DSM options makes it necessary to conduct an electric rate calculation in order to really determine which option, DSM or Supply, is the best economic choice from a customer's perspective. The importance of total costs is lessened in these calculations because total costs are merely one input into the calculation of electric rates.

Any economic analyses of competing DSM and Supply options must apply this principle to get a complete and meaningful answer to the question of which resource option is the most economic option. However, there is no need to perform this electric rate analysis when evaluating only Supply options because none of the competing Supply options will directly change the number of kWh of sales over which the utility's costs are recovered (i.e., information that will be used in an electric rate calculation). Therefore, once one knows the Supply option that is projected to result in the lowest system costs, one also knows that this Supply option is projected to result in the lowest electric rates. One could perform an additional step of calculating electric rates for all of the Supply options, but it is not needed in such cases because the cost analysis has already determined the Supply option which will result in the lowest electric rates.

This is obviously not the case when DSM options are compared to Supply options because DSM options do result in a decreased number of kWh of sales over which the utility's costs are recovered. Consequently, the additional step of calculating the projected electric rates that will result from each of the competing resource options is absolutely necessary to determine which resource option will result in the lowest electric rates for the utility's customers; i.e., which resource option is the economic choice.

Fundamental Principle #4 of Electric Utility Resource Planning: "Always Ask Yourself: 'Compared to What?'"

In all aspects of resource planning, one must always ask the question "compared to what?" when one is analyzing a particular resource option.

In the main body of the book, we introduced this Fundamental Principle in the course of discussing system air emissions that would result from adding various Supply or DSM options. The fact that resource option A is projected to either lower the number of kWh served by a utility (such as with DSM options), or to produce kWh without burning fossil fuel (such as with renewable energy options), is no guarantee that resource option A will actually result in lower system air emissions (or lower system fuel usage, etc.) for a specific utility. The key point is "what is resource option A being compared to?" Although resource option A may reduce air emissions from what the emissions otherwise would be *if no other resource option were added or considered,* there may be a resource option B that would result in even lower system air emissions. (In such a case, the selection of resource option A would actually increase system emissions compared to the selection of resource option B.)

Therefore, in order to know which resource option will really result in lower system air emissions (or lower system fuel use, lower electric rates, etc.), all resource options which are applicable for a specific utility must be evaluated.

In regard to the application of Fundamental Principle #4 in non-economic analyses, this principle can be thought of as the counterpart of Fundamental Principle #2 that addresses economic analysis. Both principles point out that a comparison of the characteristics of individual resource options can only provide an incomplete and, therefore, meaningless "answer" to the question of which resource option is the best selection.

In regard to the issue of system air emissions and resource option selection, the characteristics of individual resource options (Supply or DSM) are only one of many factors that must be considered. The utility system's marginal fuels, the emission rates of the utility's existing and new generating units, and the different capacity factors at which resource options would actually operate on the specific utility system in question, have to be accounted for before a complete picture emerges. Such an analysis may show that the options initially considered as preferable will result in higher system air emissions, for one or more types of air emissions, than would be the case if another resource option were selected. The same can be said about system fuel usage as well.

Appendix B: Glossary of Terms

A number of terms commonly used in discussing electric utilities and resource planning for electric utilities are mentioned throughout the book. An attempt has been made to explain those terms when the term is introduced. However, for handy reference purposes, a number of the terms that individuals outside of the utility industry often find unfamiliar are listed and explained in the list below.

Air Emissions: As used in the book, this term refers to certain gases that are emitted into the air as a result of burning fossil fuels in electric utility generating units. Three air emissions are used in the book when discussing utility system operation and resource planning: sulfur dioxide (SO_2), nitrogen oxides (NO_x), and carbon dioxide (CO_2).

Availability: Availability refers to the percentage of the annual hours per year that a generating unit is projected to be able to produce electricity after accounting for two factors: (i) the projected number of hours each year that the unit is out-of-service for planned or scheduled maintenance, and (ii) the projected number of hours each year that the unit is expected to be out-of-service for unplanned maintenance (i.e., the unit is "broken"). As a consequence of these two factors, the projected availability for a generating unit is typically less than 100%.

Baseload Generating Unit: A baseload generating unit is a generating unit (power plant) that is projected to operate much of the time during a year. For example, a generating unit that is projected to operate more than 70% (or so) of the hours in a year is generally considered as a baseload unit. Such a generating unit operates that many hours of the year because it will be among the least expensive-to-operate generating units on the utility system. (Please refer to the explanations/definitions of "Intermediate Generating Unit" and "Peaking Generating Unit".)

British Thermal Unit (BTU): This term refers to the amount of heat required to raise the temperature of one pound of water one degree Fahrenheit at sea level. The term is used frequently in discussing the cost of fuel in which the cost is frequently referred to as "dollars per million BTU" ($/mmBTU). The BTU term is also used in regard to the fuel efficiency (or heat rate) of generating units in which the efficiency is referred to in terms of BTU/kWh. The BTU term is again used in regard to the amount of air emissions that result from burning fuel in which the amount of emissions is referred to in terms of pounds or tons of emission per BTU of fuel burned. (Please refer to the explanations/definitions of "mmBTU" and "Heat Rate".)

Capacity Factor: Similar to the term "Availability," the term "Capacity Factor" is a value representing a percentage of the total hours in a year. While the "Availability" value tells us the projected percentage of the annual hours that a generating unit is *capable* of operating, the "Capacity Factor" value tells us the percentage of the annual hours that a generating unit is projected to actually *operate* on a particular utility system. The Capacity Factor value is typically less than, but could be equal to, the Availability value. All else equal, the lower the operating cost of a generating unit, the higher its capacity factor will be.

Cumulative Present Value of Revenue Requirements (CPVRR): The term "Cumulative Present Value of Revenue Requirements (CPVRR)" refers to the sum of the annual present values of revenue requirements for a multi-year period. (Please see the explanation/definition of "Revenue Requirements" in this appendix and the discussion of CPVRR in Appendix C.)

Demand Response Programs: (Please refer to the explanation/definition of "Load Management Program".)

Demand Side Management (DSM): This term refers to one of two types of resource options a utility can select to meet a future resource need. These two types of resource options are: Supply options (options that generate electricity) and DSM options (options that reduce the demand for, and use of, electricity). In regard to DSM programs, there are two main types of programs. These two types of programs are referred to by various terms. In this book they will be referred to by the terms: Energy Conservation programs and Load Management programs. (Please see the explanations/definitions of "Energy Conservation Programs" and "Load Management Programs".)

Discount Rate: A discount rate is a value, expressed as a percentage (such as 8%), which one uses to calculate the worth of an amount of money that will be spent or received in one specific year relative to its worth in another year. (Please see Appendix C for a further explanation/discussion of this concept.)

Electrical Demand: This term refers to the demand for electricity by a utility's customers at any one point in time. The concept of electrical demand is critical when utilities evaluate the reliability of their utility system and when they evaluate how large a resource is needed to ensure that the utility system remains reliable in the future. Electrical demand is most often referred to in terms of megawatts (MW) when discussing the utility system and Supply options, and in terms of kilowatts (kW) when discussing individual DSM options. (Please refer to the explanations/definitions of "Megawatt" and "Kilowatt".)

Energy: Energy is the amount of electrical demand over a given period of time. The period of time most frequently used in discussing an amount of energy is one hour and the term then used in either a megawatt-hour (MWh) or a kilowatt-hour (kWh). (Please refer to the explanations/definitions of "Megawatt-hour" and "Kilowatt-hour".)

Energy Conservation Programs: An energy conservation program refers to one of two basic types of Demand Side Management (DSM) options or programs. (Please see the explanation/definition of "Demand Side Management".) A utility's energy conservation programs typically offer one-time rebates or incentives to utility customers who voluntarily install energy efficient appliances, material, or equipment in their homes or buildings. (Energy Conservation programs are also referred to as "energy efficiency" programs.)

Energy Efficiency Program: (Please refer to the explanation/description of "Energy Conservation Program".)

Environmental Compliance Costs: For the purposes of the book, this term is used to denote the projected cost of air emissions, either on a "per unit" basis or in total for a resource plan. Certain air emissions (for example, SO_2) are regulated in the United States while other air emissions (for example, CO_2) are not regulated (at the time this book is written). The environmental compliance cost value for air emissions is typically expressed in terms of $/pound or $/ton. For a given year, the annual number (pounds or tons) of an emission is first calculated, then multiplied by the $/pound or $/ton cost value applicable for that year, to derive the utility's total environmental compliance cost for that year for that specific emission type. The total cost over a year (or over multiple years) is also referred to as the environmental compliance cost.

Equivalent Capacity Factor: This term represents a type of "Capacity Factor" value for DSM programs. Just as the term Capacity Factor refers to the percentage of the hours in a year in which a generating unit is projected to operate, an "Equivalent Capacity Factor" represents the projected percentage of hours in the year that DSM programs will "operate"; i.e., will impact the electricity demand and supply of a utility system. This percentage value is calculated by dividing the projected annual energy (kWh) reduction from the DSM program by the projected peak hour demand (kW) reduction from the DSM program. This value is typically used for illustrative or comparison purposes when discussing two or more resource options.

Fossil Fuels: As used in this book, this term refers collectively to three different types of fuel: oil, natural gas, and coal.

Gigawatt (GW): A Gigawatt (GW) is a measure of electrical demand and represents 1,000 MW or 1,000,000 kW.

Gigawatt-hour (GWh): A Gigawatt-hour (GWh) is a measure of electrical energy and represents 1,000 MWh or 1,000,000 kWh. The amount of electricity that an electric utility produces or sells over the course of a year is typically referred to in terms of GWh.

Heat Rate: This term refers to the efficiency with which an electrical generating unit produces electricity by utilizing fuel. Heat rate is measured by the ratio of how much fuel must be used to produce one kWh of electricity. The heat rate ratio is presented as "BTU/kWh" in which the amount of fuel used is measured in British Thermal Units (BTU). The lower the value of this ratio (i.e., the lower the amount of BTUs of fuel needed to produce one kWh of electricity), the more efficient the generating unit is. (Please refer to the explanation/definition of "British Thermal Unit".)

Integrated Resource Planning (IRP): Integrated resource planning (IRP) is an analytical approach in which both types of resource options, Supply and DSM, are analyzed on a level playing field (i.e., there is no bias or predisposition for or against any type of resource option). For each resource option, an IRP analysis accounts for all known cost impacts on the utility system that are passed on to its customers through the utility's electric rates. In addition, non-economic impacts to the utility system from the resource options are typically evaluated as well. In this way, IRP analyses foster an open competition among resource options.

Intermediate Generating Unit: An intermediate generating unit is a unit that is projected to operate on a utility system less frequently than a baseload unit, but more frequently than a peaking unit. For example, a generating unit that is projected to operate in a range between 15% and 70% of the hours in a year, might be referred to as an intermediate unit. The cost of operating an intermediate generating unit falls roughly in the middle of a range of operating costs for all of the generating units on the utility system. (Please see the explanations/definitions of "Baseload Generating Unit" and "Peaking Generating Unit".)

Kilowatt (kW): A kilowatt (kW) is a measure of electrical demand and 1 kW equals 1,000 watts (W).

Kilowatt-hour (kWh): A kilowatt-hour (kWh) is a measure of electrical energy and represents the equivalent of 1 kW of electricity being produced or consumed over 1 hour of time. For example, 1 kWh represents the amount of electricity used to light a 100 Watt lightbulb for 10 hours. A kWh is also a standard unit of electricity upon which electricity rates and bills are based.

Levelized Cost: This term refers to a proxy cost value. A levelized (or constant) cost value is the value that, when present valued each year and then summed, will result in the same cumulative present value as the cumulative present value of the sum of another group of non-constant annual cost values. (Please see Appendix C for a further explanation/discussion of this concept.)

Load Duration Curve: (Please refer to the explanation/description of "Load Shape".)

Load Management Program: A load management program refers to one of two types of Demand Side Management (DSM) options or programs.

(Please see the explanation/definition of "Demand Side Management".) A utility's load management programs typically fall into two categories. One category of load management program offers monthly incentives or credits on the electric bill to utility customers who voluntarily allow the utility to remotely control the operation of selected appliances or electrical equipment during times of high electrical demand on the utility system. The other category of load management programs typically offers electric rates in which the rate values differ either by the time of day or by the electrical peak load of the utility system. Electric rates such as these are designed to encourage customers to alter their electrical usage so that their usage is lower during times of high load on the utility system. (Load Management programs are also referred to as "demand response" programs.)

Load Shape: A Load Shape is a graph or chart that shows the total electrical demand from customers on a utility system for a selected period of time. In this book, a Peak Day Load Shape is used to show the hourly electrical demand for a 24 hour period. In addition, an Annual Load Shape (or Annual Load Duration Curve) is also used to show the electrical demand for all hours over the course of a year ranging from highest load to lowest load.

Loss of Load Probability (LOLP): The term "Loss of Load Probability (LOLP)" refers to a reliability analysis approach that is designed to determine the probability that a utility will not have enough generating resources to meet projected demand for electricity at any point in a year. This type of analysis is one of two basic types of "reliability analyses" that a utility may utilize as part of their resource planning process. (Please see the explanations/definitions of "Reliability Analysis" and "Reserve Margin".)

Megawatt (MW): A Megawatt (MW) is a measure of electrical demand and represents 1,000 kW. This measurement is commonly used in a variety of references to utility systems including the peak output of an electric generating unit and the amount of resources a utility needs to add to maintain the reliability of its system.

Megawatt-hour (MWh): A Megawatt-hour (MWh) is a measure of electrical energy and represents 1,000 kWh.

mmBTU: This term is a short-hand abbreviation for one million BTUs. (Please refer to the explanation/definition of "British Thermal Unit (BTU)".)

Peaking Generating Unit: A peaking generating unit is a unit that is projected to operate only infrequently on a utility system. For example, a generating unit that is projected to operate 15% or less of the hours in a year is often referred to as a peaking unit. Peaking units are typically the generating units on a utility system with the highest operating costs. (Please see the explanations/definitions of "Baseload Generating Unit" and "Intermediate Generating Unit".)

Present Value: This term refers to the concept of the time value of money; i.e., the relative value of money that is spent or received in a year (or years) different than the present year or some other year. Present valuing refers to a calculation method by which one can determine the relative value of money at different points in time. (Please see Appendix C for a further explanation/discussion of this concept.)

Reliability Analysis: This type of analysis is used to answer two questions in regard to the continued ability of a utility system to reliably deliver electric service to its customers. The two questions are (i) "When does the utility need to add new resources?" and (ii) "What is the magnitude of the resources the utility needs to add?" Reliability analyses are typically conducted using one or both of two criteria: a deterministic criterion such as a reserve margin criterion and/or a probabilistic criterion such as a Loss of Load Probability criterion. (Please see the explanations/definitions of "Reserve Margin" and "Loss of Load Probability".)

Reserve Margin: The term "Reserve Margin" refers to a reliability analysis calculation that is designed to show how much generating capacity a utility has compared to the projected peak summer and winter hourly loads for the utility. This type of analysis is one of two basic types of reliability analyses, along with Loss of Load Probability analyses, that a utility may perform as part of their resource planning process. (Please see the explanations/definitions of "Reliability Analysis" and "Loss of Load Probability".)

Resource Need: The term "Resource Need" refers to the timing and magnitude of additional resources that are projected to be needed by a utility in the future in order to maintain reliable electric service. The resource need is determined by reliability analyses. (Please see the explanation/definition of "Reliability Analysis".)

Resource Option: "Resource Option" is a term that refers to options that the utility system can choose from in order to meet a current or future resource need. There are two basic types of resource options: a Supply option (such as a new generating unit or a power purchase from another party) that allows the utility to provide more electricity, and a DSM option (such as an energy conservation program or a load management program) that reduces the demand for, and usage of, electricity. (Please see the explanations/definitions of "Supply Option" and "Demand Side Management".)

Resource Planning: As used in this book, the term refers to the analyses that a utility performs to answer three fundamental questions regarding its projected resource need: (i) "When does the utility need to add new resources?", and (ii) "What is the magnitude of the resources the utility needs to add?", and (iii) "What is the best resource option(s) with which to meet the projected resource need?"

Revenue Requirements: The term "Revenue Requirements" refers to the amount of revenues that are needed by a traditional regulated utility so that the utility can pay its bills and earn an authorized return for its investors. (Please refer to Appendix C for a further discussion of this concept.)

Supply Options: This term refers to one of two types of resource options a utility can select to meet a future resource need. These two types of resource options are: Supply options (options that generate electricity) and DSM options (options that reduce the demand for, and usage of, electricity). Supply options can either be a generating unit owned by the utility or a purchase of power from another party's generating unit.

Watt (W): This term refers to electrical demand. It is also the basic measure of the amount of electricity used in electrical equipment or devices (such as a 100 watt light bulb).

Appendix C: Mini-Lesson #1— Concepts of Revenue Requirements, Present Valuing of Costs and Discount Rates, Cumulative Present Value of Revenue Requirements, and Levelized Costs

When discussing resource planning for electric utilities, a number of economic and financial terms are invariably used. Among the most commonly used terms are the following:

1. Revenue requirements;
2. Present valuing of costs and discount rates;
3. Cumulative present value of revenue requirements; and,
4. Levelized costs.

Each of these terms will be discussed in order to provide a basic understanding of the terms and how they are used throughout the book.

Revenue Requirements

The term "revenue requirements" is actually a very simple term. For a traditional regulated utility, it means the amount of revenues that are needed so that the utility can pay its bills and earn an authorized return for its investors. A list of the type of bills an electric utility has would include (but not be limited to): purchase of fuels to use in power plants, purchase of new meters and transformers with which to serve new homes and new commercial buildings, replacement of old or damaged poles, salaries for its staff, maintenance for its power plants and service trucks, payments to financial lending institutions for monies that have been borrowed to pay for large capital investments, tax payments, etc. In addition, utilities have financial obligations to their investors/stockowners and these utilities strive to pay monies to these investors in order to give the investors a reasonable return on the money they have invested in the utility.

In other words, the term "revenue requirements" is simply the utility's total cost of doing business. In this book, we refer both to revenue requirements over the course of a year (annual revenue requirements) and over all of the years addressed in our analyses (cumulative revenue requirements).

These revenue requirements are "recovered" by a utility through the monthly bills its customers pay. These bills are based on electric rates that are set by a utility's regulatory body. These electric rates are basically set at a level(s) that, when applied to the utility's forecasted annual sales, result in the utility recovering the amount of money (i.e., revenue requirements) that the regulatory body believes is necessary for the utility to pay its projected bills and its investors.

Present Valuing of Costs and Discount Rates

The term "present valuing" simply refers to a way to look at amounts of money received or spent in different years that allows one to consider these monies on a comparable basis. Most of us have heard the expression "the time value of money". This expression refers to the fact that one's view of a dollar available today is likely different than one's view of the same dollar that would be available in a future year.

One can spend a dollar today or invest it to receive more dollars in the future. The expected rate of return on this investment (after accounting for risk) is used to "present value" the future dollars so that a comparison to the dollar today can be made. In other words, the term, or concept, of "present valuing" is simply a way to compare the value of dollars that are available in different years on a common basis.

This comparison works by first selecting a specific point in time, then determining the values of monies in each different year relative to that specific point in time. This specific point in time is usually the current year (but it may be a future year). In the example we will soon discuss, we will be present valuing monies back to the Current Year (just as we did in all of our analyses of our hypothetical utility system throughout the book).

The way that one then determines the values of monies in each different year (after accounting for risk) is by applying something called a "discount rate" to the amount of monies in each different year. For purposes of discussing the example that follows, we will assume a discount rate of 8% per year.*

* Discount rates can be developed in several ways. For utilities and other large companies, a discount rate is typically determined by its current after tax cost of capital (which is developed using the percent of debt and equity instruments by which the utility raises capital, plus the respective costs of the debt and equity instruments after accounting for taxes). For individual utility customers, particularly residential customers, their discount rates would be developed in different ways. However, for purposes of this discussion, how one develops a discount rate is not as important as how a discount rate is utilized.

TABLE C.1

Present Valuing Example: Costs in 1 Year (Discount Rate = 8.00%)

(1)	(2)	(3)	(4)
			= (2) × (3)
Year	**Annual Discount Factor**	**Expenditure (Nominal $)**	**Expenditure (Present Value $)**
Current Year	1.0000	$0	$0
1	0.9259	$0	$0
2	0.8573	$0	$0
3	0.7938	$0	$0
4	0.7350	$0	$0
5	0.6806	$1,000,000	$680,583

(Note that this is the same discount rate we have used throughout the book in all of our discussions regarding economic analyses of our hypothetical utility system.)

Let us now look at a simple example of how the present value of an expenditure of $1 million, 5 years in the future is calculated. We do so in Table C.1.

In this example, we see in Column (1) that we are starting with the Current Year, then extending our view 5 years into the future. In Column (2), we have used an annual discount rate of 8% to calculate annual discount factors for each of the 6 years addressed in the table (i.e., the Current Year plus 5 more years). Because we are present valuing back to the Current Year, there is no discounting of any expenditure that might have been incurred in the Current Year. Consequently, the annual discount factor for the Current Year is a value of 1.0000.

However, the annual discount factor for the next year, Year 1, is 0.9259. This value is derived by dividing the previous year's annual discount factor (which is 1.0000 in this case) by a value of (1 + the discount rate), or 1.08 in this example. In other words, 1.0000/1.08 = 0.9259.

Then, for Year 2, the same calculation approach is used and the annual discount factor for Year 2 is: 0.9259/1.08 = 0.8573. This calculation is then performed for all subsequent years through 5 years after the Current Year. In that last year, the annual discount factor is 0.6806.*

In Column (3) of the table, we present an assumption of annual expenditures. This assumption is that there will be no expenditures until 5 years after the Current Year. In this last year of the example, an expenditure of $1,000,000 is made. (The term used in the subtitle of the column, "Nominal $,"

* This example presents, and uses, an annual discount factor, shown in Column (2), in its calculation. Note that the use of present valuing formulae is also common in these types of calculations. In such cases, the annual discount factors are accounted for in formulaic calculations.

TABLE C.2

Present Valuing Example: Costs in 6 Years (Discount Rate = 8.00%)

(1)	(2)	(3)	(4)
			= (2) × (3)
Year	**Annual Discount Factor**	**Expenditure (Nominal $)**	**Expenditure (Present Value $)**
Current Year	1.0000	$1,000,000	$1,000,000
1	0.9259	$1,000,000	$925,926
2	0.8573	$1,000,000	$857,339
3	0.7938	$1,000,000	$793,832
4	0.7350	$1,000,000	$735,030
5	0.6806	$1,000,000	$680,583

merely indicates the actual expenditures that are made in each year before making an attempt to account for the time value of money by present valuing the annual costs.)

In Column (4), the annual discount factors in Column (2) are multiplied by the annual expenditures in Column (3). The resulting product of this multiplication is the present value of the annual expenditures back to the Current Year. We see that the present value of the $1,000,000 expenditure 5 years in the future is $680,583 in relation to the Current Year.*

The interpretation of this calculation is that, for whatever company or individual might be represented in this example, an amount of money equaling $1,000,000 spent 5 years from the Current Year is equivalent to $680,583 in the Current Year.

If the situation had been changed and, for example, there had been expenditures in each of the 6 years of $1,000,000, then the respective present value amount for each of the annual expenditures would be as presented in Column (4) of Table C.2.

As one might expect from our discussion so far, the present value amount of annual expenditure of $1,000,000 becomes less (by 8%, the discount rate amount) for each year one moves further away from the Current Year.

Cumulative Present Value of Revenue Requirements

We now combine the two terms or concepts of revenue requirements and present valuing to discuss the term "cumulative present value of revenue

* The fact that the $680,583 value represents a present value number is often abbreviated as $680,583 (NPV$) in which the "NPV" stands for net present value.

TABLE C.3

Cumulative Present Value of Revenue Requirements (CPVRR) Example (Discount Rate = 8.00%)

(1)	(2)	(3)	(4)
			= (2) × (3)
Year	Annual Discount Factor	Revenue Requirements (Nominal $)	Revenue Requirements (Present Value $)
Current Year	1.0000	$1,000,000	$1,000,000
1	0.9259	$1,000,000	$925,926
2	0.8573	$1,000,000	$857,339
3	0.7938	$1,000,000	$793,832
4	0.7350	$1,000,000	$735,030
5	0.6806	$1,000,000	$680,583
	Sums =	**$6,000,000**	**$4,992,710**
		CPVRR =	**$4,992,710**

requirements" or its abbreviation, "CPVRR". The term simply means the sum of the annual revenue requirements for a utility when viewed from a present value perspective. A simple example should help illustrate this.

We will use Table C.2 as a starting point, then make a couple of changes. First, we change the column titles for Columns (3) and (4) from "Expenditures" to "Revenue Requirements" so that the example denotes the annual required revenues for a utility. Second, we sum the annual nominal and present value revenue requirements in these same two columns and show those sums at the bottom of each column. The result is presented in Table C.3.

Column (3) of this table shows that this utility has annual revenue requirements of $1,000,000.* When these nominal annual revenue requirements are summed over the 6 year period, the total nominal annual revenue requirements over the 6 year period is $6,000,000. Column (4) presents the associated present value amount of the required revenues for each year. The sum of these present valued amounts is $4,992,710. As shown on the last line of the table, this sum of the annual present valued revenue requirements is called the cumulative present value of revenue requirements or CPVRR.

At this point, before we move on to discuss our next financial/economic term; it may be useful to remind ourselves of how we use a projected CPVRR value for a utility in resource planning work for that utility. Let's assume that the utility is examining two Supply options. We have seen from the discussions in the book that the introduction of a particular Supply option onto a utility system will have numerous cost impacts on the system.

* In reality, it is highly unlikely that any utility will have exactly the same amount of required revenues (or revenue requirements) from one year to the next. The annual revenue requirements are assumed to be identical in the example solely to simplify our discussion.

The cost impacts to the utility system from either of the Supply options are likely to be different from year-to-year. Consequently, the utility's projected annual revenue requirements if the first Supply option is selected are likely to be different from the projected annual revenue requirements if the second Supply option is selected. Converting the two different "streams" of annual revenue requirements into annual present valued revenue requirements, then summing the present valued costs, results in two projected CPVRR values for the utility system, one for each of the two Supply options.

The Supply option that results in the utility's lowest projected CPVRR value is the more economic Supply option.*

Levelized Costs

The last term or concept we will discuss is that of "levelized cost". As discussed in various places in the book, levelized values can be used in utility resource planning in two ways. First, levelized costs can be used in preliminary economic screening analyses of Supply options (if the Supply options are identical, or at least very similar, in regard to four key characteristics) in an analytical approach called a "screening curve" approach.† This form of preliminary economic screening analysis produces a single levelized cost value, usually in terms of $/MWh or cents/kWh. Second, levelized values can be used in the process of developing a levelized electric rate in the final (or system) economic analyses. The result is a single levelized rate value, usually in terms of cents/kWh.

Perhaps the best way to explain how a levelized cost value is developed is to show an example of the calculation. We will do so by using Table C.3 as a starting point. We will then change the values in Column (3) and add two new columns, Columns (5) and (6). The result is Table C.4.

From looking at the values in Column (3) in this table, it is obvious that we have changed the projected annual revenue requirements from the identical value of $1,000,000 each year that was assumed previously. We have kept

* And we know from our discussions in the book that the Supply option with the lowest projected CPVRR value will also be the Supply option that provides the lowest present valued electric rates for the utility's customers. This is because the CPVRR costs will be divided by the same number of kWh of sales by the utility. We also know that if either of the Supply options is to be compared to a DSM option, the economic analysis must be carried out a step further to actually determine the projected electric rates for both options because the kWh sales values over which the costs will be recovered will differ between the Supply and DSM options.

† Please see Chapters 3 and 5, plus Appendix D for a further discussion of the use of a screening curve approach in preliminary economic screening analyses, and of the severe limitations to using such an analytical approach.

TABLE C.4

Levelized Cost Example (Discount Rate = 8.00%)

(1)	(2)	(3)	(4)	(5)	(6)
			= (2) × (3)		= (2) × (5)
Year	Annual Discount Factor	Actual Revenue Requirements (Nominal $)	Actual Revenue Requirements (Present Value $)	Levelized Revenue Requirements (Nominal $)	Levelized Revenue Requirements (Present Value $)
Current Year	1.0000	$1,000,000	$1,000,000	**$1,258,782**	$1,258,782
1	0.9259	$1,100,000	$1,018,519	$1,258,782	$1,165,539
2	0.8573	$1,210,000	$1,037,380	$1,258,782	$1,079,202
3	0.7938	$1,331,000	$1,056,591	$1,258,782	$999,261
4	0.7350	$1,464,100	$1,076,157	$1,258,782	$925,242
5	0.6806	$1,610,510	$1,096,086	$1,258,782	$856,706
		CPVRR =	**$6,284,732**		**$6,284,732**

the $1,000,000 value for the Current Year, then we have escalated the annual revenue requirement by 10% each year.

This automatically changes the present valued annual revenue requirements presented in Column (4). The new projected CPVRR value for these changed annual revenue requirement stream is $6,284,732 as shown at the bottom of Column (4).

We then determine a single, constant annual revenue requirement value that when applied, without escalation, for each year, results in the same $6,284,732 CPVRR value as determined in Column (4). This value can be determined through a simple iterative process: (i) plug a number in to Column (5) and see whether the resulting CPVRR value in Column (6) is too low or too high compared to the $6,284,732 value at the bottom of Column (4); then (ii) continue to alter the original plugged-in number until the CPVRR values at the bottom of Columns (4) and (6) match. (One can also more efficiently use the "goal seek" function that is commonly found on commercially available spreadsheets.)

The concept behind a levelized cost is to find a single cost value (i.e., a cost value that remains constant or "level") that, when present valued and summed for all years, results in the same projected CPVRR value as the actual annual revenue requirements (which typically do change each year).*

Therefore, in our example we wish to find a levelized annual cost value that is projected to result in the same $6,284,732 CPVRR value as the original annual revenue requirements. We do so through the use of two new

* And, as demonstrated in Chapters 5 and 6 of the book, the same basic approach can also be used to compute a levelized electric rate.

columns in this table. Column (5) presents the levelized annual cost value and Column (6) is used to calculate the associated annual present values, then to sum up these values to produce a CPVRR that corresponds to the levelized annual cost.

As shown in Columns (5) and (6), the levelized annual cost value in this example is $1,258,782 because it produces a CPVRR value that is identical to that for the original annual revenue requirements, $6,284,732.

In this way, or by using the more sophisticated goal seek formula that performs the same basic calculation, but does not require the use of the two additional columns that are shown here, one calculates levelized cost values that can be used in preliminary economic screening analyses of Supply options (in the relatively few cases where it is appropriate to do so; i.e., where all four key characteristics, as explained in the book, of the competing Supply options are identical or very similar).

In Appendix D, we further discuss the limitations of a screening curve approach when performing preliminary economic screening analyses.

Appendix D: Mini-Lesson #2— Further Discussion of the Limitations of a Screening Curve Analytic Approach

The topic of economic analyses of resource options is discussed in numerous places in this book. In those discussions, particularly in Chapters 3, 5, and 6, it was pointed out that there are two basic types or stages of economic analyses: preliminary economic screening analyses and final (or system) economic analyses.

For preliminary economic screening analysis of Supply options, the most commonly used approach is a "screening curve" approach. (As previously mentioned in the main body of the book, some parties attempt, unfortunately, to apply a screening curve analysis approach to DSM options.) The screening curve approach is also referred to as a "levelized cost of electricity (LCOE)" approach. This analysis approach is simple to perform using only a spreadsheet. In comparison, a final (or system) economic analysis uses a much more comprehensive and accurate approach using sophisticated computer models.

Assuming the analysis is being applied to a Supply option (generating unit), the basic differences between these two types of economic analyses can be summarized as follows:

1. Preliminary economic screening analyses using a "screening curve" approach:
   a. Takes the perspective of a generating unit sitting alone in a field, completely unconnected to the utility system for which it will be a part of (if the generating unit in question is actually selected by the utility);
   b. Does not account for a number of key impacts to the utility system which will result if this generating unit is actually placed into the utility system;
   c. Can be useful, but only under *very limited* circumstances, to quickly eliminate or "screen out" the least economic Supply options from a list of Supply options that are identical or very similar in regard to four key characteristics, but does not provide meaningful results if the Supply options being compared are dissimilar in one or more of the four key characteristics; and,
   d. Should *never* be used to make a final decision regarding the selection of a resource option.

2. Final (or system) economic analyses:
   a. Takes the perspective of the entire utility system, including the generating unit in question (if the generating unit in question is actually selected by the utility);
   b. Does account for all key impacts to the utility system which will result if this generating unit is actually placed into the utility system;
   c. Always provides meaningful results for a resource option decision; and,
   d. Should *always* be used to make a final decision regarding the selection of a resource option.

Chapter 5 presented this summary level information, plus additional text that explained the severe limitations of a "screening curve" analysis approach in more detail.

We return to that discussion in this mini-lesson (which will actually turn into a bit longer than a mini-discussion). We shall first discuss the limitations of a screening curve analysis approach qualitatively. This qualitative discussion is a "lead in" to a quantitative discussion that uses an example of a screening curve analysis of a new CC generating unit. This quantitative example shows how misleading the results of a screening curve analysis can be.

Qualitative Discussion

As previously mentioned, the usefulness of a screening curve analysis approach is actually very limited. This approach can be used in a meaningful way to compare the economics of two competing resource options that are identical, or at least very similar, in regard to the following four key characteristics: (i) capacity (MW), (ii) annual capacity factors, (iii) the percentage of the option's capacity (MW) that can be considered as firm capacity at the utility's system peak hours, and (iv) the projected life of the option. If two resource options are identical, or very similar, in regard to each of these four key characteristics, then a screening curve analysis can be meaningful in performing preliminary analyses to "screen out" the less attractive of these two very similar options. (This leads to the common terminology of this type of analysis as a "screening curve" analysis.)

However, a screening curve analytical approach that attempts to compare resource options that are not identical, or very similar, in all of these four key characteristics will produce incomplete results that are of little value. Indeed, the less comparable these four characteristics are for the resource options being analyzed, the less meaningful are the results. For example, because a DSM measure and a CC generating unit are about as different in terms of resource options as one can get, a screening curve approach that attempts to compare these types of resource options provides meaningless results.

The reason that a screening curve approach can only provide meaningful results under very limited circumstances is because a typical screening curve analysis approach does not address numerous economic impacts that these resource options will have on the utility system as a whole. Instead, a screening curve approach merely looks at the cost of operating the individual option itself. One can think of a screening curve analysis as examining the costs of a resource option if it were placed out in an open field by itself and operated without its operation having any impact on the utility system. The numerous impacts an individual resource option has on the utility system—for example, how it impacts the operation of all the other generating units on the system—is ignored in a typical screening curve approach.

However, the system impacts of any resource option are very large and can result in significant system costs, or savings in system costs, that must be accounted for in order to have a complete picture of the total net system costs with any resource option. Any analytical approach, such as a screening curve approach, that ignores system cost impacts can only provide an incomplete, and, therefore incorrect, result.

This is true regardless of whether only one resource option is being evaluated, or more than one resource option is being evaluated and the evaluation results for each resource option are being compared to each other.

Limitations When Analyzing One Resource Option

We shall first look at the limitations one runs into when using a screening curve approach to analyze a single resource option. Let's assume that the resource option in question is a CC unit. In a screening curve analysis, one assumes that this generating unit will operate at a particular capacity factor (or range of capacity factors). For purposes of this discussion, we will assume the generating unit operates 90% of the hours in a year.* Then, using the generating unit's capacity and heat rate, plus the projected cost of the fuel the generating unit would burn, the annual fuel cost of operating the generating unit for 90% of the hours in a year is calculated. This calculation is then repeated for each year addressed in the screening curve analysis.

In a screening curve analysis, the unit's annual fuel costs—which will be very large for a generating unit with a 90% capacity factor—are added to all of the other costs (capital, O&M, etc.) of building and operating this individual generating unit. The present value total of these costs is then

* In the book, we used an 80% capacity factor for the CC options our hypothetical utility system was considering. We now use a 90% capacity factor assumption for this immediate discussion just to see if you are paying attention. However, for your comfort, we shall return to an assumed 80% capacity factor when we get to the quantitative discussion.

used to develop the levelized $/MWh or cents/kWh cost for this generating unit.

However, the screening curve analysis approach does not take into account the fact that this new generating unit would not operate on a utility system at 90% of the hours in a year if it was not cheaper to operate this new unit than to operate other existing generating units on the system. In other words, for almost every hour the new generating unit operates, the MWh it produces displace more expensive MWh that would otherwise have been produced by the utility's existing generating units. Whatever the annual fuel cost is of operating this new generating unit 90% of the hours in a year, the utility will save an even greater amount of system fuel costs by reducing the operation of more expensive existing generating units during these hours.

For example, let's assume that the new generating unit's annual fuel cost would be $100 million, but that the operation of this new unit will also result in an annual savings of $110 million in fuel costs from reduced operation of the system's more expensive existing units. A typical screening curve analysis will include the $100 million cost value for operating the individual unit, but will ignore the $110 million in system fuel savings that will also occur from not having to operate more expensive generating units.

For this reason, a typical screening curve analysis approach utilizes an incomplete set of information and, therefore, is an incorrect way to thoroughly analyze resource options. A complete analytical approach would take into account the total system fuel cost impact of a net system fuel savings of $10 million (=$110 million in system fuel savings – $100 million in unit fuel cost) instead of only the fuel expense of the individual CC unit. Consequently, a typical screening curve analysis will grossly overstate the actual net system fuel cost impact of the new generating unit.

In similar fashion, other system cost impacts, such as environmental compliance costs and variable O&M, are not accounted for in typical screening curve analyses because this analytical approach does not take into account the fact that the new generating unit will reduce the operating hours of the utility's existing generating units. Nor does a screening curve approach account for the impact the resource option will have in regard to meeting the utility's future resource needs. Therefore, the screening curve approach utilizes incomplete information for a number of cost categories, thus providing incomplete, and incorrect, results.

This discussion shows how a screening curve analytical approach utilizes incomplete information, thus leading to an incomplete accounting of all system cost impacts that would result from analyzing even a single new resource option. One might ask: "Is a screening curve approach even more problematic when attempting to compare two or more different types of resource options?"

The answer is "yes." We next offer a qualitative discussion that looks at problems that occur when attempting to analyze several different types of resource options using a screening curve approach.

Limitations When Analyzing More than One Resource Option

Now we shall assume that a screening curve approach is used in an attempt to compare the economics of three utility generating options:

1. Combined cycle Option A (1,000 MW)
2. Combined cycle Option B (1,000 MW)
3. Combined cycle Option C (500 MW)

Just as we did in Chapter 5, let's assume that the first comparison attempted is of two virtually identical CC units, CC Options A and B, in which the four key characteristics of these two CC units are essentially identical. In fact, let's assume that the two CC options are identical in all respects except one: the capital cost of CC Option A is lower by $1 million than the capital cost of CC Option B.

In this comparison, even though a screening curve analysis will not provide an accurate system net cost value (as per the just concluded discussion), a screening curve comparison will provide meaningful results in regard to a comparison of these two 1,000 MW CC options, CC Option A and CC Option B. This is because the impacts to the operation of existing generating units on the system will be identical from two CC units that are the same in regard to: capacity (1,000 MW), capacity factor, the amount of firm capacity (1,000 MW) each unit will provide, and the life of the two units, a screening curve analysis will provide a meaningful comparison of these two options.

In other words, these two Supply options are identical in regard to the four key characteristics that must be identical, or very similar, in order for a screening curve analysis approach to be able to provide meaningful results. What this means is that, even though the results will not be accurate from a system cost perspective for either of the two options, the results will be "off" by the same amount and in the same direction. This allows a screening curve approach to provide a meaningful comparison between these two very similar options. As would be expected, a screening-curve comparison would show that CC Option A results in a slightly lower $/MWh (or cents/kWh) value for CC Option A compared to CC Option B due to its $1 million lower capital costs.

As this example shows, a screening curve analytical approach can produce meaningful results in a case in which the four key characteristics of resource options are identical or very similar. However, as the on-going discussion will show, once these factors for competing resource options are no longer comparable, a typical screening curve approach cannot produce meaningful results. We now look at identical types of generating units that differ only by their size: CC Option A (1,000 MW) and CC Option C (500 MW).

Now at least one of the four key characteristics of resource options (that must be identical, or very similar, in order for a screening curve approach to provide meaningful results) differ significantly between CC Option A and CC Option C. This is the capacity of the two options: 1,000 MW for CC Option

A and 500 MW for Option C. Even if one were to assume that each of the other three key characteristics for the two units were identical (capacity factor, percentage of capacity that is firm capacity, and life of the units), the significant difference in capacity offered by the two options would cause a screening curve approach to yield incomplete, and therefore incorrect, results.

The capacity difference between these options would result in at least two system impacts that are not captured by a typical screening curve approach. The first of these is the impact of each of the two CC options on the utility's future resource needs. The 1,000 MW of CC Option A will address the utility's future resource needs twice as much as will the 500 MW of CC Option C. In other words, CC Option A will avoid/defer future resource additions to a greater extent than will CC Option C.

This will show up in a system cost analysis in the form of different system capital, fuel, O&M, environmental compliance, etc. costs beginning at in the next future year in which the utility would again need new resources if it selects the smaller CC Option. For example, assume that the selection of the smaller CC Option C would meet the utility's resource needs for 3 years, but the selection of the larger CC Option A would meet the utility's resource needs for 6 years. Therefore, starting 4 years in the future from the in-service date of either CC option, the utility's projected costs would be different depending upon whether the larger CC Option A, or the smaller CC Option C, had been selected. In the case of a selection of CC Option C, the utility would need to add another resource after 3 years. However, in the case of a selection of CC Option A, no other resource additions would be needed until after 6 years.

In addition, even prior to that point in the future when other new resources are needed, the 500 MW difference in capacity between the two CC options will result in different system fuel cost, variable O&M, and environmental compliance cost impacts for the utility system. In other words, the operation of the utility's existing generating units is impacted to a greater extent by the larger CC Option A than by the smaller CC Option C. This is because, assuming all else equal, the addition of a 1,000 MW generating unit will affect the operation of the utility's existing generating units more than the addition of a 500 MW unit will.

None of these system economic impacts that are driven by the difference in the capacity of two competing resource options are captured in a typical screening curve approach. The earlier discussion pointed out that a screening curve approach applied to even a single new resource option will omit a variety of significant system cost information that is necessary to develop a complete cost perspective of the one resource option. Now we see that an attempt to use a screening curve approach to compare the economics of two resource options that differ significantly only in one of the four key characteristics (i.e., in their capacity) will omit an even greater amount of important system cost information. Therefore, the use of a screening curve approach is definitely flawed when used to compare two new resource options that differ in even one of the four key characteristics.

Limitations When Attempting to Analyze Fundamentally Different Types of Resource Options (i.e., Supply and DSM Options)

Because the previous examples discussed only Supply options, one might ask the following question: "Do similar problems exist if one were to attempt to compare DSM options to Supply options using a screening curve approach?" In other words, does the fact that Supply and DSM options are fundamentally different in various characteristics also lead to limitations in the usefulness of a screening curve approach to compare such options?

The answer is a resounding "yes." All of the problems inherent in using a screening curve approach that omits the system cost impacts discussed earlier are equally applicable whether Supply or DSM options are being addressed. For discussion purposes, let's assume that the utility is also considering a DSM option which will reduce peak load by 100 MW.

In this example, the system impacts of the lower amount of DSM (100 MW) on future resource needs compared to the 1,000 MW CC Option A would not be captured in typical screening curve analyses. This would lead to the same type of incomplete and incorrect analysis discussed previously regarding a comparison of a 500 MW CC unit and a 1,000 MW CC unit. Even if one were to adjust the 100 MW of demand reduction from DSM to account for the fact that 100 MW of DSM would be equivalent to 120 MW of Supply option capacity (assuming the utility had a 20% reserve margin criterion), 120 MW of one option will have much different impacts to the utility system than will a 1,000 MW Supply option.

In addition, DSM options vary widely in terms of their actual contribution during system peak hours. In other words, would a DSM option(s) really contribute 100 MW of demand reduction at the utility's peak hour? The use of a screening curve approach that produces results in terms of $/MWh or cents/kWh makes it easy to gloss over what the real contribution of specific DSM options may be in regard to reducing the utility's demand at its peak hour.

Many DSM programs reliably reduce demand during a utility's summer and winter peak hours such as load control, building envelope improvement, and heating/ventilation/air conditioning (HVAC) programs to name a few. However, other DSM programs may contribute little or no demand reduction at the utility's summer peak hour, at its winter peak hour, or at either peak hour. A streetlight program that addressed lighting systems that operate only at night would be an example of such a program because utility systems usually do not experience their system peak hours at night.

Furthermore, at the time this book is written, attempts by various parties to analyze DSM options using a screening curve analytical approach have lumped a wide variety of DSM options together regardless of the capability of these DSM options to lower demand at a utility's peak hour. No distinction is typically made regarding what impacts the various DSM options will really have in regard to reducing demand at the utility's peak hour. This is simply another reason why attempting to apply a screening curve analysis approach to compare DSM and Supply options can only provide incomplete, and therefore meaningless, results.

Quantitative Discussion

This discussion to this point has been a qualitative one. We shall next take a quantitative look at the limitations of a typical screening curve analysis approach. We first take a look at what a typical screening curve analysis might show for a new CC unit. As is true with all typical screening curve analyses, a number of system impacts that would occur if the new CC unit were to be placed on the utility system are simply not accounted for by this type of preliminary analysis approach for the new CC unit.

Therefore, we shall show how significantly the results of a typical screening curve approach can change even if one were to account for only two of the many system impacts that are unaccounted for.

Table D.1 is an example of what the summary page of a typical screening curve analysis might be for a 1,000 MW CC unit that is going in-service 5 years from the Current Year. In this analysis, it is assumed that the CC unit will have an economic life of 25 years and that the discount rate is 8% (the same as has been used in the economic analyses throughout the book).

In Table D.1, the projected annual costs by the different cost categories, assuming the generating unit operates at a capacity factor of 100%, are shown in Columns (5) through (14). The net present value (NPV) total costs by category are shown on the next-to-last row on the table. The final row on the table shows the same NPV costs after converting them to a $/kW value (again assuming a 100% capacity factor). These values are then used in the summary box that appears in Columns (1) through (3).

Column (1) lists possible capacity factors the CC unit might operate at in increments of 5%. Column (2) presents the projected levelized costs over the 25-year period on a $/kW value for the associated capacity factor level. Column (3) then presents the projected levelized cost values in terms of $/MWh for the associated capacity factor level.

We see from the highlighted row in Column (3) that a typical screening curve analysis projects that the levelized cost for this CC unit at an 80% capacity factor (i.e., the same capacity factor that we assumed for the CC options in the economic analyses for our utility system throughout this book) is $95/MWh. (Or, converting this value to a cents/kWh value, the projected cost is 9.5 cents/kWh.) This value is highlighted in the box containing the values for Columns (1) through (3).* The results of this analysis tell us that this CC unit is projected to produce electricity, assuming it operates at an 80% capacity factor, at 9.5 cents/kWh if the CC unit is assumed to operate without any connection to the utility system as a whole.

However, as we now know, a typical screening-curve calculation does not account for a number of system cost impacts that would actually occur if the CC unit were to be placed in-service on a utility system.

We shall now see how much this result from a typical screening curve analysis might change if only a couple of these system impacts were to be accounted for. We will account for only two of the many system impacts using the results from Table D.1 as a starting point.

The conservative assumption that we shall use is that both the system net fuel cost savings, and the system net environmental compliance cost savings, will be 10% of the combined cycle unit's costs in those categories. For example, the fuel cost value for the new CC unit for in-service year 1 shown in Column (13) in Table D.1 is $865,447 (in $000). The new assumption used is that the utility system would actually realize a net savings of 10% of the fuel cost used by the new CC unit due to the new CC unit reducing the operation of existing, more expensive generating units. Thus, there will be a savings in fuel costs for the remaining generating units on the utility system of: 1.10 × $865,447 ($000) = $951,992 ($000). (Again, this fuel cost savings is the result of reduced operation of the other generating units on the utility system due to the operation of the new CC unit.)

Consequently, a net system fuel *savings* of $86,545 ($000) (=$951,992 in fuel savings from reduced operation of the utility's existing generating units – $865,447 in fuel cost from operating the new generating unit) would occur. A similar calculation is made for all years for the fuel costs and for the environmental compliance costs. None of the values in the other columns in Table D.1 are changed as a result of these assumptions for system fuel and environmental compliance cost savings.

The results of these assumption changes are presented in Table D.2. The resulting fuel cost impacts are presented in Column (12) and the resulting environmental compliance cost impacts are presented in Columns (9), (10), and (11).

Even accounting for only two of the many system impacts that would occur from adding a new fuel-efficient CC unit to the utility system, the modified

* This might be a good time to point out another problem with screening curve analyses: there are different ways to calculate the actual levelized number for a given capacity factor after one has calculated the annual costs. For example, one can present value back to the current year or to the in-service year. That choice will give widely differing values that can make comparing results from one person's screening curve analysis to results from another person's screening curve analysis even less meaningful (if such a thing is possible with a screening curve analytical approach.) The author has identified at least four different ways in which levelized costs in screening curve analyses are currently being calculated at the time this book is written. Comparing results that have used different calculation methods is a meaningless exercise.

TABLE D.1

(See color insert.) Typical Screening Curve Results for a CC Unit: With No System Impacts

(1)	(2)	(3)	(4)	(5)	(6)	(7)	(8)	(9)	(10)	(11)	(12)	(13)	(14)
Capacity Factor (%)	Levelized Cost ($/kW)	Levelized Cost ($/MWh)	In-Service Year	Capital $000	Fixed O&M $000	Capital Repl $000	Firm Gas Transportation $000	NO_x Emission $000	SO_2 Emission $000	CO_2 Emission $000	Fuel Costs $000	Variable O&M $000	Total $000
0	172	0		0	0	0	0	0	0	0	0	0	0
5	203	464		0	0	0	0	0	0	0	0	0	0
10	234	267		0	0	0	0	0	0	0	0	0	0
15	265	202		0	0	0	0	0	0	0	0	0	0
20	296	169	1	258,093	8,106	13,319	161,056	694	610	103,768	865,447	14,556	1,425,649
25	327	149	2	248,821	8,309	13,652	161,056	760	667	112,272	912,227	14,920	1,472,683
30	357	136	3	238,528	8,516	13,994	161,056	832	731	121,135	930,501	15,293	1,490,586
35	388	127	4	228,618	8,729	14,343	161,056	911	800	136,580	949,141	15,675	1,515,855
40	419	120	5	219,061	8,947	14,702	161,056	998	877	146,358	968,154	16,067	1,536,221
45	450	114	6	209,831	9,171	15,070	161,056	1,093	960	163,062	987,547	16,469	1,564,259
50	481	110	7	200,889	9,400	15,446	161,056	1,198	1,052	180,509	1,007,328	16,881	1,593,760
55	512	106	8	192,194	9,635	15,832	161,056	1,022	1,028	191,875	1,027,505	17,303	1,617,450
60	543	103	9	183,630	9,876	16,228	161,056	872	1,005	210,720	1,048,085	17,735	1,649,207
65	574	101	10	175,085	10,123	16,634	161,056	745	982	230,387	1,069,076	18,179	1,682,266
70	604	99	11	166,541	10,376	17,050	161,056	636	960	258,285	1,090,487	18,633	1,724,024
75	635	97	12	157,997	10,636	17,476	161,056	543	939	279,871	1,112,327	19,099	1,759,943
80	666	95	13	149,455	10,902	17,913	161,056	407	911	304,610	1,134,603	19,576	1,799,433

85	697	94	14	140,914	11,174	18,361	161,056	264	881	331,183	1,157,325	20,066	1,841,223
90	728	92	15	132,374	11,454	18,820	161,056	113	849	359,684	1,180,501	20,567	1,885,418
95	759	91	16	123,939	11,740	19,290	161,056	0	815	390,214	1,204,140	21,082	1,932,276
100	790	90	17	115,716	12,033	19,773	161,056	0	778	422,875	1,228,252	21,609	1,982,091
			18	107,598	12,334	20,267	161,056	0	740	457,776	1,252,847	22,149	2,034,765
			19	99,481	12,643	20,774	161,056	0	698	495,030	1,277,933	22,703	2,090,316
			20	91,365	12,959	21,293	161,056	0	655	534,754	1,303,521	23,270	2,148,872
			21	83,933	13,283	21,825	161,056	0	608	577,072	1,329,621	23,852	2,211,249
			22	77,866	13,615	22,371	161,056	0	559	622,110	1,356,242	24,448	2,278,268
			23	72,484	13,955	22,930	161,056	0	507	670,002	1,383,396	25,059	2,349,390
			24	67,102	14,304	23,503	161,056	0	453	720,887	1,411,093	25,686	2,424,085
			25	61,722	14,661	24,091	161,056	0	395	774,909	1,439,344	26,328	2,502,507
NPV total costs =				982,398	53,769	88,350	860,046	3,396	4,350	1,357,990	5,658,207	96,554	9,105,059
NPV $/kW at 100% Capacity Factor =				85	5	8	75	0	0	118	491	8	790

TABLE D.2

(See color insert.) Modified Screening Curve Results for a CC Unit: With Two System Impacts

(1)	(2)	(3)	(4)	(5)	(6)	(7)	(8)	(9)	(10)	(11)	(12)	(13)	(14)
Capacity Factor (%)	Levelized Cost ($/ kW)	Levelized Cost ($/ MWh)	In-Service Year	Capital $000	Fixed O&M $000	Capital Repl $000	Firm Gas Transportation $000	NO_x Emission $000	SO_2 Emission $000	CO_2 Emission $000	Fuel Costs $000	Variable O&M $000	Total $000
0	172	0		0	0	0	0	0	0	0	0	0	0
5	170	387		0	0	0	0	0	0	0	0	0	0
10	167	191		0	0	0	0	0	0	0	0	0	0
15	164	125		0	0	0	0	0	0	0	0	0	0
20	162	92	1	258,093	8,106	13,319	161,056	(69)	(61)	(10,377)	(86,545)	14,556	358,079
25	159	73	2	248,821	8,309	13,652	161,056	(76)	(67)	(11,227)	(91,223)	14,920	344,165
30	156	60	3	238,528	8,516	13,994	161,056	(83)	(73)	(12,114)	(93,050)	15,293	332,067
35	154	50	4	228,618	8,729	14,343	161,056	(91)	(80)	(13,658)	(94,914)	15,675	319,679
40	151	43	5	219,061	8,947	14,702	161,056	(100)	(88)	(14,636)	(96,815)	16,067	308,195
45	148	38	6	209,831	9,171	15,070	161,056	(109)	(96)	(16,306)	(98,755)	16,469	296,330
50	146	33	7	200,889	9,400	15,446	161,056	(120)	(105)	(18,051)	(100,733)	16,881	284,664
55	143	30	8	192,194	9,635	15,832	161,056	(102)	(103)	(19,187)	(102,750)	17,303	273,878
60	141	27	9	183,630	9,876	16,228	161,056	(87)	(101)	(21,072)	(104,808)	17,735	262,457
65	138	24	10	175,085	10,123	16,634	161,056	(74)	(98)	(23,039)	(106,908)	18,179	250,957
70	135	22	11	166,541	10,376	17,050	161,056	(64)	(96)	(25,829)	(109,049)	18,633	238,619
75	133	20	12	157,997	10,636	17,476	161,056	(54)	(94)	(27,987)	(111,233)	19,099	226,896
80	130	19	13	149,455	10,902	17,913	161,056	(41)	(91)	(30,461)	(113,460)	19,576	214,849

85	127	17	14	140,914	11,174	18,361	161,056	(26)	(88)	(33,118)	(115,732)	20,066	202,605
90	125	16	15	132,374	11,454	18,820	161,056	(11)	(85)	(35,968)	(118,050)	20,567	190,156
95	122	15	16	123,939	11,740	19,290	161,056	0	(81)	(39,021)	(120,414)	21,082	177,590
100	120	14	17	115,716	12,033	19,773	161,056	0	(78)	(42,287)	(122,825)	21,609	164,996
			18	107,598	12,334	20,267	161,056	0	(74)	(45,778)	(125,285)	22,149	152,267
			19	99,481	12,643	20,774	161,056	0	(70)	(49,503)	(127,793)	22,703	139,289
			20	91,365	12,959	21,293	161,056	0	(65)	(53,475)	(130,352)	23,270	126,050
			21	83,933	13,283	21,825	161,056	0	(61)	(57,707)	(132,962)	23,852	113,218
			22	77,866	13,615	22,371	161,056	0	(56)	(62,211)	(135,624)	24,448	101,465
			23	72,484	13,955	22,930	161,056	0	(51)	(67,000)	(138,340)	25,059	90,094
			24	67,102	14,304	23,503	161,056	0	(45)	(72,089)	(141,109)	25,686	78,408
			25	61,722	14,661	24,091	161,056	0	(39)	(77,491)	(143,934)	26,328	66,394
NPV total costs =				982,398	53,769	88,350	860,046	(340)	(435)	(135,799)	(565,821)	96,554	1,378,722
NPV $/kW at 100% Capacity Factor =				85	5	8	75	(0)	(0)	(12)	(49)	8	120

screening curve's levelized cost value for the CC unit at an 80% capacity factor has now dropped dramatically from $95/MWh, or 9.5 cents/kWh, to $19/MWh, or 1.9 cents/kWh. The adjustment begins to provide a more complete picture of the actual net system cost impact when the CC unit is connected to the utility system.

Stated another way, the typical screening curve result presented in Table D.1 projected a CC unit cost value that is five times higher than what a projection of the net system costs from the new CC unit would be if only two of many system cost impacts had been accounted for.

The Moral of the Story

The moral of the story is that, by leaving out system cost impacts, typical screening curve analyses are based on very incomplete information and can provide very misleading results. Just how misleading these results can be is demonstrated by the just concluded quantitative example.

As previously stated, a screening curve approach may be useful to screen out resource options (but only if the resource options are identical, or nearly so, in regard to four key characteristics). Therefore, this approach may be useful in certain preliminary economic screening analyses, but only in those limited cases in which the resource options being compared are identical, or virtually identical, in those four key characteristics.

However, the quantitative example just discussed clearly points out the fallacy of using a screening curve analysis result and jumping to conclusions about which resource option is best for a particular utility system. Resource option decisions should only be made based on a full economic analysis that accounts for all known system cost impacts.

If an individual or organization attempts to justify a resource option selection solely with the results of a screening curve analysis, the individual or organization attempting to use such an analysis as justification either: (i) does not understand how utility systems work and/or the significant limitations inherent in a screening curve approach, or (ii) knows better, but is hoping to convince someone else who is not aware of the severe limitations of screening curve analyses. In either case, such an attempt does not speak highly of that individual or organization.

Appendix E: Mini-Lesson #3—Further Discussion of the RIM and TRC Preliminary Cost-Effectiveness Screening Tests for DSM

The RIM and TRC preliminary cost-effectiveness screening tests for DSM options have probably been the subject of more prolonged discussion over the last several decades than any other single utility resource planning-related issue that I am aware of. This has certainly been the case in the state of Florida and I seem to continually run into cases in other states where the issue of the two preliminary screening tests has again resurfaced. We discussed the basics of these two preliminary cost-effectiveness screening tests in Chapter 6. In this mini-lesson, we shall discuss the two tests a bit more.

As evidenced by the discussion of resource planning analyses throughout this book, neither of these preliminary cost-effectiveness screening tests should be used to make a final resource option decision for a utility. Neither test is able to fully account for all of the utility system impacts that selection of a DSM option will result in. Furthermore, neither of these screening tests was designed to be used to make final resource option decisions. They were designed to serve as tools for performing *preliminary* economic analyses of DSM options. The proper role of these tests is to allow a comparison of a number of DSM options quickly against a common standard (which is usually the Supply option that the utility would build absent any incremental DSM). In this way, the least economically attractive DSM options can be identified and screened out.

As discussed in some detail in Chapter 6, the two cost-effectiveness screening tests differ in regard to whether each test fully accounts for all of the DSM-related cost impacts that will be passed on to customers in the form of electric rates. As we saw in that chapter, the RIM test does account for all of these DSM-related cost impacts, but the TRC test does not.

Consequently, the TRC test represents a much easier "test" for DSM options and, in general, fewer DSM options are screened out with the TRC test than with the RIM test.* For that reasons, proponents of DSM options have tended

* However, the fairly recent inclusion of CO_2 environmental compliance costs in DSM cost-effectiveness screening has greatly narrowed the gap between the number of DSM measures that will pass RIM versus those that pass TRC in cost-effectiveness screening analyses. For example, in a DSM filing before the Florida Public Service Commission that occurred just prior to the writing of this book, 56% as many DSM measures passed RIM as passed TRC if no CO_2 costs were included in the analyses. However, that ratio jumped significantly to 90% if CO_2 compliance costs were included in the analyses.

to support the TRC test, even though it clearly is fundamentally flawed in its role as an economic tool because it does not account for all of the DSM-related cost impacts that are passed on to a utility's customers.

Over the years, some of the TRC test supporters have fallen into the practice of making ill-informed statements about the RIM test in hopes of casting aspersions on the RIM test despite the fact that it is the only cost-effectiveness test that does fully account for all DSM-related costs. A few of these incorrect statements have been repeated so often over the years that the incorrect statements have, unfortunately, seemed to be taken for granted by some.

Therefore, in order to set the record straight, I shall address three of these incorrect and misleading statements about the RIM test. I shall characterize those three statements as follows: (1) "the RIM test is the 'most restrictive' cost-effectiveness test"; (2) "Supply options are not evaluated using the RIM test"; and (3) "energy conservation or efficiency programs cannot pass the RIM test." Each of these misleading and incorrect comments will be discussed in turn.

The RIM Test is the "Most Restrictive" Cost-Effectiveness Test

This misleading statement has actually been intended as a disparaging remark regarding the RIM test by individuals/organizations that seem to want the outcome of any utility resource planning analysis to end up with either more, or only, DSM options being selected. However, from a utility resource planning perspective that seeks to evaluate all types of resource options on a level playing field, this statement is actually a ringing endorsement for the RIM test.

As we discussed in Chapter 6, when comparing the RIM and TRC tests, the benefits calculations are identical for both tests. The two tests differ only in regard to DSM-related costs that are included in each calculation. Because the RIM test accounts for all DSM-related costs and cost impacts that are passed on to a utility's customers, and the TRC test omits two significant costs (incentive payments to participants and unrecovered revenue requirements), the RIM test will naturally present a tougher hurdle for DSM options to pass. In this sense, it is definitely "more restrictive" than the TRC test.

However, because a real objective in integrated resource planning is to evaluate resource options with a full accounting of all cost impacts associated with the resources options that will be passed on to the utility's customers, the phrase "more restrictive" in regard to the RIM test actually means: "most accountable," "most complete," or "most informative." After all, when comparing different Supply options such as new generating units, a utility ensures that it accounts for all of the costs of new generating units that will be passed on to its customers. It only makes sense to do the same for DSM options as well.

In other words, the meant-to-be-disparaging phrase "most restrictive" when applied to the RIM test is actually an unintended compliment which acknowledges that the RIM test is the only DSM cost-effectiveness test that fully accounts for all of the DSM-related costs and cost impacts that will be passed on to customers.

"Supply Options Are Not Evaluated Using the RIM Test"

This statement is intended to give the impression that somehow utilities are evaluating DSM options with a "more restrictive" cost-effectiveness test than is used to evaluate Supply options, thus giving an unfair advantage to Supply options in economic comparisons with DSM options. However, the statement is both incorrect and misleading.

First, the statement is incorrect. When using the RIM test, both the DSM option and the competing Supply option are evaluated. Therefore, when using the RIM test, one is evaluating a Supply option at the same time one is evaluating a DSM option. (Note that this is also the case when a TRC test is used to compare a Supply option to a DSM option.)

Second, the statement is also misleading. When making this statement, the point that the one making the statement is probably trying to convey is: when only Supply options are evaluated against each other, the RIM test is not used. This better phrased statement is a correct statement, but an irrelevant one that makes the statement misleading. When evaluating only Supply options, neither the RIM nor the TRC test is used.

As previously discussed, Supply options are typically large resource options ranging from a few dozen MW to over 1,000 MW in size. There are typically only a relatively small number of Supply options that are suitable for a given utility to consider at a given time. Due to its size, each Supply option will generally have a noticeable impact on the utility system if it is chosen. Therefore, Supply options readily lend themselves to analyses of resource plans in which one or more of the competing Supply options are incorporated into resource plans. The resource plans are then analyzed using sophisticated computer models. Stated another way, evaluations of Supply options often forego any preliminary economic screening analyses and proceed directly with the final (or system) economic analyses.

On the other hand, the impact of an individual DSM option per participating customer is much smaller than the impact of a single Supply option. The majority of DSM options, especially for residential and small commercial customers, may have demand reduction values close to 1 kW per installation.

In addition, the utility may have hundreds, or even thousands, of DSM measures that are potentially suitable for it to consider. Because of the small nature of individual DSM measures, they do not lend themselves well to direct analysis of resource plans (such as used in final or system economic analyses) because the small size of the DSM measures would result in small system impacts that would be very difficult (if not impossible) to accurately judge. Nor would it be practical to even attempt to evaluate hundreds of DSM measures individually in resource plan analyses due to the time it takes to set up and perform these analyses.

Consequently, several DSM preliminary cost-effectiveness screening tests have been created so that preliminary economic screening analyses of individual DSM measures could be carried out quickly. In this way, large numbers of DSM measures can be "screened" using these tests. Then the best DSM measures can be combined into DSM portfolios for the much more meaningful analyses of resource plans with portfolios of these DSM options versus resource plans with either different DSM portfolios or with only Supply options.

But, let's take another look at how two Supply options are actually compared. There are two key characteristics of such an evaluation. The first key characteristic is that the evaluation is performed with a full accounting of all costs for the two Supply options. The second key characteristic is that an evaluation between two Supply options is simultaneously an evaluation of both system costs and system average electric rates.

In regard to the second point, because the utility system will be serving the same amount of kWh regardless of which Supply option is selected, the typical analytical approach is to evaluate the cumulative present value of revenue requirements (CPVRR); i.e., the total costs, for a resource plan featuring each Supply option. However, this also represents a comparison from an electric rate perspective. The system costs for each Supply option represent the numerator, and the identical number of kWh served represents the denominator, in a system average electric rate calculation of cents/kWh. Due to the fact that the denominator does not change when analyzing competing Supply options, the Supply option with the lowest system cost will also result in the lowest system average electric rate. Consequently, the evaluation of only Supply options is simultaneously an evaluation of both costs and system average electric rates.

Now let's return to the RIM and TRC preliminary economic screening tests. By design, the TRC test does not address electric rate impacts, but the RIM (or Rate Impact Measure) test is focused on electric rates. Furthermore, the TRC test omits two significant DSM-related costs: incentive payments and unrecovered revenue requirements.

Therefore, the TRC test is definitely not evaluating a DSM option versus a Supply option in a manner that is consistent with how two Supply options are compared; i.e., with a full accounting of costs related to the resource option that will be passed on to a utility's customers and from an electric rate perspective.

However, the RIM test both accounts for all DSM-related costs and takes an electric rate perspective. Consequently, even though the RIM test itself is not used when two Supply options are evaluated, the RIM test does evaluate a DSM option versus a Supply option in a manner that is consistent with how two Supply options are evaluated.

Energy Conservation or Efficiency Programs "Cannot Pass the RIM Test"

This statement is also incorrect and misleading. The gist of the statement is that, although load management type DSM programs that have relatively low kWh-per-kW reduction ratios can pass the RIM test, energy conservation or energy efficiency type DSM programs that have much larger kWh-per-kW reduction ratios cannot pass the RIM test.

This statement is simply incorrect. The state of Florida has used the RIM test (in combination with the Participant test) for at least 15 years at the time this book is being written and the electric utilities in that state have offered a large number of energy conservation programs that passed the RIM screening test. Using FPL's DSM efforts as an example, the utility has achieved approximately 4,300 MW of Summer MW reduction through 2010 from a combination of load management and energy conservation programs. Of that total, approximately 2,500 MW, or almost 60% of the total, has been achieved with energy conservation programs.

Therefore, the statement that "use of the RIM test does not support the implementation of large quantities of energy conservation DSM programs" has been proven by actual results to be incorrect.

Furthermore, the unspoken premise of the statement, that many more energy conservation programs pass the TRC test than pass the RIM test, is not only irrelevant, it is becoming even less meaningful over time. This unspoken premise is irrelevant for two reasons. First, the premise is irrelevant because large numbers of energy conservation DSM programs can pass the RIM test as evidenced by DSM efforts in Florida. Second, the premise is also irrelevant because one *would expect* the use of a "test" such as TRC that does not account for all DSM-related costs to "pass" more DSM programs of any type. Having more DSM programs of all types "pass" an incomplete cost-effectiveness screening test is not only expected, it is not much of an accomplishment.

The premise is also becoming less meaningful over time as mentioned in an earlier footnote. The introduction of environmental compliance costs (such as projected costs for CO_2 compliance) into DSM preliminary cost-effectiveness screening in recent years has increased the benefits of DSM programs that feature relatively large kWh-per-kW reduction ratios. These increased benefits occur in both the RIM and TRC tests. However, because the TRC test does not account for all DSM-related costs, the benefit-to-cost ratios under the TRC test are generally much larger to start with than are the ratios under the RIM test that does account for all DSM-related costs.

We saw exactly those results in Chapter 6 with Table 6.6 that examined the RIM and TRC test results for two DSM options. The TRC test benefit-to-cost results for the two DSM options ranged from 2.40 to 2.69, while the RIM test benefit-to-cost ratios ranged from 0.78 to 1.07. But why is that important for the discussion at hand?

The reason this is important is that the introduction of environmental compliance costs will increase the benefits for a specific DSM option by exactly the same amount under either the RIM or TRC tests. However, because RIM test benefit-to-cost ratios generally are much closer to the 1.00 benefit-to-cost ratio "pass/fail" criterion than are TRC test ratios, this increase in benefits will result in *greater numbers* of DSM options moving from "fail" to "pass" under RIM than will be the case with TRC.

As the previous footnote mentioned, this is exactly what occurred in a recent preliminary economic analysis of a large number of DSM measures in Florida just prior to the writing of this book. FPL found that when performing preliminary economic screening analyses of the DSM measures using the RIM and TRC tests, 166 measures passed the RIM test, and 296 measures passed the TRC test, when the assumption was that there were zero CO_2 compliance costs. This equates to 56% as many measures passed the RIM test as passed the TRC test.

However, when the assumption was changed to include a projection of CO_2 compliance costs, the picture changed significantly. The number of measures now passing the TRC test increased only slightly from 296 to 309. But the number of measures now passing the RIM test increased greatly from 166 to 279. This equates to 90% as many measures passed the RIM test as passed the TRC test once CO_2 compliance costs were accounted for in both tests.

Consequently, the introduction of CO_2 compliance costs (and/or other environmental compliance costs) in DSM preliminary cost-effectiveness screening is likely to result in the number of DSM options passing the RIM test more closely matching the number of DSM options "passing" the TRC test than in previous years when no such CO_2 compliance costs were accounted for.

A Final Thought on RIM versus TRC

Although I believe that the RIM test is the correct DSM preliminary cost-effectiveness screening test to use, I also believe that the RIM versus TRC debate has diverted attention away from a more important issue.

Recall that both the RIM and TRC tests are designed as *preliminary* economic analysis screening tools. As such, the two tests are intended to be used to screen out the economically inferior DSM options. They are definitely *not* intended to be used to make final resource option decisions. Final resource option decisions should only be made after analyses have examined all of the important system impacts of resource plans that feature specific resource options.

Because RIM and TRC tests are merely preliminary economic analysis screening tools, these screening tests could, in theory, be scrapped. A utility *could* attempt to include all DSM options in various resource plans that would then compete in final (or system) economic analyses. However, that would not be practical due to the sheer volume of potential DSM measures, and the corresponding volume of resource plans, that would have to be evaluated. Therefore, for practical reasons, utilities will continue to use these preliminary economic screening tools.

So why mention the theoretical possibility of scrapping the use of the RIM and TRC tests? This was done to help point out that preliminary economic screening should *not* be the primary issue when discussing resource planning involving DSM options (although, unfortunately, it often is). The most important issue is whether the utility, and the regulatory body that governs the utility, have a thorough set of analysis results of different resource plans (featuring different resource options) that account for both system economic and system non-economic impacts.

A far more productive use of time in regulatory hearings involving DSM options would be to focus on the electric rate, system fuel use, system emissions, etc., impacts of varying levels of DSM versus Supply options, rather than focus on heated debates regarding the two *preliminary* economic screening tools of RIM and TRC.

Another key issue is how much DSM is *actually needed* to meet the utility's projected resource needs. As we have previously discussed in the main body of the book, a utility should not be adding an amount of DSM that exceeds its near-term resource needs. To do so will definitely increase the electric rate impact of the DSM portfolio due to the earlier-than-needed introduction of DSM costs, combined with the greater kWh reduction values from the additional DSM further lowering the number of kWh over which the utility recovers its revenue requirements. In addition to raising near-term electric rates, the additional DSM may result in the entire DSM portfolio being non-cost-effective.

In my opinion, regulatory bodies, interveners, and utilities spend too much time discussing the results of these two *preliminary* economic screening tests and too little time discussing both the amount of new resources that are needed, and the variety of system impacts that occur as a result of the potential resource options being evaluated. Stated another way, it is almost irrelevant whether the RIM or TRC test is used solely to identify a number of DSM options that will then be evaluated in a final (or system) economic analysis.* (However, this assumes that one does *not* incorrectly stop the analytical process at the conclusion of this preliminary economic screening step in the IRP process.)

What is important is that the final (or system) economic analyses correctly takes into account the actual resource need(s) of the utility, uses the correct electric rate perspective in regard to the economics of the various resource plans, and accounts for all of the system impacts, both economic and non-economic of the resource plans. Conversely, a utility or regulatory body that prematurely stops the process at the end of the preliminary economic screening of DSM options, and decides that the utility should perform all of the DSM that passed either of the RIM or TRC preliminary economic screening tests without accounting for the utility's resource needs or all system impacts, is almost certainly doing the utility's customers a disservice. This is because such a decision would be based on preliminary, and therefore very incomplete, information regarding all of the electric rate and other impacts to the utility system that would be passed on to the utility's customers.

* Because the TRC test does not include all relevant DSM costs and, therefore, represents a much lower "test" to pass, more DSM options will be identified with TRC than with RIM. In a true final (or system) economic analysis, many/all of these "additional" TRC-based DSM options will likely be eliminated. Therefore, the use of the TRC test in the preliminary economic screening is not as efficient a process as would be the case if the RIM test were used in the preliminary economic screening.

Appendix F: Mini-Lesson #4— How Can a Resource Option Result in Lower Costs, but Increase Electric Rates?

When a discussion of utility resource options reaches the point at which DSM options are compared to Supply options, one often encounters the statement that certain DSM options (i.e., DSM options that do not pass the Rate Impact Measure or RIM preliminary economic screening test) can result in lower utility system costs, but will also result in higher electric rates.

For some, the first reaction to hearing this statement is (and I quote): "Huh?" To them, it seems counterintuitive that something that lowers costs could also result in increased electric rates. (Their next thought in regard to an explanation of this statement may be something along the lines that "Those pesky utility people are up to something!")

However, the explanation of why this statement is true is a lot less exciting and is based on the simple fractions that we all learned in grade school. We shall now walk through an explanation of how this statement can be true using another hypothetical utility system.

Let's assume that this new hypothetical utility system currently has sales of 50,000 GWh (or 50,000,000,000 kWh once we remember that 1 GWh = 1,000,000 kWh). Let's also assume that the utility has total costs (or revenue requirements) of $5 billion dollars for the current year.

Now, recalling that an electric rate is (basically) a simple fraction in which the utility's costs are divided by the utility's sales, we have $5 billion in costs divided by 50,000,000,000 kWh of sales or $0.10/kWh. Because electric rates are typically expressed in terms of cents/kWh, we will state that the electric rate for the current year is 10.00 cents/kWh. This calculation is presented in Table F.1.

We assume (simplistically) that this particular hypothetical utility is growing both in load and costs at 2% per year. Therefore, because of the growth in load, the utility projects that it will have to add a resource option in the future. We assume this point will occur 5 years in the future. The utility is considering two resource options that will result in acceptable system reliability. The first resource option (a Supply option) will have an annual net cost (after accounting for capital and operating costs of the new Supply option as well as system fuel and environmental compliance cost savings) of $100 million dollars. The second resource option (a DSM option) will actually lower the utility's total costs by $10 million due to reduced system fuel and environmental cost savings being greater than the cost to implement the DSM option. However, the second option will also lower the utility's annual sales (GWh) by 3%.

TABLE F.1

How Can a Resource Option Result in Lower Costs, but Increase Electric Rates?

I. Current Situation (for a Hypothetical Electric Utility)	
(a) Current total costs (revenue requirements) =	$5,000,000,000
(b) Current total sales (GWh) =	50,000
(c) Current electric rates (cents/kWh) =	**10.00**

The utility is, naturally, interested in determining the impact that each resource option will have on the electric rates that will be charged to its customers. The utility knows that its costs will have increased by 2% for each of 5 years to $5,520,404,016 (before accounting for the addition of one of the resource options) and that its sales will have increased by 2% for each of 5 years to 55,204 GWh (again, before accounting for the addition of one of the resource options). Using those values as starting points, the utility first calculates the electric rate if Resource Option #1 (the Supply option) is selected. The result is presented in Table F.2 (which is an expansion of Table F.1).

In this scenario, we see in the table's section II, rows (a2) and (a3), that the utility's total revenue requirements are increased by $100 million if Resource Option #1 is selected. We also see in rows (b2) and (b3) that the number of GWh sold with which the utility's total revenue requirements are collected is not affected by the selection of Resource Option #1. The end result, after dividing the value on row (a3) by the value on row (b3) and converting from $/GWh to cents/kWh, is that the utility's projected electric rate for this year if Resource Option #1 is selected is 10.18 cents/kWh.

TABLE F.2

How Can a Resource Option Result in Lower Costs, but Increase Electric Rates?

I. Current Situation (for a Hypothetical Electric Utility)	
(a) Current total costs (revenue requirements) =	$5,000,000,000
(b) Current total sales (GWh) =	50,000
(c) Current electric rates (cents/kWh) =	**10.00**
II. Future Scenario #1	
(a1) Projected costs w/o resource option =	$5,520,404,016
(a2) Projected cost of potential resource option #1 =	$100,000,000
(a3) Projected total future costs (revenue requirements) =	$5,620,404,016
(b1) Projected sales w/o resource option (GWh) =	55,204
(b2) Impact on sales from resource option #1 (GWh) =	0
(b3) Projected total future sales (GWh) =	55,204
(c) Projected future electric rates (cents/kWh) =	**10.18**

Table F.3 now expands the picture to add similar information if the utility were to select Resource Option #2 (the DSM option) that lowers the projected system costs by $10 million and decreases the number of GWh served by 3% or 1,656 GWh (=55,204 GWh × 0.03).

In this scenario, we see in the table's section III, rows (a2) and (a3), that the utility's total revenue requirements are indeed decreased by $10 million if Resource Option #2 is selected. We also see in rows (b2) and (b3) that the number of GWh sold with which the utility's total revenue requirements are collected is decreased by 3% due to Resource Option #2. The end result is that the utility's projected electric rate for this year if Resource Option #2 is selected is 10.29 cents/kWh, i.e., a higher electric rate than if Resource Option #1 is selected.

By comparing the results of the two scenarios, it is clear that although Resource Option #2 results in significantly lower costs ($5,510 million vs. $5,620 million), it also results in higher electric rates (10.29 cents/kWh vs. 10.18 cents/kWh.)

Therefore, assuming all else equal, the utility's customers would prefer that Resource Option #1 were selected (even though the utility's total costs

TABLE F.3

How Can a Resource Option Result in Lower Costs, but Increase Electric Rates?

I. Current Situation (for a Hypothetical Electric Utility)	
(a) Current total costs (revenue requirements) =	$5,000,000,000
(b) Current total sales (GWh) =	50,000
(c) Current electric rates (cents/kWh) =	**10.00**
II. Future Scenario #1	
(a1) Projected costs w/o resource option =	$5,520,404,016
(a2) Projected cost of potential resource option #1 =	$100,000,000
(a3) Projected total future costs (revenue requirements) =	$5,620,404,016
(b1) Projected sales w/o resource option (GWh) =	55,204
(b2) Impact on sales from resource option #1 (GWh) =	0
(b3) Projected total future sales (GWh) =	55,204
(c) Projected future electric rates (cents/kWh) =	**10.18**
III. Future Scenario #2	
(a1) Projected costs w/o resource option =	$5,520,404,016
(a2) Projected cost of potential resource option #2 =	($10,000,000)
(a3) Projected total future costs (revenue requirements) =	$5,510,404,016
(b1) Projected sales w/o resource option (GWh) =	55,204
(b2) Impact on sales from resource option #2 (GWh) =	(1,656)
(b3) Projected total future sales (GWh) =	53,548
(c) Projected future electric rates (cents/kWh) =	**10.29**

would be lower if Resource Option #2 were selected). After all, a customer's monthly bill is directly driven by the electric rate that applies to the customer. Moreover, a customer's monthly bill is only partially or indirectly driven by the utility's total costs. This is because the utility's total costs are only one part of the calculation that determines the actual electric rate as shown by this example. The number of kWh with which those costs are recovered is the other part.

Returning to our grade school work in simple fractions in which we have a numerator (the value that is the top part of the fraction) and a denominator (the value that is the bottom part of the fraction), we can explain what is happening in the aforementioned example between the two future scenarios. Using the scenario with Resource Option #1 as a starting point, we have a numerator (utility costs) of $5,620,404,016 and denominator (GWh of sales) of 55,204 GWh that yields an electric rate of 10.18 cents/kWh.

The scenario with Resource Option #2 lowers the numerator to $5,510,404,016, but also lowers the denominator to 53,548 GWh. The reason that this yields a higher electric rate of 10.29 cents/kWh is that the denominator is lowered by a greater percentage than the numerator is lowered. In this example, the differences between Resource Option #1's and Resource Option #2's "fractions" (i.e., electric rates) were that the numerator (cost) value was lowered by approximately 2%, while the denominator (sales) value was lowered by 3%, if Resource Option #2 is selected. Consequently, the resulting ratio (cents/kWh) increased.

Hopefully, the concept of how a resource option can lower costs, but increase electric rates is no longer counterintuitive.*

---

* This should be especially true for those of you who were paying attention in grade school when fractions were taught (even if the only reason you concentrated on fractions was to avoid thinking about what would likely happen to you during the next recess period when the dodge ball game resumed).

Appendix G: Mini-Lesson #5—How Can a Resource Option That Produces Emissions Lower a Utility's Total System Emissions? ("The Taxi Cab Example")

In Chapter 7, as part of our non-economic analyses of our hypothetical utility system, we evaluated each of the five resource plans in terms of projected system air emissions. The result was that a resource plan (Supply Only Plan 1 (CC)) that added a new generating unit resulted in the lowest projected system emissions for SO_2 and NO_X. Furthermore, the new generating unit that would be added, a 500 MW CC unit, was the largest generating unit that was evaluated and this unit was projected to operate on our utility system 80% of the hours in the year.

The result of that evaluation may have been surprising, and even counterintuitive, to some readers. The question such a reader may have asked at that point is framed as the main title of this appendix: "How can a resource option that produces emissions lower a utility system's total emissions?" After all, a commonly held belief is that energy conservation efforts *have* to result in lower air emissions.* Because the new CC unit was compared not only to two other Supply options, but also to two DSM options, the system emission outcome may be even more surprising or counterintuitive.

A reader may have been tempted to ask one or both of the following questions:

- "Is this some trick that utility companies utilize in their analyses?"; and,
- "Is this something that is peculiar to electric utility companies; i.e., could a similar result occur for another type of business?"

The answer to the first question is "no" and the answer to the second question is "yes."

In order to demonstrate why these are the correct answers, we will examine the same air emission question using another type of business that is also reliant upon using fossil fuel. This business is one that most people are more familiar with than they are with electric utilities: a taxi company.

* At this point, the reader should recall Fundamental Principle #4 of Electric Utility Resource Planning which states that one must always keep in mind the question of: "compared to what?"

The Taxi Cab Company Analogy

The analogy of using a taxi cab company as a proxy for an electric utility system for the purpose of examining air emissions, while not perfect, is useful. This is because both electric utilities and taxi companies share at least three fundamental operating characteristics. These three shared operating characteristics are:

1. Both power plants and taxis use fossil fuel when operating;
2. Air emissions result from the use of this fuel; and,
3. All else equal, the greater amount of fuel that is used, the more air emissions which result.

We shall now take a look at a taxi company and how its total air emissions may change as it considers its "resource options." The taxi company analogy will be kept very simple in order to keep the calculations simple. However, this simplicity does not change the underlying principles that will emerge.

We shall start by listing certain assumptions that describe the operation of our taxi company.* These assumptions are:

- The taxi company has one cab which gets 20 miles per gallon;
- This cab operates 24 hours each day and drives an average of 20 miles each hour; and,
- For each gallon of fuel that the cab uses, one "unit" (e.g., 1 lb) of air emission is released.

Therefore, our cab is driven 480 miles per day (480 miles = 24 hours × 20 miles each hour). Our cab also consumes 24 gallons of fuel per day (24 gallons = 480 miles/20 miles per gallon). From this information, we can now calculate the current daily emissions for our taxi company:

$$\begin{aligned} &24 \text{ gallons of fuel consumed per day} \\ &\times 1 \text{ unit of air emission per gallon} \\ &= 24 \text{ emission units per day.} \end{aligned}$$

The current situation in regard to daily mileage, fuel usage, and emissions for our taxi company is summarized in Table G.1.

* You read this correctly; this is now "our" taxi company. When you picked up this book, I bet you never thought that you'd own (or at least share on paper) an electric utility company and a taxi company. No need to worry; not a word of this will be breathed to the U.S. Internal Revenue Service.

TABLE G.1

Taxi Company Analogy: Current Situation

	Miles Driven	Gallons of Fuel Consumed	Emission Units
Current Situation	480	24	24

However, our taxi company is now forecasting that changes will need to be made because of growing customer demand for our taxi company's services. The forecasted change is that the demand for service will increase for 5 "peak hours" of each day and remain unchanged for all other hours. For each of these 5 peak hours, an additional 20 miles of driving is forecasted, or an additional 100 miles per day of driving. Therefore, the forecast is that the taxi company's customers will soon be requesting 580 miles of transportation per day.

The taxi company realizes that something needs to be done because its sole cab is already occupied during each hour of the day. The taxi company must somehow address this increased demand.

The conclusion that is reached is that the taxi company has two basic options:

1. The "Conservation" Option: Make all of the expected increased demand vanish by persuading the new "peak hour" prospective riders, or an equivalent number of its existing riders, to participate in "conservation" activities that result in these folks not using the taxi company's services. (For example, the taxi company could lend bicycles for these people to use or could give them bus tokens.)
2. The "Supply" Option: Purchase a second cab and use it. The new cab the taxi company is considering purchasing gets 30 miles per gallon.*

Our taxi company now needs to make a decision as to which of these two "resource options" it will choose. The company decides to base its decision solely on one criterion: "which option results in the lowest air emissions from the operation of the taxi company?"

The taxi company now analyzes its system emissions for each of the two resource options. It starts its analysis with the Conservation Option.

Because the Conservation Option removes all of the projected new miles of travel, and there are no changes made to operation of the taxi company's sole

* The assumption that the new cab is more fuel-efficient than its current cab is consistent with the fact that new generating units are typically more fuel-efficient than existing generating units. Note also that, although new car models typically have lower emission rates than prior models, we will assume that the new cab still has the same emission rate of 1 emission unit per gallon of fuel consumed as the current cab.

TABLE G.2

Taxi Company Analogy: Current Situation and Forecast with Conservation Option

	Miles Driven	Gallons of Fuel Consumed	Emission Units
1. Current Situation	480	24	24
2. Forecast w/the Conservation Option	480	24	24

cab, neither the daily miles driven nor the daily amount of fuel consumed change. Consequently, the resulting daily emission level remains unchanged as shown in Table G.2.

Our taxi company feels pretty good about the projected emission level outcome with the Conservation Option. It will have satisfied all of its customers needs for transportation (otherwise no customers would have voluntarily selected the Conservation Option) and it will have done so without increasing its daily emission levels. However, in order to satisfy itself that it has evaluated all of its options, the taxi company now evaluates its Supply Option.

The taxi company notes that, because the Supply Option (a new cab) gets 30 miles per gallon while its existing cab gets only 20 miles per gallon, it will change the way it operates. The new 30 miles per gallon cab will now be operated (or dispatched) as often as possible (24 hours per day) and the existing 20 miles per gallon cab will be operated only during the 5 "peak" hours of the day. In other words, the taxi company will operate the cab with the lower fuel cost as often as possible (as a "baseload" cab) and will operate the cab with the higher fuel cost only when needed to meet peak load (as a "peaking" cab).

The daily emission profile for the taxi company if the Supply Option is chosen is now calculated as follows:

a. The new cab will be driven 24 hours per day and will cover 20 miles each hour. This results in 480 miles driven per day (480 miles = 24 hours × 20 miles/hour). The new cab gets 30 miles per gallon, so the new cab consumes 16 gallons of fuel per day (480 miles/30 miles per gallon = 16 gallons). Because cabs emit one emission unit for each gallon of fuel consumed, the new cab operation results in 16 emission units per day.
b. The existing cab will now be operated 5 hours per day and will also cover 20 miles each hour. This results in 100 miles driven per day (100 miles = 5 hours × 20 miles/hour). The existing cab gets 20 miles per gallon, so the new cab consumes 5 gallons of fuel per day (100 miles/20 miles/gallon = 5 gallons). Because cabs emit one emission unit for each gallon of fuel consumed, the existing cab operation results in 5 emission units per day.

TABLE G.3

Taxi Company Analogy: Current Situation, and Forecasts with Conservation and Supply Options

	Miles Driven	Gallons of Fuel Consumed	Emission Units
1. Current Situation	480	24	24
2. Forecast w/the Conservation Option	480	24	24
3. Forecast w/the Supply Option	580	21	21

Consequently, the total daily emission level for the taxi company if the Supply Option is chosen is 21 emission units ($21 = 16 + 5$).

Our taxi company has now evaluated each of its available options and the results of its analysis are summarized in Table G.3.

The results show that, with its Supply Option, although the taxi company will have doubled the number of its cabs and will drive significantly more miles, its daily fuel consumption and emissions will be lower if it selects the new Supply Option (a second cab) than if it had selected its Conservation Option.

This result was probably counterintuitive to the taxi company prior to performing the analysis. Nevertheless, the Supply Option (adding another, more fuel efficient cab) emerges as a better choice for reducing air emissions than the Conservation Option.

Now let's return to why we examined the taxi company in the first place. The taxi company was used as a proxy for an electric utility company because both types of companies have certain characteristics in common as previously mentioned. In other words, the taxi company example provided an analogy to an electric utility company. But what did we learn from the analogy?

This analogy tells us several things. First, we have learned that it is clearly possible for the addition of a specific "Supply" option to result in lower system-air emissions than a specific "Conservation" option for a specific system that uses fossil fuels. Second, because it is possible for a Supply option to be a better choice in regard to system emissions than a Conservation option for one such system (i.e., the taxi company), it is logical to conclude that it may be true for other similar systems (such as an electric utility).

Third, the analogy tells us that, all else equal, the resource option that results in the lowest amount of fossil fuel used *by the system* is the resource option that results in the lowest air emissions for the system as a whole. If the taxi company had simply looked at the two resource options, noted that the Conservation Option by itself has no emissions while the Supply Option itself has emissions, it may have leapt to the conclusion that the taxi company's system emissions have to lower if it selects the Conservation Option.

As we have seen, this look solely at the resource options themselves, without accounting for all of the system impacts, would have lead it to the wrong decision.

Fourth (and finally), the analogy tells us that a number of factors must be accounted for in analyzing the system in question, including:

- The current characteristics of the system in question (such as the number of cabs, the fuel efficiency of those cabs, and the emission rates of those cabs in our taxi company example);
- These same characteristics for each resource option being considered (the characteristics for the Conservation and Supply options); and,
- How each option will be utilized on the system and how the operation of the system's existing components will be affected (how the existing cab's operation will be affected by each of the two resource options being considered).*

With this in mind, one can go back to the taxi company example and see how the outcome might have been different if even one of the key assumptions had been changed. (For example, if the new cab had only gotten 25 miles per gallon rather than 30 miles per gallon, the Conservation Option's 24 emission units would now just edge out the Supply Option's 24.2 emission units, resulting in the Conservation Option emerging as the resource option which resulted in the lowest system-air emissions.)

If one is truly interested in determining which resource option lowers system air emissions the most for a fossil fuel-based system, one *cannot* immediately select the resource option that may seem intuitively correct. All of the resource options must be carefully analyzed to account for all impacts to the specific utility system.

Back to Our Utility System

We now take a taxi cab ride back to our utility system† and take another look at system air emissions. Rather than presenting values for total system emissions for all of the years in the analysis for five resource options, we shall simplify things by looking at only 1 year's emission profile for two resource options, the CC unit and the DSM Option 2. We recall that the CC unit is the largest (500 MW) Supply Option and that DSM Option 2 is

* The reader should immediately be reminded of Fundamental Principles of Electric Utility Planning #1 and #4.

† Sorry, I couldn't resist.

the DSM option which reduces system energy use (GWh) more than DSM Option 1.

For this example, we will simplify things even further by assuming that the CC unit is only 120 MW. This allows a comparison of a comparable amount of Supply (120 MW) and DSM (100 MW) from a system reliability perspective. Other than that, the characteristics of the two resource options remain unchanged.

In order to more clearly see how the annual system air emissions are impacted by the introduction of a resource option, we shall calculate the impact in three steps*:

Step (1) = The annual emissions added by the individual resource option's operation.

Step (2) = The annual emissions reduced on the utility system by the resource option's operation.

Step (3) = The net annual emission reduction which is calculated as follows: Step (3) = Step (2) – Step (1).

In this example, we shall examine two types of air emissions: SO_2 and CO_2. We will also assume that the utility's marginal fuels will be natural gas 80% of the time and oil 20% for the year shown in this calculation.† And, to further simplify the example, we assume that both resource options will only impact the operation of the utility's existing steam-oil/gas generating units.

We will start by examining DSM Option 2 for SO_2 emissions only.

For the DSM option, Step (1) is easy. Because DSM consumes no fossil fuel during its "operation," DSM produces no SO_2 emissions. Consequently, the Step (1) value is zero.

However, Step (2) requires some calculations. We will first show the detailed calculation for the emissions saved by DSM from energy that would have been produced by steam-oil/gas power plants using an approximate assumption of an SO_2 emission rate for a steam-oil/gas power plant with a 10,000 BTU/kWh heat rate: 0.68 lb of SO_2 for each mmBTU of oil burned. (We will discuss the corresponding SO_2 emission rate if the steam-oil/gas unit is burning natural gas in a moment.)

* In reality, the emission impacts to the utility system described in Steps (1), (2), and (3) are occurring simultaneously. These impacts are discussed separately in the example in order to clarify the different impacts. Also, these three steps are somewhat different than the three steps discussed in Chapter 6. This is only because here we are examining impacts from a single resource option being added to a utility system. In Chapter 6, we were discussing impacts from one resource option addition versus another resource option. The results of the calculations in this appendix, if merged to compare the impacts of two resource options would be identical to those resulting from an analysis using the three steps discussed in Chapter 6.

† Few utilities rely this much on oil, but this oil percentage was chosen to include a meaningful amount of a fossil fuel other than natural gas in the example.

We also recall from Table 6.1 that DSM Option 2 is projected to reduce 6,000 kWh for each 1 kW reduction. This means that 100 MW of DSM Option 2 will be projected to reduce 600,000,000 kWh annually (600,000,000 = 6,000 kWh/kW × 100,000 kW/MW). We can now calculate the SO_2 emission reduction by DSM Option 2 from steam-oil/gas units when burning oil.

Step (2) Calculation for DSM Option 2 re SO_2 Emission Reductions from Oil:

a. 600,000,000 kWh reduced × 10,000 BTU/kWh for steam-oil/gas units = 6,000,000,000,000 BTU reduced or 6,000,000 mmBTU reduced (6,000,000 = 6,000,000,000 BTU/1,000,000 BTU per mmBTU).
b. 6,000,000 mmBTU reduced × 20% from steam units burning oil = 1,200,000 mmBTU reduced from oil.
c. 1,200,000 mmBTU reduced from steam-oil units × 0.68 pound (lb) of SO_2 per mmBTU = 816,000 lb of SO_2 reduced from steam-oil units.
d. 816,000 lb/2,000 lb per ton = 408 tons of SO_2 reduced from steam-oil unit operation.

Therefore, the impact of DSM Option 2 for this particular year is a net reduction in SO_2 emissions of 408 tons from steam units burning oil. A similar calculation for the SO_2 reduction from steam units burning natural gas is then made by making two substitutions: (i) substituting 80% for 20% (to represent the percentage of energy reduced from natural gas instead of from oil) and (ii) substituting 0.006 for 0.68 (to represent the SO_2 pounds per mmBTU emission rate for these steam units burning natural gas instead of oil). The result of the calculation for gas-fired steam generating units is 14 tons of SO_2 reduced from steam units burning natural gas for this particular year.

Therefore, the total SO_2 reduction from steam-oil/gas generating units is a reduction of 422 tons of SO_2 (422 = 408 + 14). The information from this three-step approach is summarized in Table G.4.

We now turn our attention to the CC unit. Because the CC unit burns fossil fuel (natural gas), and does so at an annual capacity factor of 80%, its value for Step (1) will not be zero. The calculation for Step (1) is as follows:

TABLE G.4

SO_2 Emission Reduction with DSM Option 2

Resource Option	Step (1) Annual Emissions Added by Options (tons)	Step (2) Annual System Emission Reduction due to Option (tons)	Step (3) = Step (2) – Step (1) Net Annual Emission Reduction (tons)
DSM	0	422	422

Step (1) Calculation for CC Unit re SO_2 Emissions Added:

a. 120 MW = 120,000 kW of generation running 80% of the year = 120,000 kW × 8,760 hours per year × 80% = 840,960,000 kWh generated per year.
b. 840,960,000 kWh per year × 6,600 BTU/kWh heat rate for the CC unit × 1 mmBTU/1,000,000 BTU = 555,034 mmBTU per year of gas burned by the CC unit.
c. 555,034 mmBTU of fuel × 0.006 lb of SO_2 per mmBTU of fuel = 33,302 lb of SO_2 emissions.
d. 33,302 lb/2,000 lb per ton = 17 tons of SO_2 emissions.

Therefore, the operation of the new CC unit is projected to emit 17 tons of SO_2 for this year. We now calculate the system emission reductions from the operation of the new CC unit. We again start with the system emission reductions from the reduced operation of steam units burning oil.

Step (2) Calculation for CC Unit re SO_2 Emission Reductions from Oil:

a) 840,960,000 kWh per year × 20% that replaces oil-fired generation = 168,192,000 kWh of reduced operation of existing oil-fired generation.
b) 168,192,000 kWh × 10,000 BTU/kWh for oil-fired units × 1 mmBTU per 1,000,000 BTU = 1,681,920 mmBTU of oil usage reduced.
c) 1,681,920 mmBTU of oil usage reduced × 0.68 lb of SO_2 per mmBTU of oil burned = 1,143,706 lb of SO_2 reduced.
d) 1,143,706 lb of SO_2 reduced × 1 ton per 2,000 lb = 572 tons of SO_2 reduced.

The CC unit is projected to lower system SO_2 emissions in this year from existing oil-fired steam generating units by 572 tons. Using the same two substitutions as mentioned earlier in the DSM discussion (80% instead of 20%, and 0.006 for 0.68), the same calculation for the CC unit's impact on system SO_2 emissions from reducing the operation of existing gas-fired steam generating units is 20 tons. Consequently, the total system SO_2 emission reduction from the operation of the new CC unit is 592 tons (592 = 572 + 20).

The full picture for this particular year of the projected impact of both of these resource options on our utility system's projected SO_2 emissions is presented in Table G.5.

The result is that the CC option is projected to reduce system SO_2 emissions for this year by a greater amount, 575 tons versus 422 tons, compared to the DSM option. This is consistent with the result from the multi-year analysis presented in Chapter 7 in which the CC unit resulted in lower system SO_2 emissions than the DSM option.

TABLE G.5

SO_2 Emission Reduction with DSM Option 2 and CC Unit A

Resource Option	Step (1) Annual Emissions Added by Options (tons)	Step (2) Annual System Emission Reduction due to Option (tons)	Step (3) = Step (2) – Step (1) Net Annual Emission Reduction (tons)
DSM	0	422	422
CC	17	592	575

The key drivers in this analysis that led to this result are: (i) the fuel efficiency of the new CC unit compared to the existing steam generating units on the utility system; (ii) the very low SO_2 emission rate of the new CC unit; and (iii) the fact that the new CC unit's capacity factor is greater than the equivalent capacity factor of the DSM option (80% versus 68%, respectively).

However, in Chapter 7, the results for the multi-year analysis also showed that DSM Option 2 reduced CO_2 emissions more than CC unit did. Using the simplified three-step approach we just utilized for analyzing SO_2 emissions for 1 year, we can now shed light on why we get the opposite result for CO_2 emissions.

Table G.6 presents the results of this analysis for system CO_2 emissions.

The results of this table show that the projected CO_2 system emission reduction for our utility system from this particular DSM option is greater (366,000 tons) than the projected reduction from the new CC unit (207,717 tons). The same calculation formulae were used, so what is the reason for the completely different outcome for CO_2 emissions compared to SO_2 emissions?

A clue is given by the projected emissions from the CC unit shown in Step (1) of the two analyses. In the SO_2 analysis, the CC unit is projected to emit 17 tons of SO_2. However, in the CO_2 analysis, the CC unit is projected to emit more than 305,000 tons of CO_2. Because the only difference between these two calculations is the emission rate (pounds of emission per mmBTU of fuel burned), one would suspect that there must be huge differences in the CC unit's emission rates for the two types of emissions.

One would be correct in this suspicion. We previously mentioned that the CC unit's SO_2 emission rate was 0.006 lb per mmBTU. The CC unit's CO_2 emission rate is 110 lb per mmBTU, or more than 18,000 times greater for CO_2

TABLE G.6

CO_2 Emission Reduction with DSM Option 2 and CC Unit A

Resource Option	Step (1) Annual Emissions Added by Options (tons)	Step (2) Annual System Emission Reduction due to Option (tons)	Step (3) = Step (2) – Step (1) Net Annual Emission Reduction (tons)
DSM	0	366,000	366,000
CC	305,268	512,986	207,717

than for SO_2. Consequently, the CC unit emerges from Step (1) with a much larger increased emission "hurdle" to overcome for CO_2 than for SO_2.

The CC unit is then not able to overcome this hurdle because the difference in CO_2 emission rates in Step (2) between the existing oil units (170 lb per mmBTU) and the CC unit in Step (1) (110 lb per mmBTU) is much smaller than difference between the corresponding SO_2 emission rates of 0.68 and 0.006, respectively.*

Conclusions

This discussion started by questioning the seemingly counterintuitive result we saw in Chapter 7 in which the selection of a new generating unit that would operate most of the hours in the year resulted in the lowest system air emissions for two of the three types of air emissions, even when compared to resource options whose individual operations produced no air emissions (i.e., the DSM and PV options). In order to explain how such a result could occur, we used the analogy of a taxi company.

This analogy allowed us to see how it was possible for a system that consumes fossil fuels to add a new resource (that also consumes fossil fuels) and actually end up with lower system emissions. This simple example established the fact that such an outcome, which we first saw in Chapter 7, was definitely possible for a system that consumes fossil fuels.

We then returned to our utility system and looked at a one year only calculation of how a new CC unit and a DSM option would impact our utility system in regard to two air emissions. We broke the calculations down in a three-step format to better understand what would actually happen on our utility system.

By doing this, we more clearly see how a specific new generating unit such as a CC unit can result in lower system SO_2 emissions than a specific DSM option, and also how the opposite result can occur in regard to system CO_2 emissions.

In turn, this leads us back to the conclusion that one cannot rely upon one's intuition regarding which types of resource options will result in the lowest system air emissions for a specific utility system. As pointed out in Fundamental Principles of Electric Utility Planning #1 and #4, such a leap to an intuitive conclusion is very likely to be proven incorrect once a complete analysis of a specific utility system is performed. We have now seen this in

* In simple terms, the reason for this is that the carbon contents of oil and natural gas are fairly similar to each other while the sulfur contents of those fuels are different by orders of magnitude. Consequently, the CO_2 emission rates of 170 and 110 lb per mmBTU differ by less than a factor of 2, but the SO_2 emission rates of 0.68 and 0.006 lb per mmBTU differ by a factor of more than 100.

three ways: the multi-year utility system analysis in Chapter 7; the taxi company analogy; and the just concluded one year analysis of a CC unit versus a DSM option.

There should now be no doubt that one needs to do a complete analysis of a utility system in order to accurately determine how different types of resource options will impact the air-emission profile of a specific utility system.

Index

F

K

L

M

N

O

P

W

For Product Safety Concerns and Information please contact our EU representative GPSR@taylorandfrancis.com Taylor & Francis Verlag GmbH, Kaufingerstraße 24, 80331 München, Germany

T - #0016 - 160425 - C16 - 234/156/19 [21] - CB - 9781439884072 - Gloss Lamination